IEE CONTROL ENGINEERING SERIES 47

Series Editors: Professor P. J. Antsaklis
Professor D. P. Atherton
Professor K. Warwick

A history of control engineering 1930-1955

Other volumes in this series:

Volume 1	**Multivariable control theory** J. M. Layton	
Volume 2	**Elevator traffic analysis, design and control** G. C. Barney and S. M. dos Santos	
Volume 3	**Transducers in digital systems** G. A. Woolvet	
Volume 4	**Supervisory remote control systems** R. E. Young	
Volume 5	**Structure of interconnected systems** H. Nicholson	
Volume 6	**Power system control** M. J. H. Sterling	
Volume 7	**Feedback and multivariable systems** D. H. Owens	
Volume 8	**A history of control engineering, 1800-1930** S. Bennett	
Volume 9	**Modern approaches to control system design** N. Munro (Editor)	
Volume 10	**Control of time delay systems** J. E. Marshall	
Volume 11	**Biological systems, modelling and control** D. A. Linkens	
Volume 12	**Modelling of dynamical systems—1** H. Nicholson (Editor)	
Volume 13	**Modelling of dynamical systems—2** H. Nicholson (Editor)	
Volume 14	**Optimal relay and saturating control system synthesis** E. P. Ryan	
Volume 15	**Self-tuning and adaptive control: theory and application** C. J. Harris and S. A. Billings (Editors)	
Volume 16	**Systems modelling and optimisation** P. Nash	
Volume 17	**Control in hazardous environments** R. E. Young	
Volume 18	**Applied control theory** J. R. Leigh	
Volume 19	**Stepping motors: a guide to modern theory and practice** P. P. Acarnley	
Volume 20	**Design of modern control systems** D. J. Bell, P. A. Cook and N. Munro (Editors)	
Volume 21	**Computer control of industrial processes** S. Bennett and D. A. Linkens (Editors)	
Volume 22	**Digital signal processing** N. B. Jones (Editor)	
Volume 23	**Robotic technology** A. Pugh (Editor)	
Volume 24	**Real-time computer control** S. Bennett and D. A. Linkens (Editors)	
Volume 25	**Nonlinear system design** S. A. Billings, J. O. Gray and D. H. Owens (Editors)	
Volume 26	**Measurement and instrumentation for control** M. G. Mylroi and G. Calvert (Editors)	
Volume 27	**Process dynamics estimation and control** A. Johnson	
Volume 28	**Robots and automated manufacture** J. Billingsley (Editor)	
Volume 29	**Industrial digital control systems** K. Warwick and D. Rees (Editors)	
Volume 30	**Electromagnetic suspension—dynamics and control** P. K. Sinha	
Volume 31	**Modelling and control of fermentation processes** J. R. Leigh (Editor)	
Volume 32	**Multivariable control for industrial applications** J. O'Reilly (Editor)	
Volume 33	**Temperature measurement and control** J. R. Leigh	
Volume 34	**Singular perturbation methodology in control systems** D. S. Naidu	
Volume 35	**Implementation of self-tuning controllers** K. Warwick (Editor)	
Volume 36	**Robot control** K. Warwick and A. Pugh (Editors)	
Volume 37	**Industrial digital control systems (revised edition)** K. Warwick and D. Rees (Editors)	
Volume 38	**Parallel processing in control** P. J. Fleming (Editor)	
Volume 39	**Continuous time controller design** R. Balasubramanian	
Volume 40	**Deterministic control of uncertain systems** A. S. I. Zinober (Editor)	
Volume 41	**Computer control of real-time processes** S. Bennett and G. S. Virk (Editors)	
Volume 42	**Digital signal processing: principles, devices and applications** N. B. Jones and J. D. McK. Watson (Editors)	
Volume 43	**Trends in information technology** D. A. Linkens and R. I. Nicolson (Editors)	
Volume 44	**Knowledge-based systems for industrial control** J. McGhee, M. J. Grimble and A. Mowforth (Editors)	
Volume 45	**Control theory—a guided tour** J. R. Leigh	
Volume 46	**Neural networks for control and systems** K. Warwick, G. W. Irwin and K. J. Hunt (Editors)	

A history of control engineering 1930-1955

S. Bennett

Peter Peregrinus Ltd. on behalf of the Institution of Electrical Engineers

Published by: Peter Peregrinus Ltd., on behalf of the
Institution of Electrical Engineers, London, United Kingdom

© 1993: Peter Peregrinus Ltd.

Apart from any fair dealing for the purposes of research or private study, or criticism or review, as permitted under the Copyright, Designs and Patents Act, 1988, this publication may be reproduced, stored or transmitted, in any forms or by any means, only with the prior permission in writing of the publishers, or in the case of reprographic reproduction in accordance with the terms of licences issued by the Copyright Licensing Agency. Inquiries concerning reproduction outside those terms should be sent to the publishers at the undermentioned address:

Peter Peregrinus Ltd.,
The Institution of Electrical Engineers,
Michael Faraday House,
Six Hills Way, Stevenage,
Herts. SG1 2AY, United Kingdom

While the author and the publishers believe that the information and guidance given in this work is correct, all parties must rely upon their own skill and judgment when making use of it. Neither the author nor the publishers assume any liability to anyone for any loss or damage caused by any error or omission in the work, whether such error or omission is the result of negligence or any other cause. Any and all such liability is disclaimed.

The moral right of the author to be identified as author of this work has been asserted by him/her in accordance with the Copyright, Designs and Patents Act 1988.

British Library Cataloguing in Publication Data

A CIP catalogue record for this book
is available from the British Library

ISBN 0 86341 280 7 (Casebound)
ISBN 0 86341 299 8 (Paperback)

Printed in Great Britain by Redwood Books, Trowbridge, Wiltshire

Contents

		Page
Preface		vii

1 Control technology in the 1930s — 1
  1.1 Introduction — 1
  1.2 Electric power industry — 2
  1.3 Sectional drives — 7
  1.4 Amplidyne, Metadyne and military applications — 10
  1.5 Stabilisation of ships — 15
  1.6 Automatic steering and auto-pilots — 18
  1.7 Summary — 20
  1.8 Notes and references — 23

2 Process control: technology and theory — 28
  2.1 Introduction — 28
  2.2 Measuring and recording instruments — 29
  2.3 Introduction of automatic controllers — 31
  2.4 Problems of on-off control — 34
  2.5 Wide band proportional control — 39
  2.6 Derivative control action — 46
  2.7 Process control theory — 49
  2.8 Process simulators — 55
  2.9 Tuning of three term controllers — 60
  2.10 Summary — 61
  2.11 Notes and references — 64

3 The electronic negative feedback amplifier — 70
  3.1 The repeater amplifier — 71
  3.2 The invention of the negative feedback amplifier — 73
  3.3 The practical carrier amplifier — 76
  3.4 Nyquist and the stability criterion — 81
  3.5 Hendrik W. Bode — 84
  3.6 Wireless communication systems — 86
  3.7 Summary — 89
  3.8 Notes and references — 92

vi *Contents*

4 Theory and design of servomechanisms — **97**
    4.1 Network analyser — 97
    4.2 Product integraph — 98
    4.3 Differential analyser — 103
    4.4 Cinema integraph — 104
    4.5 High speed servomechanism — 106
    4.6 Theory of servomechanisms — 108
    4.7 Summary — 110
    4.8 Notes and references — 112

5 Wartime: problems and organisations — **115**
    5.1 Anti-aircraft fire control — 115
    5.2 Organisation in the UK — 117
    5.3 Organisation in the USA — 120
    5.4 Relationships between civilian scientists and engineers, and military government personnel — 124
    5.5 The systems approach — 125
    5.6 Notes and references — 127

6 Development of design techniques for servomechanisms 1939-1945 — **130**
    6.1 Remote power controls for heavy AA guns — 130
    6.2 British approach to design — 133
    6.3 Development of design methods in the USA — 136
    6.4 Albert C. Hall and frequency response design methods — 140
    6.5 The Radiation Laboratory and automatic tracking — 143
    6.6 Frederick C. Williams — 146
    6.7 A.L. Whiteley, the inverse Nyquist technique and 'Standard Forms' — 148
    6.8 The practical problems arising from non-linearities — 153
    6.9 Relay and pulsed servomechanisms — 155
    6.10 Notes and references — 157

7 Smoothing and prediction: 1939-1945 — **164**
    7.1 Manual tracking — 165
    7.2 Smoothing circuits for predictors — 168
    7.3 Gun predictor developments — 170
    7.4 Background to Wiener's work — 175
    7.5 The Wiener predictor — 176
    7.6 Notes and references — 181

8 The classical years: 1945-1955 — **186**
    8.1 Technical publications on developments during the Second World War — 189
    8.2 The root locus technique — 196
    8.3 Analogue simulation — 198
    8.4 The transition to modern control — 200
    8.5 Notes and references — 205

Bibliography — **208**

Selected technical publications — **217**

Index — **245**

Preface

In 1930 the behaviour of the many devices and systems incorporating feedback loops was poorly understood: at the most fundamental level there was still confusion between positive and negative feedback. The crucial role of time delays—both time constants and transport lags—was not understood and engineers attempting to build control systems lacked both the theoretical and practical tools necessary to construct such systems. Engineers turned to feedback control only when they found themselves at an impasse. The lack of knowledge and understanding of the effect of feedback in systems resulted in reverse salients developing in many areas of engineering and, as a consequence of this, there was greater interest in the problems of control. Gradually, engineers began to realise that feedback control was not just a last resort solution, but an essential part of complex engineering systems. The changes taking place were described in an article published in 1930 in the newly published journal, *Instruments*:

> Great as has been the extension of temperature control in industry, it is in its infancy, and its adolescence waits upon the growth and manifestation of a greater interest on the part of precisely that group of industrial engineers who would reap its greatest benefits. Let us not forget that there is a limit, after all, to the mind-reading ability of the instrument maker and to the power of advertising and salesmanship. The further development of automatic control and of controllers, demands first, a greater temperature consciousness on the part of prospective users of temperature controllers in industry; second, the granting of a freer entree to instrument men by the former ... and, third, a more receptive attitude on the part of the equipment manufacturer. It is high time for builders and manufacturers of furnaces, kilns, melting pots and tanks, lehrs, ovens, boilers, kettles, stills, towers, vats and of all other diversified industrial equipment, to realise that some among them are holding up the procession. It is high time for all to follow the example of a few who make 'controllability' a part of design routine.

In the first part of this book I give a brief overview of the general history of feedback from the period 1920 to 1939. During this time developments occurred in the field of automatic control devices, in the understanding of their behaviour and the discovery of analysis and design techniques. Their application was in areas as diverse as the process industries, amplifiers for the telephone system, the distribution system for electrical power, automatic stabilisation controls for aircraft, analogue computing machines, electrical drives for the paper making and steel industries, and many others. If we examine three areas in detail — process control, the electronic negative feedback amplifier and servomechanisms — it is apparent that progress was prevented by common problems, but recognition of this commonality and the development of appropriate abstractions was hindered by the lack of a common language with which to describe the problems. Gradually, during the 1930s, appropriate concepts began to emerge in each of these areas and by the end of the decade two distinct approaches to the analysis and design of control systems had emerged: a time domain approach based on modelling the system using linear differential equations, and a frequency domain approach based on plotting the amplitude and phase relationship between the input and output signals on some form of gain and phase plot.

The second part of the book deals with the developments that took place during the Second World War and its immediate aftermath. The exigencies of war brought together engineers with diverse experience in using feedback control to work on one problem: the design of a system to automatically track the course being flown by and aim anti-aircraft guns at an aircraft. The nature of the problem, the equipment, and the computational tools available were such that the frequency response approach to analysis and design was used and developed. Consequently, in the years after the war frequency response methods dominated control theory. This dominance persisted to the end of the period covered in this book.

The body of knowledge developed during the years 1930 to 1955 acquired the name 'classical control theory' in the early 1960s. This was to distinguish it from the so called 'modern control theory', a name that began to be used during the early 1960s for the new time domain approaches to control system design. Both approaches were based on the assumption that for the purposes of analysis and design, real systems can be represented by deterministic mathematical models. However, as emerged clearly during the wartime period, such models were an inadequate representation of the complex systems that exist in the real world: real systems are nonlinear, subject to disturbances and have to respond to noisy, non-deterministic inputs. From the mid-1950s onwards, the research frontier shifted to the study of stochastic and nonlinear control systems.

I have concentrated on telling the story of the technical developments and this is therefore a traditional, narrative history. I leave it to others to delve into the complex relationships between the technology and its social and economic consequences. However, at both the beginning and end of the period covered by this book, it is impossible to ignore the serious concerns being expressed about the impact of automation. There is a similarity between the concerns and arguments propounded in the early 1930s and those of the mid-1950s, covering unemployment, economic growth, removal of degrading and onerous work, and de-skilling. The editorials in *Instruments* during the 1930s, repeatedly argued the case for automation. Typical comments stated that it 'reduces the sum total of waste and drudgery', it enables 'people to have more and better foods, clothes, houses, books, cars and goods of all kinds, and also to have more and more leisure' and 'employers to pay high wages'. Others argued cogently that the problem lay not in the use of technology but in the distribution of its profits. The US Secretary of Agriculture in an address to the American Association for the

Advancement of Science, said in December 1993 'the man who invented our labor saving machinery, the scientists who developed improved varieties and cultural methods, would have been bitterly disappointed had they seen how our social order was to make a mockery of their handiwork ... [they were seeking to] free mankind from the fear of scarcity, from the grind of monotonous all-absorbing toil, and from the terrors of economic insecurity. Things have not worked out that way'. Some engineers were sure that the economic conditions of the 1930s were not an inevitable consequence of technological change or even of market forces — 'economic conditions are due to economic practices, and economic practices arise from the rules man makes' was one comment in 1938, in an article in the *Transactions of the American Society of Mechanical Engineers*. The point had been made even more forcefully in an editorial in *Industrial and Engineering Chemistry* in 1933: 'in conquering the problems of production, we have practically neglected the difficulties of proper distribution and that in applying science to industry, we have brought complete leisure to a few, when we should have made possible more leisure to all. What lies before us is the task of finding an equitable way of distributing to labour a greater proportion of the wealth and the goods which technology helps to create'. A diversity of views has persisted and these same issues remain debated but unresolved.

During the wartime years the concerns were more immediate and meant initially, a total commitment to produce the best possible weapons as quickly as possible, followed by increasing concern about the destructive powers of the weapons being developed, and then during the later years of the war, an awareness that the discoveries being made and the techniques being developed would have immense social and economic consequences.

I began work on this book in 1979, immediately following the publication of my book covering the period 1800 to 1930, and during its preparation I have received help from many people and many organisations. I have received financial support from the University of Sheffield, the Smithsonian Institution Fellowships and Grants programme and the Royal Society. The University of Sheffield granted me study leave for four months in 1982 which, thanks to the assistance of Otto Mayr, I spent as a visiting scholar in the Division of Mechanisms, Department of the History of Science and Technology, National Museum of American History, Smithsonian Institution, Washington DC. I returned to the National Museum of American History for a year in 1988-89 when I was awarded a senior postdoctoral research fellowship of the Smithsonian Institution. The University of Sheffield granted me special leave to take the fellowship and Carlene Stephens, supervisor of the Division of Engineering & Industry (successor to the Division of Mechanisms) made the arrangements for my stay. I am indebted to her and to Peter Liebhold, Steven Lubar, Jeffry K. Stine, Bill Worthington, Kay Youngflesh and Bonny Lilienfeld of the Division for their support and encouragement and their advice and tolerance. I also benefited from discussions with other staff and visitors of the museum and from the regular seminar programme: I wish to thank Peggy Caldwell, W. Bernard Carlson, Harold Closter, Nanci Edwards, Barney Finn, Paul Forman, Louis Hutchens, Uta Merzbach, Arthur Mollela, Jane Morley, Harry Rubenstein, David Shayt and Debbie Warner. A benefit of being in the National Museum of American History was the excellent support received from the Librarian and her staff and in particular from Jim Roan.

Preface

I am grateful for the assistance received from the staff of the Public Record Office, London; the National Archives Record Service, Washington DC; the Archive Center, MIT; The Naval Operational Archives, Navy Yard, Washington DC; Marconi Radar Systems and General Electric Company (Schenectady).

In carrying out this work I have had the privilege of meeting and interviewing a number of engineers who have made major contributions to control engineering. During 1987, I met with J.F. Coales, H.L. Hazen, A. Porter, A. Tustin, A.L. Whiteley, F.C. Williams and during 1982, Charles Beck, H. Chestnut, Admiral E.B. Hopper, Carlton W. Miller, N.B. Nichols, R.S. Phillips, Denny D. Pidhayny, Y. Takahashi, J.G. Ziegler and W.R. Evans (sadly unable to speak as a consequence of a stroke but nevertheless able to convey some of the excitement and frustrations of his invention of the root locus technique). I am grateful to all of them for willingly giving up their time and in many cases offering me hospitality. I am also grateful to Gordon S. Brown, Charles Concordia, K.A. Hayes, E.I. Jury, E.C.L. White and K.L. Wildes who have written to me providing either new information or corrections to previous published work.

Over the years I have received help and encouragement from many people including Dean Allard, David Allison, Derek Atherton, Chris Bissell, Tom Fuller, Derek Linkens, Otto Mayr, David Noble, Harry Nicolson, Annraoi de Paor (Harry Power), K.F. Raby, T.P. Speed, D.B. Welbourn and J.H. Westcott and from many friends and colleagues who from time-to-time have enquired, sometimes tentatively, about progress with the book. I am also grateful to John St Aubyn of Peter Peregrinus who has waited patiently for me to deliver the manuscript.

Finally, I wish to thank my wife Maureen Whitebrook who has constantly encouraged and supported me.

Stuart Bennett
Sheffield

End of Chapter Notes and References

Full citations are given in most cases. However, where a technical paper is mentioned simply as an example and no direct reference is made to it, only the author and date are given. A full citation can be found in the Bibliography at the end of the book.

Abbreviations:
NARS: National Archives Record Service, Record Group 227, Records of OSRD

AVIA, ADM, AIR, CAB, DSIR, PREM, SUPP, WO all indicate document series in the Public Record Office, London.

Trade catalogues: unless otherwise indicated all trade catalogues cited are to be found in the Columbia Collection of the National Museum of American History, Smithsonian Institution.

Chapter 1
Control technology in the 1930s

1.1 Introduction

Improved management, the introduction of new machinery, and new production processes brought about a rapid growth in output and productivity in manufacturing industries during the late 1920s.[1] Managers and engineers used information gathered from plant instruments to plan operations, and they used technology more and more to ensure their plans were carried out. They preferred continuous to batch processing and, whenever possible, used remote, semi-automatic, and in some cases, automatic control. Adoption of remote, semi-automatic, or automatic control was consistent with the growth of a managerial bureaucracy that fervently believed in bringing all aspects of production and distribution of products under managerial control. This bureaucracy had evolved over a period during which major changes in methods of distribution of goods and in their production, took place. Central to the changes was the creation of a mass market for standardised products. The mass production of standard products required close control at all stages of production; such control, managers believed, was too important to be left to the workers.

From the turn of the century onwards recorders were developed to 'help the manager or superintendent to secure and maintain uniform operating conditions', and, as instrument manufacturers claimed, to 'give an absolute check on the efficiency of employees'.[2] Recorders were used to measure the performance of machinery, to check on hours worked and also to ensure that working practices determined by managers—for this was the era of Taylorism and the Scientific Management movement—were carried out. One instrument manufacturer, the Bristol Company, in its 1912 Catalogue, gave as an example the use of a recorder for ensuring 'that the men keep the furnace uniformly full, and do not stop filling longer than twenty minutes... It is not a good plan to let the men rest too long... It is a better plan for the men to fill a short time, then rest a short time, and we want them to do this and this chart shows us at a glance whether or not they do it'.[3]

The standardisation of products, components and procedures not only made it easier to observe and control what workers were doing but also encouraged managers to consider automatic operation, for with automatic operation the uncertainty involved in human operations could be removed.

The change to mass production and to the use of assembly line and continuous production techniques was accompanied by a change in the social organisation of work. The semi-autonomous skilled craftworker began to be replaced by semi-skilled and unskilled factory hands who 'operated' automatic and semi-automatic machinery.

The desire to replace measurement and control actions previously carried out by skilled operatives, and to use new machinery and production techniques (the control of which was beyond unaided human capability), placed enormous demands on engineers concerned with instrumentation and control.

At the beginning of the 1920s, 'control' was, with a few exceptions, thought of as the switching on, or off, of devices—motors, pumps, valves—either directly or through relays. The exceptions included the specialist applications of the mechanical governor for speed control, electrical voltage regulators, thermostats (although these tended to be designed as high gain devices giving on-off action), and ship steering engines. A common characteristic of nearly all control devices was that the measuring and actuating elements were combined in one physical element. By the end of the decade it was becoming clear that improved control could be obtained if measurement and actuation were separated by a unit that could both amplify, and in some manner modify, the signal produced by the measuring device.

There were, of course, many control systems in use before 1930 that depended on complex manipulation of the signal obtained from the measuring device. The Tirrill voltage regulator is one such device and Elmer Sperry invented many devices that used complex mechanical linkages to modify the control signal. However, theoretical analysis of complex mechanical linkages and vibrating relays is difficult, and it was only with the development of pneumatic and electronic devices that full understanding of the signal manipulation being performed, and its connection with theoretical ideas, was established.

The major developments in techniques and in understanding took place independently in several different areas: process control (pneumatic devices), communications (the negative feedback amplifier), and computing mechanisms (theory of servomechanisms). These developments are examined in detail in the next three chapters. At the same time, slow advances, without any major breakthroughs, were being made in several other areas: electrical power generation, electrical devices, steering and stability controls for aircraft and ships, and gunnery fire-control systems both (land and sea).

1.2 Electric power industry

The growing availability of electricity was a key element in promoting both industrial change and the development of automatic control. For example, between 1919 and 1927 use of electric motors in the USA more than doubled, from 9.2 million to 19.1 million. This growth was accompanied by increased generating efficiency: during this period the amount of coal needed to produce one kilo-watt hour of electricity was reduced from 3.20 pounds to 1.84 pounds.[4] The increase in efficiency resulted from improved boiler design, gains from increases in scale, and from the use of instruments and controllers.

Before 1930, the electric power industry was the main user of instruments and controllers, and its demands led to improvements and advances in such devices. The Bailey Meter Company and the Smoot Engineering Corporation in the USA, Elliott Brothers and George Kent in England, and Siemens in Germany, were leaders in this field and supplied complete boiler control systems. These systems controlled both steam pressure and water level by manipulating the firing rate, forced and induced draft, and water supply. Full automatic control was expected to provide constant pressure without hunting in the presence of variable load, to maintain complete combustion with minimum losses, to maintain correct heat transfer, and to provide load sharing between boilers.[5]

To satisfy users who wanted supplies at a constant voltage and constant frequency the generating companies also had to invest in voltage and frequency controllers. Frequency was controlled using the established technology of the mechanical governor, and improvements in this area were mainly confined to actuators used to follow the governor motion. However, the apparatus in use for voltage regulation, although adequate, was not entirely satisfactory. The common method of voltage regulation used a control signal derived from the output voltage to change the resistance of the field circuit of the exciter generator (the generator supplying current to the field of the main alternator). By 1930, two basic techniques were widely used:

(i) the control signal operated a motor that changed the setting of a rheostat;
(ii) the control signal moved the rheostat directly.

The first method provided floating control and the second provided proportional control.[6] Motor-operated, floating-action systems did not respond rapidly to sudden, large changes in voltage and, in practice, they were fitted with a subsidiary on-off controller that switched part of the field resistance in and out of the circuit in the event of large changes in voltage.[7]

The 'Tirrill' regulator was the most widely used proportional-action regulator, and Figure 1.1 shows a schematic diagram of such a regulator from the early 1930s. Its operation is complex since it includes both load and load power factor compensation. However, the basic principle is simple: vibrating contacts are used to switch a resistance (exciter field rheostat in Figure 1.1) in and out of the field circuit of the dc exciter of the generator that supplies field current to the alternator. If there is no error between the set voltage and actual voltage, then the 'in' and 'out' periods are of equal length. If there is an error, then one period is shortened and the other lengthened, thus changing the average value of the field resistance and hence the current supplied to the alternator field circuit.

Vibrating regulators of the Tirrill type are unsuited to conditions of extreme vibration, for example on moving vehicles, or for applications in which the speed of the prime mover for the generator fluctuates widely. For such applications carbon-pile regulators were developed. Figure 1.2 shows a General Electric carbon pile regulator (circa 1932). The output voltage from the alternator (ac generator) is rectified and applied to a torque motor that compresses a stack of carbon plates — the carbon pile — placed in the field circuit of the exciter. The resistance of the carbon-plate stack is a function of the pressure applied to the plates. The force generated by the torque motor causes the field resistance of the exciter to change and hence a change in current flowing in the ac generator field-circuit.

The General Electric Company and the Westinghouse Company were main suppliers of voltage-regulators, and they developed a wide selection of types of

4 *Control technology in the 1930s*

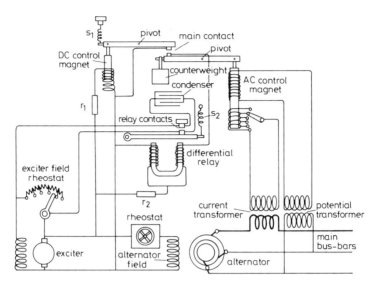

Figure 1.1 *Tirrill regulator*

      Reproduced (with partial redrawing) by permission of P.A. Borden and M.F. Behar, from *Instruments*, 1936, **9**, p.209

regulators. The mechanical, moving parts in both the Tirrill and carbon-pile regulators led to reliability problems, and there was much interest in techniques that would avoid mechanical moving parts, as these offered the potential of improved reliability and reduced maintenance.

Saturable reactors were used as the basis of some designs, but the main thrust of research during the 1930s was towards electronic regulators based on the use of the thyratron valve. The thyratron can be used to rectify an ac supply. The average voltage of the dc output is controlled by adjusting the 'angle of conduction' of the thyratron by applying a variable dc voltage to the grid of the valve. Maximum voltage output is obtained when the thyratron conducts for the full half-cycle of the ac supply — 180° angle of conduction — and is reduced as the angle of conduction is reduced. By applying the rectified ac output from the alternator, through a suitable circuit, to the grids of a pair of thyratrons supplied from the alternator, a dc voltage proportional to generated ac voltage can be applied to the exciter generator field, hence giving a proportional feedback control system.

The importance of the thyratron based method of voltage was, as A.W. Hull of the General Electric Company explained in 1929, 'the absence of mechanical inertia which made the method remarkably free from hunting or overshooting of voltage with sudden changes of load'.[8] During the following years, many people in many different application areas, discovered and commented on the benefits that accrued from the absence of inertia in electronic control systems.[9]

Electronic voltage regulators were also developed by the Westinghouse Company. The adoption of electronic voltage regulators did not entirely eliminate all stability problems. Figure 1.3, shows the circuit for the Westinghouse AT regulator of about 1933. The schematic diagram of the regulator shows clearly

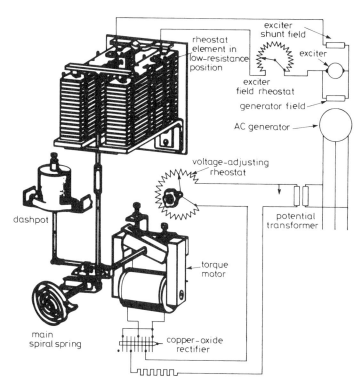

Figure 1.2 *GE Carbon Pile regulator*

Reproduced (with partial redrawing) by permission of P.A. Borden and M.F. Behar, from *Instruments*, 1936, **9**, p. 208

a subsidiary feedback circuit, labelled 'anti-hunting circuit'. The purpose of the anti-hunting circuit, F.H. Gulliksen its designer explained, is to provide 'anticipatory action...[which] is imperative in order to obtain stable operation as well as quick response action', and it does this by feedback of the exciter voltage.[10]

A.L. 'John' Whiteley, of the British Thomson-Houston Company, in an article published in 1936 on applications of the thyratron, included several circuits for electronic voltage regulators for ac and dc generators.[11] Figure 1.4 shows his circuit for an ac generator. There is a main feedback signal from the alternator (labelled *reset*) and a subsidiary feedback signal (*stabilising feed-back loop*) from the dc exciter voltage. The idea of the subsidiary circuit was taken from the Tirrill regulator with which Whiteley was familiar. Direct feedback from the dc exciter gives an improvement in the transient response; however, this is accompanied by a loss of overall gain which reduces the accuracy of the voltage regulation. Whiteley reasoned that since the subsidiary circuit is needed only for improving the transient performance, it might be possible to introduce a capacitor (marked *C*) into it, and thus remove any effect on the steady state. He recalls using operator techniques to analyse the circuits and later, probably after Nyquist's work had

6 Control technology in the 1930s

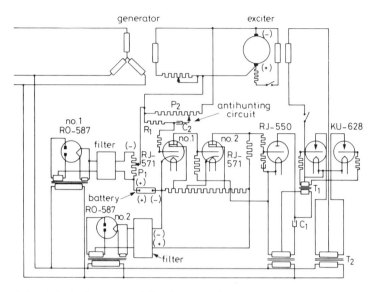

Figure 1.3 *Westinghouse type AT voltage regulator*

Reproduced (with partial redrawing) by permission of F.H. Gulliksen, from *Trans. AIEE*, 1934, **53**, p. 877

begun to spread, he obtained frequency responses for the circuits by substituting $j\omega$ for p in the transfer operators.

The success of the ac regulator (many commercial systems were installed) led Whiteley to analyse the earlier dc voltage regulator. He realised that it contained what was then known as the 'quick response' circuit — in modern terminology, a phase advance circuit. With the help of L.C. Ludbrook he investigated other passive networks with a view to their use as feedback-path stabilising circuits.[12] The circuits designed by Whiteley and those designed by Gulliksen (referred to above) produced the equivalent of proportional-plus-derivative control action.

The enthusiasm with which the thyratron was greeted emphasises the fact that a crucial problem for feedback control systems was the provision of power amplification. The simplest and best understood power amplifier was the electro-

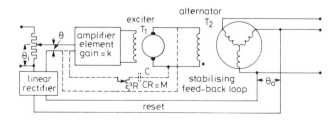

Figure 1.4 *Electronic voltage regulator*

Reproduced (with partial redrawing) by permission of A.L. Whiteley, from *J. IEE*, 1936, **78**, pp. 516-539

mechanical relay (or some variation) and it was to this device that most designers turned, unless they could avoid it by not using feedback at all.

1.3 Sectional drives

Many mechanised continuous and semi-continuous processes require material — paper, rubber, steel sheet — to be pulled through a series of driven rollers. These systems are called *sectional drives*. The control problem is difficult: the drives have to be synchronised and this requires accurate speed control of the drive motors. In paper making, if the speed of successive drives is not matched correctly the paper will tear. In steel rolling, the strip will not tear but there will be a spectacular, dangerous, and expensive 'cobble' as the strip tangles up between two sets of rollers. In the USA both the General Electric Company and the Westinghouse Company developed sectional drives for paper making and other applications. One of the first such systems was installed by Westinghouse in 1919.[13]

In the same year, the General Electric Company introduced a system with a mixture of inherent regulation and manual reset. Each individual section of the machine is fitted with a dc motor and a synchronous motor rated at approximately 20% of the power of the dc motor. The dc motor drives the section directly and each section is connected to the synchronous motor through a set of gears and a cone and pulley arrangement. The synchronous motor stators for each section are connected together electrically. The belts on the cone pulleys are set to provide the correct 'draw' (the speed of the successive motors has to be slightly greater than the previous motor in order to draw or pull the paper through the sections). The synchronous motors act to pull the speed of each drive to the correct value as the electrical interconnection forces the synchronous motors to run at the same speed. A section synchronous motor running at the correct speed generates no torque and consumes no power. Each motor is fitted with a zero-centre wattmeter and the operator adjusts the field of the dc section motor so that the wattmeter reads zero. Thus proportional control action was provided automatically and the operator was responsible for removing any remaining steady-state error.

Over the next ten years many variations, improvements, and attempts to reduce manual intervention were made. A major improvement, the Selsyn regulator, was introduced in 1929 and Figure 1.5 shows the general arrangement of the system. Each section is driven by a dc motor the speed of which can be adjusted by changing the resistance R. One of the motors acts as a master, or reference, and a small alternator fitted to this motor generates a three-phase reference signal that is supplied to the stator windings of Selsyns fitted to each section. The output of a small alternator (P) is applied to the rotor field of the Selsyn (S) for that drive. If the speed of the drive is correct then the rotating fields of the Selsyn are synchronised and no torque is generated; thus the rotor remains stationary and the field resistance, and hence the speed, of the dc drive motor are unchanged. If the speed departs from the set value then the phase difference in the rotating fields generates a torque causing the Selsyn rotor to turn and change the resistance (R). The rotor continues to turn until the stator and rotor are again synchronised and the torque drops to zero. The system provides floating (integral) control action. It cannot compensate for instantaneous load changes without the danger of hunting and it is restricted to operating in a narrow speed range.[14]

8 *Control technology in the 1930s*

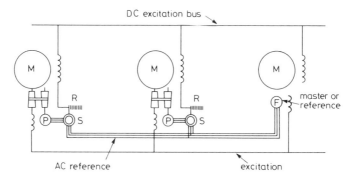

Figure 1.5 *Synchronous regulator – differential type*
(General Electric 1929)
Reproduced (with partial redrawing) by permission of H.W. Rogers, from unpublished report of the General Electric Company

During the period 1922–3 and again in 1928–9, the General Electric Company conducted experimental work on electronic speed-regulators for use in paper making machinery, and on both occasions the work was postponed because of uncertainty about the life of electronic valves. Interest revived in 1932. J. Liston reported in a review of electrical developments for 1932 that the 'modernisation of industrial plants, [was] in line with the program of the National Committee on Industrial Rehabilitation' and with the growing confidence in electronic components. He particularly noted that many of the new control devices made use of the 'unique characteristics of the electronic tube'.[15] The majority of the new control devices made use of the thyratron valve. Applications included dimmers for theatre lights, control of ac generators, dc motor control, register control for cutting and folding printed papers, constant tension devices for wire drawing, and an automatic screwdown mechanism for a steel slabbing mill.[16] However, it was not until 1937 that the General Electric Company introduced a completely electrical speed regulator incorporating an electronic amplifier, and such systems did not become standard until the 1940s.

Figure 1.6 shows the 1937 system. Each section is operated by a separate motor-generator set instead of a dc motor drawing current from a common bus. The drive motor (M) has constant field excitation and its speed is controlled by the armature excitation supplied from the generator (G). A tachogenerator (dc pilot generator P), driven by the motor, produces a voltage proportional to the speed and this is compared with a reference voltage (RV). The difference between the two is amplified by the electronic amplifier (EA) and the output is further amplified by Amplidyne generator (A) which supplies the armature of the section drive motor. The relative speeds of each section can be set by adjusting the draw rheostat (DR) which sets the proportion of the tachogenerator voltage supplied to the electronic amplifier. According to Charles Beck, an engineer with GE, this regulator was, compared with the previous design, fast, accurate and reliable. 'The high amplification factor inherent in the electronic Amplidyne system makes it very accurate', he argued, and 'the absence of inertia makes it very fast; and

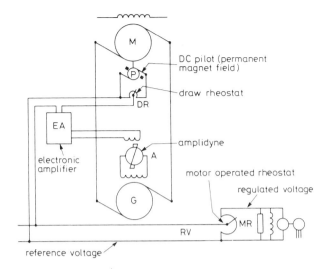

Figure 1.6 *General Electric speed regulator for sectional drives*

Reproduced (with partial redrawing) by permission of H.W. Rogers, from unpublished report of the General Electric Company

the elimination of cone pulleys, belts, carbon piles, and other mechanical parts, make it more reliable'.[17]

As noted earlier, the General Electric Company used the carbon pile in voltage regulators as an alternative to the vibrating relay regulator. The Westinghouse Company also used the carbon pile extensively, and they continued to develop systems based on its use, including speed regulators for sectional drives, into the 1940s.[18]

Similar systems were being developed in the UK and A.L. Whiteley described one such system which serves to illustrate the difficulties faced by engineers in the early 1930s. Newne and Pearson, publishers of 'Home Chat', 'Tit-bits' and 'Radio Times', wanted to combine preprinted photogravure material with the normal black and white type. This required precise registering (to within ± one sixteenth of an inch) between the preprinted roll and the press. Figure 1.7 shows a schematic diagram of the system. The printing press, driven by an electric motor, runs at its normal speed with no feedback control on the motor drive. An additional drive shaft from the press motor is connected via a cone and pulley drive to the web rollers which pull the paper from the preprinted roll. The paper forms a loop as it passes round a pivoted roller placed between two sets of rollers. On starting, the preprinted roll is aligned with the plate on the press drum. Should slippage occur between the paper and the press drum the alignment is lost and the length of the paper loop will change. Since the pivot arm supporting the roller is spring loaded, a change in the length of the paper loop causes the pivot arm to move. This movement operates relay contacts causing a pilot motor to adjust the setting of the cone and pulley system linking the press drive motor to the web drive rollers, and hence the speed of the web rollers relative to the press roll. The paper is thus pulled back into alignment. Analysing such a control system is not easy: Whiteley

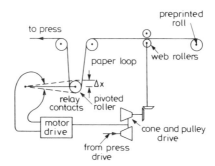

Figure 1.7 *Schematic of Whiteley's paper drive speed control system*
Adapted from sketch provided by A.L. Whiteley

was able to show that the deadspace in the relay must be greater than, or equal to, half the maximum allowable change in the length of the paper loop (Dx) for the system to be stable.

Before Whiteley's work on the printing press the British Thomson-Houston Company had installed only one closed-loop system — for sectional drives for paper making machinery — and in this system they used ac shunt commutator (Schrage) motors with brush shifting for the control of speed. Similar systems were still being installed in 1938.[19]

1.4 Amplidyne, Metadyne and military applications

The precursor to Amplidyne was the direct current, cross-field generator which had been known about for many years. Gravier in 1882, and Rosenberg in 1904 and 1907 described such generators, and in 1907 Osnos discussed a number of possible arrangements.[20] Pestarini, between 1922 and 1930, developed the theory of such machines (but dealt only with the static characteristics). He proposed the name 'Metadyne' for the general class of machines obtained by fitting additional brushes to otherwise normal generators. He visited Britain in 1930 and his ideas were taken up by the Metropolitan Vickers Company who used the Metadyne in the control systems of electric trains. Pestarini also visited the USA in 1930 but the Metadyne does not seem to have been used there.[21]

The Metadyne is a constant current device: the current flowing in the load circuit generates a flux which opposes the flux in the control circuit, hence it is a voltage-to-current amplifier. A voltage-to-voltage amplifier can be obtained by adding a compensating winding to nullify the effects of the load current flux. This was the modification made by the General Electric Company engineers led by Alexanderson, and they gave the name 'Amplidyne' to the fully compensated machine.[22]

The development costs for the Amplidyne were predominantly underwritten by military contracts, mainly with the US Navy. The contracts were for mechanical computing mechanisms, the redesign of Selsyn units, hydraulic and electrical servomechanisms, and vertical stabilisers (gyroscopic devices) for the aiming and firing of guns.[23]

Navies throughout the world had been interested in automatic methods of fire control since the beginning of the twentieth century. They wanted solutions to two problems: the stabilisation of the gun platform against the pitching and rolling of the ship, and the correction of the target angles (azimuth and elevation) to compensate for the movement of the target during the time of flight of the shell. Although several schemes for stabilised gun platforms were proposed during the nineteenth century there was little urgency while the traditional 'broadside' technique remained the normal battle tactic. Using this tactic the guns could be held at some fixed elevation and fired as the motion of the ship brought them to the correct elevation.[24] However, as the range of guns increased, tactics changed, and the broadside fell out of favour. The need for compensation for the movement of the ship became urgent. As the range of the guns increased, the flight time of a shell meant that the distance travelled by a moving target between the firing of a shell and its impact had to be taken into account when aiming the gun.[25] For the simple case with the target moving directly towards a stationary gun with velocity v, an implicit equation of the form

$$y = x - v \cdot T(y) \tag{1.1}$$

where x is the present distance of the target from the gun and $T(y)$ is the time of flight of the shell to a future position of the target y, has to be solved to find the predicted distance of the target from the gun in order to determine the range setting for the gun.[26] In practice, the problem is more complicated. The gunner has to deal not only with the change in the relative positions of the target and gun, but also has to consider the wind speed and direction, the temperature and density of the air, and specific characteristics of the gun itself. Mechanical calculators were developed to assist with this task.[27] During the First World War, Elmer Sperry had designed, as part of an overall fire-control system for ships, a 'battle tracer' that was, in essence, a mechanical computer for the solution of equation 1.1.[28] Similar 'predictors' (called 'directors' in the USA) were developed in the UK by the Vickers Company and by the Admiralty Research Laboratory.[29]

Fire-control for gunnery involves sighting, range finding, data transmission, calculation, and position control. The present position and velocity of the target has to be found, transmitted to, and entered into the mechanical calculator. The output — computed elevation and azimuth angles — has to be conveyed to the guns and used to position the gun barrel. In the early naval fire-control systems the plotting room staff obtained oral reports of the target position (azimuth and range) from the observers in the look-out station. They plotted the course and calculated the velocity of the target and then entered this data into the mechanical computer. They read the computed azimuth and elevation angles and reported them orally to the gun crew, who then aimed the guns. Gradually, this slow, error prone method gave way to one in which electrical devices were used to transmit data between observers and the plotting room, and between plotting room and gun crew. At first, simple, direct current step-by-step transmitters (a crude and simple type of stepping motor that had been developed for use in torpedoes) were used. These transmitters were not very accurate and were slow: in 1901 performance was quoted as a resolution of 7° and a maximum slewing rate of 30 steps/s; by 1922 slewing rates had increased to 300 steps/s.[30] They were succeeded by alternating current devices. The German navy was using an alternating current synchronous transmitter before the First World War[31], as was the General Electronic Company for its searchlight control systems designed for the USA coastal defence system.[32] GE had developed synchronous transmitter and receiver units (synchros) for use in remote control systems in power stations.[33] They also used them for remote control operations on the Panama Canal[34] and in a wide variety of other industrial problems. A similar device, the Autosyn, was developed for use in aircraft instruments in 1932.[35]

Observers could provide more accurate measurements of target positions if they were isolated from the motion of the ship. Schemes ranging from stabilised telescope mountings to complete ship stabilisation were proposed, many of which were tried out. By 1940, the typical solution adopted was the stabilised 'director tower', which was a stabilised platform holding both observers and the director equipment.

The aiming data from the early gun directors was displayed to the gun crew as a target pointer moving on a dial. The operator manipulated the power controls of the gun to align the gun elevation (or azimuth) pointer with this target pointer. In some early systems the gun aiming data was computed relative to the ship's axes and the guns were fired automatically when the ship passed through the horizontal plane. In later systems, the pitch and roll motion (derived from the ship's gyroscopes) was added to the aiming data and the operator attempted to keep the guns on target as the ship moved — a difficult task.

The potential of automatic gun positioning systems for improved fire control was obvious to many people. What was needed was an automatic follow-up system — a position control servomechanism — that could follow accurately, with zero velocity error, a signal that combined target position with a correction for the pitching and rolling motion of the ship.

In the UK the remote-power-control (RPC) group of the Admiralty Research Laboratory (ARL), working under C.V. Drysdale, tackled the simpler problem of searchlight control: simpler because less power was needed to move the smaller and lighter searchlights. Their system illustrates some of the problems of automatic position control, and also provides some possible solutions.[36] It provided follow-up control from the optical sight to the searchlight, and also stabilised both the searchlight and the optical sight against roll and yaw of the ship. Figure 1.8 shows the general arrangement of the stabiliser, the optical sight, and the transmitter box. The Magslips transmitters (devices similar to the Selsyn) conveyed position information to the hydraulic (oil) motors on the searchlights (Figure 1.9 and Figure 1.10). The motor speed for both the gyroscope follow-up motor and the searchlight positioning motors is proportional to the appropriate positional errors. In trials carried out in 1928, the RPC group found that the amplifier gains for which the system was stable, there was an unacceptable large velocity error which resulted in the searchlights lagging behind a moving target.

J.M. Ford and H. Clausen proposed adding a second motor, 'the advancer motor', (B in Figure 1.9) and this was introduced into the later systems. The input to motor A is also fed to motor B and the outputs of both A and B are added to the movement of the tracking telescope and the searchlight motors. A feedback loop around motor B gradually returns the input of B to zero. The overall effect

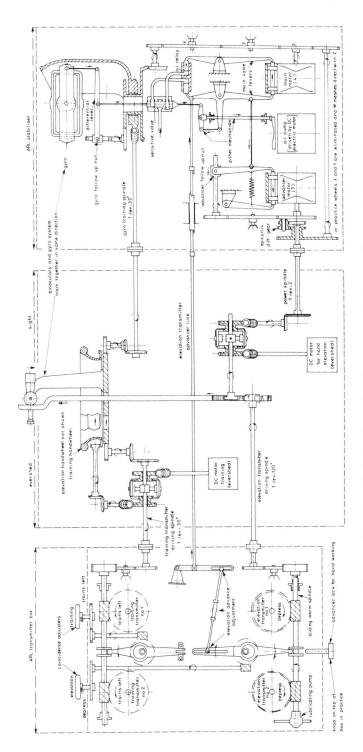

Figure 1.8 *The Admiralty Research Laboratory searchlight control system 1928–1930*
Reproduced (with partial redrawing) by permission of J.O.H. Gairdner, from *J. IEE*, **94**, 1947, (IIA), p. 210

14 *Control technology in the 1930s*

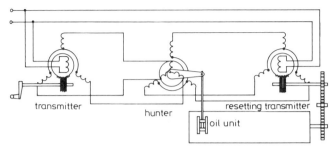

Figure 1.9 *Magslip hunter arrangement*

Reproduced (with partial redrawing) by permission of J. Bell, from *J. IEE*, 1947, **94**, (IIA), p. 266

is to add an angular displacement proportional to the angular velocity of the searchlights, to the searchlight movement thus provided a feed-forward signal to compensate for the velocity lag.[37] (If y = angle of searchlight then $y = ky + y\,dt$).

The ARL group also met the problems common to mechanical servomechanisms — backlash and friction — and they stressed the importance of the 'elimination of backlash in the valve linkwork...for precise operation'. They incorporated a dither mechanism on the 'sensitive oil relay' (the pilot valve) to reduce the effects of friction and to increase the permissible manufacturing tolerance for the valve.[38] The searchlight stabilisation system went into service on *HMS Champion* in 1929, and further sets were installed on *HMS Exeter* in 1930, and subsequently in other ships.[39]

Land based forces also were interested in gun directors and power controls. Their major interest was in their use in anti-aircraft gun control systems and, in the late 1920s, several countries began to develop such systems.[40]

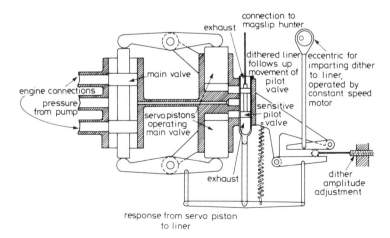

Figure 1.10 *The Admiralty Research Laboratory oil-servo relay*

Reproduced (with partial redrawing) by permission of J.O.H. Gairdner, from *J. IEE*, **94**, (IIA), p. 212

1.5 Stabilisation of ships

Interest in the general problem of the stabilisation of ships which began in the early part of the century continued into the 1930s. Naval forces were concerned about the ability of the crew to operate effectively in rough seas, and operators of passenger ships wished to increase the comfort of passengers and to avoid the need to slow down in rough seas.[41] Table 1.1 shows a summary of schemes for ship stabilisation developed and applied up to 1949. Excluding bilge keels, which were invariably used in conjunction with the other methods, the schemes divide into four groups: moving weights, anti-roll tanks, gyroscopes, and fins, and in each case both passive and active versions of the method were used. The methods remaining in use today are active fins for medium-to-high speed ships (passenger ships and destroyers); and anti-roll tanks for low speed ships.

Gyroscopic stabilisation methods were extensively investigated between 1910 and 1930, particularly by Elmer Sperry. These methods are effective for small ships, but there was a gradual recognition that the methods were impractical for large ships. One large passenger ship, the *Conti di Savoia* (41,000 tons), was fitted with three gyroscopes, each 13 ft in diameter running at 900 rpm and with a total weight of 650 tons, and its performance was monitored and widely reported.[42] However, in the 1930s attention turned to the use of anti-roll tanks.

Passive anti-roll tanks were proposed as early as 1875, and in 1911 H. Frahm, a German naval engineer, revived the method.[43] He used sealed tanks on each side of the ship which were partially filled with water and connected at the top and bottom; a restriction was placed in the top connection. The water in the tanks resonates in anti-phase with the rolling of the ship and by adjusting the size of the restriction the resonating frequency can be altered. The system worked well on ships with poor inherent stability (low metacentric height) when operating in seas with a regular wave pattern. Reductions in roll of up to 50% were reported. However, in choppy seas the results were poor, and for ships with good inherent stability there was virtually no improvement. The method was abandoned immediately after the First World War in favour of active and semi-active anti-roll tanks.

Active anti-roll tanks were investigated in both Germany[44] and the USA during the 1930s. Nicolas Minorsky carried out extensive investigations of the system for the US Navy between 1934 and 1940.[45] The principle of the active tank method is that the level of water in tanks placed on each side of the ship should be kept proportional to the rate of roll of the ship. In a typical system a variable delivery pump moves water between the tanks, and since the pump controls the rate of flow of the water, its speed is made proportional to the roll acceleration of the ship.

Minorsky, in his active-tank stabilisation systems as he had done for the steering of ships, added first and second order derivative terms to the basic control signal, thus obtaining:

$$e = m\frac{d^2h}{dt^2} + n\frac{d^3h}{dt^3} + p\frac{d^4h}{dt^4} - ka \tag{1.2}$$

where h is the angle of roll, a is the angle of the pump vanes, and e is the difference between the velocity of roll and the level of water in the tank. His experimental

Table 1.1 Chronoligical list of actual stabilisation installations

Date circa	Designer	Ship	Stabiliser	Control type	Control method
1870		Almost all	Bilge keels	Passive	None
1880	Bridge (England)	Colussus, Edinburgh	Tanks (slosh)	Passive	?
1883	Watts (England)	Inflexible	Tanks (slosh)	Passive	Tuning and damping
1891	Thornycroft (England)	Cecile	Weight	Active	Position of weight proportional to effective wave-slope and roll angle
1906	Schlicke (Germany)	See-bar and others	Gyro	Passive	Tuning and damping
1909	Cremieux (France)	Channel steamer	Weight	Passive	Tuning and damping
1910	Frahm (Germany)	Ypiranga, Europa, and many others	Tanks (U-tube)	Passive	Tuning and damping
1912	Frahm (Germany)	Deutschland, Hamburg, and many others	Tanks (sea-ducted)	Passive	Damping
1915	Sperry (USA)	Worden, Conte de Savoia, and many others	Gyro	Active	Precessional velocity proportional to roll velocity
1924	Fieux (France)	French destroyer	Gyro	Passive	Tuning and damping
1925	Motora (Japan)	Mutsu Maru and others	Fins (variable angle)	Active	Manual control and ?
1929	Hort (Germany)	Fuchs, Rossarol	Weight	Active	Velocity of weight proportional to roll angle
1930	Hort (Germany)	Konige Louise	Tanks (sea-ducted)	Active	Velocity of fluid proportional to roll angle
1963	Kefeli (Italy)	Aviso Etourdi	Fins (variable area)	Active	Area of fins proportional to roll velocity

Table 1.1 *Continued*

1935	Hort (Germany)	*Prinz Eugen* and others	Tanks (U-tube)	Active	Velocity of fluid proportional to roll angle
1936	Denny-Brown (England)	*Isle of Sark, Queen Elizabeth*, and many others	Fins (variable angle)	Active	Angle of fins proportional to roll velocity
1938	Dutch Engineers (Holland)	?	Fins (hydrofoil keels)	Passive	None
1939	Minorsky (USA)	*Hamilton*	Tanks (U-tube)	Active	Velocity of fluid proportional to roll acceleration
1949	U.S. Navy (USA)	*Peregrine*	Tanks (U-tube)	Active	Velocity of fluid proportional to roll acceleration

18 *Control technology in the 1930s*

system foundered because vibrations caused by water hammer in the vane-shifting mechanism of the pump were picked up by the roll accelerometer and gave rise to unstable oscillations of the system. Before he could make modifications the ship being used for the experiments was recalled for active service (in 1940) and further work was abandoned.[46]

A third method of ship stabilisation, the active fin system, was developed in the UK between 1918 and 1939 by the collaborative effort of Brown Brothers, William Denny & Brothers and the Admiralty Research Laboratory. The active fin system operates on the same principle as the aircraft wing. Hydrodynamic action caused by the forward motion of the ship through the water generates vertical forces on the fins: if the profile and orientation of each of the fins is identical, then the vertical forces generated balance, and there is no effect on the roll of the ship. If, however, the orientation of each fin is changed or the profile is changed (for example, by the use of the equivalent of flaps or ailerons) then the vertical forces do not balance and a couple opposing the roll of the ship is produced.[47]

In the early Denny-Brown system the direction of roll was detected using a gyroscope and an on-off controller was used to position the fins. In later versions the control signal was obtained by combining signals from a position and a rate gyroscope and the on-off controller replaced by an oil-hydraulic servo (the ARL unit referred to above was used).[48] Active fin stabilisers are widely used on high speed ships but have the disadvantage that stabilisation is lost when the ship is stationary or moving at slow speed. Development of anti-roll fins continued after the 1939–45 war at the Admiralty Experiment Works, Gosport.

1.6 Automatic steering and auto-pilots

Automatic steering mechanisms based on the use of the gyroscopic compass also received attention during the 1920s and 1930s. Nicolas Minorsky, in a major theoretical paper published in 1922, proposed the method of control now known as PID (proportional + integral + derivative), or three term control.[49] Tests on a practical implementation of the system were carried out during the late 1920s but work was abandoned in 1930 when the US Navy withdrew its support because '...the operating personnel at sea were very definitely and strenuously opposed to automatic steering'.[50] Minorsky sold his patent rights to the Bendix Corporation.[51]

The major supplier of automatic steering gear for commercial use was the Sperry Company. The Sperry system was described and analysed by Minorsky in an article published in *The Engineer* in 1937 in which he showed that the anticipator, referred to by Rawlings of the Sperry Company as the 'dodge', introduced a term proportional to the yaw acceleration of the ship thus giving a control signal containing proportional and acceleration terms.[52] This is an effective combination and automatic steering systems based on this principle, manufactured by the Sperry Company, became widely used.

In one of those public events that draws attention to technical changes, Wiley Post flew a Lockheed Vega 5B around the world in July 1933. Post credited the success of the flight to 'Mechanical Mike', the new Sperry A2 autopilot that was installed in the aircraft. The *New York Times* called the flight 'a revelation of the new art of flying' adding that:

> By winning a victory with the use of gyrostats... Post definitely ushers in a new stage of long-distance aviation. The days when human skill alone, an almost birdlike sense of direction, enabled a flyer to hold his course for long hours through a starless night or over a fog are over. Commercial flying in the future will be automatic.[53]

Some commercial flying was already 'automatic': Eastern Airlines had fitted the previous version of the Sperry autopilot to their Curtiss Condors in 1932, and in 1934 United Airlines fitted the A2-autopilot to their Boeing 247 aircraft. Other companies, including Honeywell and Bendix, entered the market, and there were similar extensive developments in Germany with autopilots being developed by the Askania, the Siemens, and the Anschütz companies.

The A2-autopilot was a mixed pneumatic and hydraulic system: pneumatic pickoffs were used on the gyroscopes and the follow-up servomechanism was hydraulic. Subsequent developments in the USA moved rapidly to all-electric operation for gyroscope drive, pickoff, power amplification and actuation. Investigations covered the use of pneumatic, hydraulic and electrical schemes as well as mixed systems.

Directional guidance (autopilot) systems attracted the attention of the popular press. However, the success of long distance flights depended much more on autostabilisers which relieved the pilot of having to make constant adjustments to the controls to maintain the aircraft in straight and level flight. Development of autostabilisers began around 1910 and, in his extensive report of 1936, F. Haus listed 22 longitudinal and 12 lateral stabilising systems developed since 1912.[54]

Improvements to autopilots and autostabilisers during the 1930s were made on the basis of experiment and intuition; there was no theoretical underpinning. D. McRuer and D. Graham have argued that this remained so until the late 1940s and that the 'hands-off' flight between Stephenville, Newfoundland, and Brize Norton, England, which took place on September 22 1947 was a 'triumph of the tinker/inventor'.[55] This is not to say that there were no theoretical advances in the understanding of aircraft stability and control during the period, but that there was a disjunction between theory and practice. For example, in the UK, Bryan (1904, 1911) and later Bairstow and Melvill Jones (1913) contributed to a developing understanding of aircraft stability. Their work was extended by Gates (1924), Garner (1926) and Cowley (1928) of Royal Aircraft Establishment who began to take into account the effect of automatic feedback control; both Garner and Cowley made provision in their theoretical treatment for the time lag in the control actuator.

Furthermore, autopilot designers were not unaware of the theoretical developments. Meridith and Cooke, working at the Royal Aircraft Establishment, described in 1937 both the theoretical and practical developments that were taking place. The puzzle is the apparent lack of use of theoretical work. There were, of course, computational difficulties in pursuing calculations of dynamic stability — with or without considering the effect of the presence of feedback control — but, as McRuer and Duncan comment, designers of airplanes continued to disdain dynamic stability analysis even after the computational effort was reduced through the introduction of operator methods and design charts.[56] It is also surprising that Melvill Jones (1935) in a major survey of aircraft stability and control did not cover feedback control.

20   *Control technology in the 1930s*

W. Oppelt suggests more was known than was published partly because of military secrecy and partly because of an unwillingness to disclose techniques to commercial competitors.[57] Commercial organisations were willing to describe the actual devices (once patent protection had been granted) because it was in their interests to do so, but the release of information on design methods could be of assistance to competitors.[58] While both military and commercial secrecy may provide a reason for the slow spread of theoretical ideas, it is not the sole reason for aircraft builders eschewing dynamic stability calculations. A more plausible and compelling reason is that they did not need to make such calculations: the problems, even the difficult problems, could be solved by trial and error, intuition, and experience. Performance requirements were such that there was no need to study the interaction between the aerodynamics and control surface feedback loops; it was sufficient simply to add the feedback controllers. This position did not change until after the Second World War — McRuer and Graham suggest 1947–1948 as a watershed.[59]

1.7 Summary

As at any time of rapid technical change the understanding of the problems and the adoption of new ideas, devices, and techniques varied greatly, both between and within various application areas, and there is no single definitive path of innovation. The experience of A.L. Whiteley when he joined the Industrial Engineering department of the British Thomson-Houston Company in September 1930 epitomises the general state of industrial control technology in the early 1930s. He recalls that the department designed electrical drive systems for the steel and paper manufacture, for printing, and for mine winders, lifts, cranes, excavators and machine tools. Design was done by 'control gear engineers' and amounted to no more than the layout of the system using standard relays and contactors. Occasionally, but with great reluctance, some design and development work might be done on a new relay or contactor. Control was largely open loop and there was heavy reliance on the inherent properties of the device or system: automatic control frequently meant little more than remote control not involving feedback. Remote control was employed widely in the electricity distribution system for the operation of sub-stations. It was used experimentally in aircraft, and a number of radio-controlled aircraft were built, particularly for military use. The General Electric Company exploited their Selsyn system in a wide range of remote control applications including theatre lighting control; the system they designed and installed in the Rockefeller Center Theater, New York, even allows for a range of preset sequences. It is arguable that remote control was a necessary stage before the widespread adoption of automatic control. Through remote control applications both instruments and actuators were developed, and engineers became familiar with and confident in handling conversion between electrical systems and systems involving mechanical, pneumatic and hydraulic devices.

The major concern was with regulation, the maintenance of some process at a steady value — speed of a motor, voltage, frequency, flow, pressure — and feedback was avoided by a reliance on the inherent properties of the device (this sometimes involved modifications to the design of the device to improve its natural regulation). Generators were designed to provide constant voltage or constant current output.

If the regulation was not adequate for the application then attempts were made to modify the generator to improve its inherent regulatory properties: the use of closed loop feedback control was seen as a last resort.

Most electrical drives relied on inherent speed regulation. Variable speed drives were avoided whenever possible; for example there was heavy reliance on the use of induction motors, the process, at whatever disadvantage, being trimmed to suit. If the trimming required was too great then the speed was changed by the use of gears. In applications which used feedback the normal approach was to use an on-off controller; continuous action control devices were seen as a last resort. The major reason for this was the difficulty in amplifying electrical signals and that of converting an electrical signal into a form in which it could operate an actuator. The simplest and most reliable power amplifier available was the electro-mechanical relay (or variations of it in the form of contactor mechanisms). It is therefore not surprising that the thyratron and the photoelectric cell were seen to have wide application. The photoelectric cell, the 'magic eye', was seen as adding sight to automatic devices and bringing with it 'a vast range of future economic effects'. Its potential in modern industrial manufacturing — seen as the use of continuous processes, automatic operation, registering devices, and controlling devices — was thought to be enormous.[60] Whiteley recalled being told that the electric eye should have wide applications and was charged with finding such applications.[61] He had little to guide him except the patent information from the General Electric Company (USA) (the British Thomson-Houston Company's parent company) and from this source he learned of the photoelectric relay and the thyratron. During the 1930s he developed for the company several applications for the thyratron, particular for the control of electric arc welding.[62]

The attraction of the photoelectric cell (together with other newly emerging electronic devices) was the speed of operation and the absence of mechanical inertia factors common. It was natural for the photoelectric cell to be thought of as a replacement for the ubiquitous relay used both as a sequence control device and as an amplifier as this was the largest market. M.E. Behar commented that engineers 'pounced eagerly on this means of actuating...valves...in quick and obedient response to the feeble impulses of sensitive primary elements. But let us...not be carried away...for serious problems remain to be solved...you can get wonderful on-and-off action...but you can't get exactly proportional amplification'.[63] The device could, however, be used other than as an on/off controller and it was widely used for 'talkies'. The edge of the film carried the sound track in the form of a varying width opaque strip, through which light was passed, falling on a photoelectric cell; the intensity of the light received varied with the width of the opaque strip and hence modulated the output of the photocell which was amplified and passed to a loudspeaker. The ability of the photoelectric cell to provide an electric signal proportional to the intensity of light falling on it led to proposals for its use in closed loop control systems. A.J. McMaster suggested that it could be used with an optical radiation pyrometer for temperature control, and two Westinghouse engineers, C.A. Styer and E.H. Vedder made several proposals, including its use for the control of chemical processes, where the reaction involved a change in colour or opacity.[64]

The Metadyne and Amplidyne provided the control engineer with a power amplifier which, coupled with the rapidly developing electronic amplifier, enabled the construction of a wide range of electro-mechanical control systems. Control of hunting was still a problem. The compensating winding poses stability problems if the devices are used to amplify ac signals, and under certain circumstances it may result in positive feedback to the control circuit and hence instability.[65] Solutions to stability problems were still largely based on the trial and error application of dashpots, although other methods were being tried.

At the beginning of the 1930s engineers tackling technical problems for which the adoption of closed-loop (feedback) control was a possible solution were not helped by the absence of clearly defined concepts and a shared language through which concepts, ideas and designs could be exchanged. It is arguable that one of the major of the achievements of both H.L. Hazen and H.S. Black was to clarify the concepts and language of automatic feedback control. F.D. Waldhauer claims that 'Black's diagram and equations were central because they established a *language* with which to talk about feedback systems'.[66]

The word feedback illustrates the difficulty: it was introduced by radio engineers and their usage implied positive feedback; it was only after the publication of Black's paper on the negative feedback amplifier that the significance of the distinction between positive and negative feedback was clearly understood. Similarly, 'stability' to the radio engineer implied stable oscillations of a frequency generator, whereas to a mechanical engineer dealing with governors it implied the lack of oscillation.

Similarly there was confusion between the effects of open-loop and closed-loop control: they were both subsumed in the term automatic control and frequently the differences were not recognised. There was perhaps a greater willingness to accept and commend open-loop control than closed-loop control. In January 1931, the *Electrical Review*, commenting on an open loop, pre-programmed mechanism for setting the roll gap in a reversing, steel rolling mill, expressed the view that:

> automatic screw-down is an example of eliminating the uncertain human element from industry, and it may be an indication that the operation of steel mills entirely automatically is a future possibility.[67]

However, in March 1931, in discussing the fully automatic boilers fitted to the new Kirkstall power station in Leeds, it asked what was the point in having fully automatic control since one operator was still required. An operator provided with the instruments and remote control, it suggested, could achieve all that the fully automatic system achieved. It conceded, however, that a less intelligent operator with less training could perhaps be used with a fully automatic plant — an argument that was to be repeated many times — and that such a plant might be justified when low grade coal with great variation in quality was to be used. Clearly, automatic control is useful in difficult cases and it can reduce the effect of disturbances. However, the leader writer's preferred solution was to prevent the disturbances; provide a supply of coal with constant qualities — the quality of the coal being supplied to Kirkstall was high and consistent — and automatic control is not required.[68] This argument, keep the environment and the input constant, is a seductive argument, but it results in solutions that simply transfer the problem elsewhere. This is fine, if the transferred problem is simpler to solve. However, in many cases the failure to use feedback led to complex and expensive systems, and an example illustrating the problems that can be caused is given in Chapter 3. It concerns the attempts to be stable oscillators for wireless transmission.

Practising engineers were naturally reluctant to employ feedback. Not only was it a new concept which was unsupported by an established body of analysis and design techniques, but also its adoption in many applications depended on using the new, untested technology of electronic valves. It was many years before the electronic valve and electronic amplifier became accepted as reliable, robust devices, simple and cheap enough for general industrial use.

Some analysis could be done by applying Heaviside's operational calculus and the Routh-Hurwitz stability criteria. The Routh-Hurwitz criteria provided a test for stability but it did not provide an indication of how stable the system was or how to modify the system. Fritz Johnson of the General Electric Company recollects, 'we could solve for the roots of the characteristic equations in the systems, but if they were unstable, how to make them stable was something we simply didn't know how to handle. It was a matter of experiment. It was really amazing that we were able to get things to go, I guess'.[69]

However, during the 1930s the common problems of closed loop feedback systems were recognised, and attempts were made to identify appropriate mathematical models for such systems which could form the basis for analysis and design and which would help to unify the handling of control problems. The major contributions to this are described in the next three chapters.

1.8 Notes and references

1 That productivity increased dramatically during the first thirty years of the twentieth century is not in dispute: explaining why and how is more difficult and contentious. Economic analysis based on measurements of capital inputs does not fully account for the increase in productivity; it is argued that the excess productivity arose from economies of scale and 'added knowledge' (for a summary of the arguments see Rosenberg, N.: *Inside the Black Box: Technology and Economics*, (Cambridge University Press, Cambridge, 1982) pp. 24-5; Uselding, P.: 'Studies of technology in economic history', in *Research in Economic History, Supplement 1, Recent Developments in the Study of Business*, Robert E. Gollman (ed), pp. 195-202)
2 See, for example, Souvenir 25th Anniversary Brochure Panama-Pacific Exposition, San Fransisco, 1915, Bristol Company (Columbia). A similar theme ran through a long series of advertisements placed by the company in *Engineering Magazine* from 1905 through 1915. See also 'Recording and Index Thermometers', Taylor Instrument Company Brochure (1916) (Columbia)
3 'Bristol's Electric Time Recorder', The Bristol Company, Bulletin 138 (September 1912): 2 (NMAH Columbia)
4 Sobel, R., *The Age of Giant Corporations: a Microeconomic History of American Business 1914-1970* (Greenwood Press, Westport, Connecticut, 1972), p. 53. See also Woolf, Arthur G.: 'Electricity, Productivity, and Labor Saving: American Manufacturing, 1900-1929', in *Explorations in Economic History*, **21**(2) (1984), pp. 176-191
5 Young, J.M.: 'Automatic remote control of boilers', *J. Institute of Fuel,* **5** (1931-2), pp. 217-223
6 In 'floating control' there is no fixed relationship between the error signal and the actuator (rheostat in this case) position. The actuator 'floats' and can take up any position between fully open and fully closed for any given value or error. The action is purely integral.
7 For a detailed account of voltage regulators in use in the early 1930s see Borden, P.A., Behar, M.F.: 'Automatic control of voltage and current', *Instruments,* **v.9** (1936) pp. 201-210, 235-238, 259-262. See also West, C.P., Applegate, T.N.: 'A generator-voltage regulator without moving parts', *Electric Journal,* **33** (1936), pp. 181-183
8 Hull, A.W.: 'Hot-cathode thyratrons', *General Electric Review,* **32** (1929), p. 394
9 See, for example, Kjolseth, K.E.: 'An automatic electrode regulator for three-phase arc furnaces', *General Electric Review,* **37**(6) (1934), pp. 301-303
10 Gulliksen, F.H.: 'Electronic regulator for ac generators', *Trans. American Institute of Electrical Engineers,* **53** (1934), pp. 877-881, discussion pp. 1530-1

24 Control technology in the 1930s

11 Whiteley, A.L.: 'Application of the hot-cathode grid-controlled rectifier or Thyratron', *J. Institution of Electrical Engineers*, **78** (1936), pp. 516–539
12 Whiteley's work was not published until after the war: (Whiteley, A.L., 'Theory of servo systems with particular reference to stabilisation', *J. Institution of Electrical Engineers*, **93** Pt.II (1946), pp. 353–367 discussion pp. 368–372. He included the material in a series of lectures he gave to designers at British Thomson-Houston Company, Rugby, in the early part of 1944. Extensive investigations into both active and passive networks were carried out by F.C. Williams at the Telecommunications Research Establishment in connection with radar developments during the early part of the war. Information about the work was not made public until 1946. (See Williams, F.C., 1946, p. 261)
13 Sectional drives made an important contribution to the increase in productivity in the paper making industry. Although Westinghouse was the first to introduce automatic control on such machines, by 1924 the General Electric Company had captured 75% of the market. See Lorant, John H.: *The Role of Capital-Improving Innovations in American Manufacture During the 1920s* (New York, 1975), pp. 132–137, for a discussion of economic effects of introduction of machinery into the paper making industry.
14 Various versions of this system were introduced between 1923 and the late 1930s. I am indebted to Charles Beck of General Electric for providing copies of unpublished surveys of paper-making machinery developments written by him and by H.W. Rogers.
15 Liston, J.: 'Developments in the electrical industry during 1932', *General Electric Review*, **36** (1933), p. 7. The National Committee for Industrial Rehabilitation was set up by President Hoover in 1932 and headed by A.W. Robertson. The purpose of the Committee was to put 1,620,000 people back to work through a nationwide drive to modernise equipment, see New York Times, (12th September 1932), p. 17:6. The actual effect on orders and industrial confidence of this committee was probably small as it was formed at a time at which Hoover had lost public confidence.
16 The electronic control circuit for the screw down mechanism was designed by A.L. Whiteley, who was working at the General Electric Company in Schenectady on a two year British Thomson-Houston Company Fellowship.
17 Charles Beck, unpublished paper, n.d. (see note 14 above).
18 *General Electric Review*, **33** (1935), p. 39. See also Hanna, C.R., Oplinger, K.A., Mikina, S.J.: 'Recent developments in speed regulation', *Trans. American Institute of Electrical Engineers*, **56** (1940), pp. 692–700
19 For a similar type of installation in the USA the General Electric Company used dc motor-generator sets.
20 Gravier, British Patent 1211, 1882; Rosenberg, British Patent 17,423, 1904 and British Patent 28,350, 1907; Osnos, M.: 'New lighting generators of the Felten and Guilleaume-Lahmeyerwerke', *Elektrotechnische Zeitschrift*, **38** (1907), p. 917. For further details see Tustin, A.: *Direct Current Machines for Control Systems* (Spon, London, 1952), pp. 162–164
21 Pestarini, J.M.: 'The theory of the dynamic operation of the Metadyne', *Revue Générale de l'Electricité*, **27** (1930), p. 355, p. 395; 'Metadynes and their derivatives', *Revue Générale de l'Electricité*, **28** (1930), p. 813, 851, 900; *Metadyne Statics*, (Wiley, New York, 1952)
22 See papers by Alexanderson, E.F.W., Edwards, M.A., Bowman, K.K.: 'Dynamoelectric amplifier for power control', Fisher, A.: 'The design characteristics of Amplidyne generators', and Shoults, D.R., Edwards, M.A., Crever, F.E.: 'Industrial applications of Amplidyne generators', all published in the *Trans. American Institute of Electrical Engineers*, **59** (1940), pp. 937–9, 939–44, 944–9
23 The company was working on fire-control computers for the Navy, remote control of aircraft, gyroscope stabilised gunsights, radio and eventually radar systems. The depression, despite the National Industrial Recovery Act, was seriously affecting GE but 'certain parts of the company had money to spend: military applications' see Interview with Frithiof 'Fritz' Johnson, Oral History, Hall of History Committee, General Electric Co., interview date 29th August 1981
24 See, for example, Tower, B.: 'An apparatus for providing a steady platform for guns etc., at sea', *Trans. Institution of Naval Architects*, **30** (1889), pp. 345–361; *The Engineer*, **89** (1900), p. 344
25 For information on battle tactics and fire control see Sumida, Jon Testuro: 'British capital ship design and fire control in the Dreadnought era: Sir John Fisher, Arthur Hungerford Pollen, and the battle cruiser', *Journal of Modern History*, **51**(2) (1979), pp. 205–230 and Garcia y Robertson, R.: 'Failure of the heavy gun at sea', *Technology & Culture*, **28**(3) (1987), pp. 539–557

26 Hartree, D.R.: *Calculating Instruments and Machines* (Cambridge University Press, Cambridge, 1950), p. 53
27 An early reference to such calculators is in 'Coast defenses of the United States', *Scientific American*, **106** (1912), p. 438
28 Hughes, T.P.: *Elmer Sperry: Inventor and Engineer* (Johns Hopkins Press, Baltimore 1971), pp. 232-233
29 A Mr E.T. Hanson developed a plan position predicting instrument for Anti-Aircraft guns in 1922 (ADM 204/192). In 1926, suggestions for improving this instrument by adding improved techniques for transmitting the aiming data to the gun layers were made (PRO, ADM 212/51). The system was still being refined in 1936 (ADM 212/72). See also Scott, J.D.: *Vickers: a History* (Weidenfeld and Nicolson, London, 1962), p. 113
30 Bell, J.: 'Data-transmission systems', *J. Institution of Electrical Engineers*, **94** (Pt.IIA) (1947), p. 222
31 *Ibid.* p. 224. Development of these devices continued in Germany between the wars. See BIOS Report No. 864
32 Hall, T.: 'Distant electrical controls for military searchlights', *General Electric Review*, **22** (1919), pp. 718-21; see also Hughes, T.P., *op. cit.* (n. (28) above), p. 220; the Selsyn was not used by the Admiralty Research Laboratory in Britain until the mid-1930s
33 'Review of progress in the electrical engineering industry', *General Electric Review*, **30** (1927), p. 55. The General Electric Company trade name was Selsyn, similar devices were known as Magslips in the UK.
34 Corby, R.A.: 'The versatility of application of Selsyn equipment', *General Electric Review*, **32** (1930), p. 706
35 Reichel, W.A., Sylvander, R.C.: 'Autosyn application for remote indication of aircraft instruments', *J. Aeronautical Science*, **6** (1939), p. 464
36 PRO ADM 218/7 Dr A.B. Wood's papers
37 Gairdner, J.O.H.: 'Some servo mechanisms used by the Royal Navy', *J. Institution of Electrical Engineers*, **94** (Pt. IIA) (1947), pp. 209-213; see also PRO ADM 204/251, 'The A.R.L. system of stabilised searchlight control', Feb. 1930. The feed-forward scheme was devised by J.M. Ford and H. Clausen, private communication from J.F. Coales. See also Wood, 1965; (see note 39 below)
38 PRO ADM 204/251; by the 1920s the efficacy of dither in reducing the effects of friction appears to have been widely known. For example, it was used by Bush and Hazen on the Cinema Integraph (see Chapter 4), although in a later paper Hazen remarked that at best its use 'reduces one evil at the cost of another' (Hazen, H.L.: 'Theory of Servomechanisms', *J. of the Franklin Institute*, **21** (1934), p. 283). Dither was patented in 1892 by Charles Parsons, and the specification stated that 'This freedom of movement [of the control valve] is very much assisted by the reciprocating movement of all the parts, which neutralises all frictional resistance', (British Patent No. 15677, 1892)
39 Wood, A.B.: 'From board of invention and research to Royal Naval Scientific Service', *J. Royal Naval Scientific Service*, **20** (1965), p. 59
40 See, for example, Ward, Roswell H.: 'Anti-aircraft Gun Control', *Army Ordnance*, **XI** (1931), pp. 452-457. Report of French system appeared in Algrain, P.: 'Calculating machine for directing anti-aircraft fire', *Rev. de l'Armée de l'Air*, No. 58 (1934), pp. 68-613. See also *Aeron J.* (1935), p. 511
41 For accounts of the early work see Bennett, S.: *A History of Control Engineering 1800-1930*, (Peter Peregrinus, Stevenage, 1979), pp. 123-33; an excellent account of Sperry's work is given in Hughes, T.P., *op. cit.* (n.28 above)
42 A historical review of ship stabilisation is in Chadwick, J.H.: 'On the stabilisation of roll', *Trans. Naval Architects and Marine Engineers*, **63** (1955), pp. 234-280. It gives a review of findings relating to the *Conte di Savoia* and other references to the ship and its performance are: 'The gyroscopic stabilising equipment of the Lloyd Sabado liner *Conte di Savoia*', *The Engineer*, **153** (1932), pp. 32-35, 62-65; Schilovsky, P.: 'The gyroscopic stabilisation of ships', *Engineering*, **134** (1932), pp. 689-690, discussion **135** pp. 53-54, 107; Hodgkinson, R.P.: 'Stabilisation of *Conte di Savoia*', *Engineer*, **158** (4109) (Oct. 1934), p. 369
43 Frahm, H.: 'Results of trials of the anti-rolling tanks at sea', *Trans. Institution Naval Architects*, **53** (1911), pp. 183-216
44 See, for example, Hort, H.: 'Pneumatic stabilising system', *Shipbuilding & Shipping Record*, **44**(20) (Dec. 1934), p. 647; Rellstab, L.: 'Activated anti-rolling tank system', *Engineer*, **157**(4094) (June 1934), pp. 648-650

45 The work was reported in Minorsky, N.: 'Note on the angular motion of ships', *Trans. American Society of Mechanical Engineers*, **63** (1941), pp. 111–120; and Minorsky, N.: 'Experiments with activated tanks', *Trans. American Society of Mechanical Engineers*, **69** (1947), pp. 735–747
46 Minorsky carried out further investigations on active tank stabilisation systems at Stanford University after the war. For a brief account of Minorsky's career see Flugge-Lotz, I.: 'Memorial to N. Minorsky', *IEEE Trans. Automatic Control*, **AC-16** (1971), pp. 289–91
47 Passive fins were tried first but to have any appreciable effect on the rolling motion, the size of fin required was such as to make them impractical.
48 Allan, J.F.: 'The stabilisation of ships by activated fins', *Trans. Institution of Naval Architects*, **87** (1945), pp. 123–159
49 Minorsky, N.: 'Directional stability of automatically steered bodies', *J. American Society of Naval Engineers*, **34** (1922), pp. 280–309
50 Captain H.S. Howard, discussion of paper by Henderson, J.B.: 'The automatic control of the steering of ships and suggestions for its improvement', *Trans. Institution of Naval Architects*, **76** (1934), p. 30
51 For a full account of Minorsky's work on the automatic steering of ships see Bennett, S.: 'Nicolas Minorsky and the automatic steering of ships', *IEEE Control Systems Magazine* (November 1984), pp. 10–15
52 Minorsky, N.: 'The principles and practice of automatic control', *The Engineer* (1937), pp. 322–3; the Sperry automatic steering mechanism including the anticipator is described in Hughes *op. cit.* (n.28 above), pp. 280–2
53 Quoted from McRuer, D., Graham, D.: 'A historical perspective for advances in flight control systems', *AGARD Advances in Control Systems*, No. 37 (1974), pp. 2.1–2.7. This section is based largely on the above paper and the following papers: McRuer, D., Graham, D.: 'Eighty years of flight control: triumphs and pitfalls of the systems approach', *American Institute of Aeronautics and Astronautics, J. Guidance and Control*, **4** (1981), pp. 353–362 (a revised version of the 1974 paper); Oppelt, W.: 'A historical review of autopilot development, research, and theory in Germany', *American Society of Mechanical Engineers J. Dynamic Systems, Measurements, and Control*, **98** (1976), pp. 215–223; Draper, C.S.: 'Flight control', the 43rd Wilbur Wright Memorial Lecture, *J. Royal Aeronautical Society*, **59** (1955), pp. 451–477; Bollay, W.: 'Aerodynamic stability and automatic control', *J. Aeronautical Sciences*, **18** (1951), pp. 569–624
54 Haus, F.: 'Automatic Stability', *NACA Tech. Memorandum*, No. 802 (1936) (translated from French); see also Haus (1932)
55 McRuer, D., Graham, D. (1981) *op. cit.* (n.53 above), p. 357
56 McRuer, D., Graham, D. (1981) *op. cit.* (n.53 above), p. 356
57 Oppelt, W. *op. cit.* (n.53 above), p. 220
58 There is extensive patent literature on aircraft control devices in the 1930s and descriptions were published in journals such as *Scientific American, Flight, Engineering* and *The Engineer*.
59 McRuer, D., Graham, D. (1981) *op. cit.* (n.53 above), p. 358; in a discussion reported in the *J. of the Royal Aeronautical Society* (1949), p. 279, F.W. Meridith expressed the view that prediction of performance depended on the analysis of aeroplane, automatic pilot and power controls; another contributor stressed the need for power controls to be stiff, i.e. to have a high gain and low steady state error.
60 Gillian, S.C., p. 24 in Ogburn, W.F., Merriam, J.C., Elliott, E.C. 'Technological trends and national policy: including the social implications of new inventions' Report of Subcommittee on Technology to the National Resources Committee (GPO, Washington, DC, 1937)
Gillian was a former curator of Social Sciences, Museum of Science and Industry, Chicago.
61 Whiteley, A.L., private communication
62 Whiteley, A.L. (1936)
63 There were many references to its use as a relay. A typical article is 'New device uses light to control machinery', *General Electric Review*, **33**(7) (1930), p. 398, and the Engineering Index for 1930 lists 41 entries dealing with the photoelectric cell used as a relay. Behar's comments appeared in an editorial in *Instruments*, **4** (1931), p. 525
64 See, for example, McMaster, A.J.: 'Photoelectric cells in chemical technology', *Industrial and Engineering Chemistry*, **22** (1930), pp. 1070–3 and Styer, C.A., Vedder, E.H.: 'Process control with the electric eye', *Industrial and Engineering Chemistry*, **22** (1930), pp. 1062–1069

65 Alexanderson, *et al.*, *op. cit.* (n.22 above), p. 938
66 Waldhauer, F.D.: *Feedback*, (Wiley, New York, 1982), p. 4
67 'Automatic adjustment for steel-mill rolls', *The Electrical Review*, **CVIII** (Jan. 1931), p. 38
68 *The Electrical Review*, **CVIII** (March 1931), p. 537
69 Johnson, F., *op. cit.* (n.23 above), p.5

Chapter 2
Process control: technology and theory

Prior to the World War [1914-1918], European instrument manufacturers held an important position in the field of industrial instrumentation. To apply automatic control to industrial processes at that time was a daring procedure. Today America leads the world in the art of industrial measurements, and European engineers who have come to our shores for the purpose of studying our methods have been astounded at the widespread use of automatic control equipment, and the simplicity and directness in which American industry solves its automatic control problems.

H.M. Schmitt, 1937.

2.1 Introduction

During the nineteenth and early part of the twentieth century there was sporadic use of automatic controllers in the process industries — metals, power generation, petroleum, chemicals, food, textiles, paper and pulp, glass, ceramics, brewery and distillery, sugar, lumber, paint and varnish, and others — but such use did not become widespread until the mid 1920s.[1] A survey carried out for the United States government in the late 1930s found that the sales of industrial instruments had grown rapidly, in both absolute terms and relative to other forms of machinery, between 1919 and 1929. And although the absolute value of sales fell sharply during the Depression, the relative share continued to increase, changing from approximately 0.4% of machinery sales in 1919 to 1.4% in 1935, and peaking at almost 1.6% in 1933. American industry spent over $300 million on instruments between 1920 and 1936.[2] The survey covered indicators, recorders and controllers and found a steady rise in the proportion of controllers sold, from 8% of total sales in 1923 to 32% in 1935, and from about 1932 onwards between 40% and 50% of all new instruments introduced were controllers. I estimate that over 75,000 automatic controllers were sold by the American instrument companies between 1925 and 1935.[3]

In the USA several industrial instrument manufacturing companies were formed around the turn of the century, and a number of scientific instrument makers began to produce instruments suitable for industrial use. In England and also in Germany similar instrument companies were developing. In the early years of the century they concentrated on improving the measuring elements of indicators and developing recorders. By the mid 1920s accurate, reliable, and cheap indicators and recorders were readily available for the more commonly required measurements: temperature, pressure and flow.[4] During the 1920s design leadership in industrial instruments passed from Europe to the USA and by the mid 1930s there were over 600 companies in the USA manufacturing and selling industrial instruments. Of these seven dominated the market with combined sales amounting to 65% of total sales.

Early users of instruments were the power generation companies, seeking to improve the efficiency of steam generation and interested in CO_2 emissions: the automobile industry for heat treatment of parts, the dairy industry for pasteurisation of milk, paper manufacturers in measurement of relative humidity, and the chemical industry for a wide range of activities.[5] The need to measure temperatures, pressures and flows was common to a wide range of industries; of these, temperature measurement was the predominant requirement. The ability to measure a quantity is prerequisite to controlling it, and in order to understand the way in which controllers developed we need to first examine briefly the development of measuring (recording) instruments during the early years of this century.

2.2 Measuring and recording instruments

Two major classes of temperature measuring instruments emerged: those capable of measuring up to 800°F (commonly referred to as thermometric instruments), and those for higher temperatures (referred to as pyrometers). Thermometric devices are based on the expansion principle. A metal tube containing a liquid or gas is connected by a capillary tube to a bellows or some form of Bourdon tube — a slightly flattened tube bent into a curve — the free end of which moves as the pressure changes.[6] A temperature change at the bulb results in a pressure change which is converted by the bellows (or Bourdon tube) into a small mechanical movement. This movement has to be amplified using mechanical levers in order to operate indicators and recorders. Tube shapes, materials, methods of construction, and means of connection to the recording or indicating arm were investigated extensively with the aim of improving the accuracy of the measurement and pressure range. Industrial pyrometers were based either on the thermocouple principle or the change in electric resistance.[7] In both cases a bridge circuit was used to detect the small changes in potential which were converted to a mechanical movement using a moving coil galvanometer. The galvanometer was used either to move an indicator arm or as a null sensing element. Instruments based on the former are referred to as deflection instruments; those on the latter as potentiometric instruments.

For many reasons — legal, scientific study, quality control, 'policing the work force', or simply fashion — managers wanted recorders as well as indicators. Early recorders include those produced by the Bristol Company which were introduced in the 1890s and extensively promoted in management journals.[8] For thermometric instruments it was not difficult to develop helical spring movements that generated sufficient force to overcome pen drag and which did not load the transducer and hence distort the reading. Producing pyrometric recorders was more difficult. The accuracy of the measurement is dependent on the linearity and sensitivity of the galvanometer. Industrial galvanometers were produced that had greater torque and better damping than the laboratory versions but they were still unable to provide sufficient torque to prevent recorder pen drag affecting the accuracy of the measurement.[9] To avoid pen drag loading, instruments were developed in which the pointer arm was allowed to swing freely and was only brought into contact with the recording paper at intervals by means of a depressor bar. In 1905 the Cambridge Scientific Instrument Company introduced its 'Thread Recorder' based on this principle,[10] and in 1915 the Bristol and Taylor companies introduced similar recorders. They continued to manufacture them until the early 1930s.[11] Many users disliked the curved grid on the recording paper of deflection instruments (it was needed because the indicating arm necessarily swings in an arc) and for precise work preferred the more expensive potentiometric recorder.

In potentiometric instruments some means has to be used to generate a voltage to match that of the thermocouple. The most commonly used method was to change the position of a contact on a slide wire placed in one arm of a bridge circuit. The position of the contact is a measure of the thermocouple output and hence the temperature. The principle of this mechanism is simple but implementation is complex and difficult. The position of the galvanometer has to be detected without disturbing its movement and then the slide wire contact has to be accurately positioned. The instrument manufacturers were faced with solving a position control (servomechanism) problem.[12]

At the beginning of the century the only available potentiometric recorder was the Callendar recorder, manufactured by the Cambridge Scientific Instrument Company and marketed in the USA by Taylor Instruments. The Callendar recorder was originally developed for laboratory use and the first potentiometric recorder devised for industrial applications was designed by Morris E. Leeds in 1911. The Leeds and Northrup Company, first with the Leeds recorder, and then the Micromax (1930) and the Speedomax (1932), dominated this part of the market until the end of the 1930s. The success of their recorders was largely the result of the proportional action in the follow-up mechanisms where other companies used on-off action. In the original Leeds recorder the galvanometer needle was periodically clamped and its position sensed by 'mechanical fingers'. The position of the needle relative to the fingers was used to operate a follow-up mechanism that attempted to position the fingers centrally over the needle. The fingers thus followed the movement of the galvanometer. The Micromax was an improved mechanical version but the Speedomax used electronic amplification of the out-of-balance signal and thyratrons to operate the motors.[13] The Bailey Meter Company introduced its Galvatron recorder, which combined an electronic amplifier and a magnetic amplifier, in 1934 and in 1937, the C.J. Tagliabue Company introduced a recorder-controller—the Celectray—based on using a photoelectric cell to detect the galvanometer deflection, followed by use of electronic amplifier and thyratron drives.[14]

By the mid 1930s, all the major instrument companies were manufacturing potentiometric recorders and all had adopted the principle of intermittent action. The galvanometer was allowed to swing freely on its suspension but at periodic intervals a depressor bar mechanism held it fixed while its position was sensed and the follow-up mechanism corrected for any movement since the last reading. During the 1930s shaped sensors were gradually introduced whereby the size of the deviation could be determined and a correction proportional to the deviation applied.

For installations with a large number of instruments, central instrument or control rooms became popular during the 1920s. One operator or supervisor decided, on the basis of the instrument reading, if a valve or other actuator should be opened or closed. Then either someone was sent to adjust the valve or actuator, or the supervisor changed the colour of a light to indicate to the furnace operator or another worker that a change was required. Typical of such arrangements were the heat treating systems sold by the Leeds & Northrup Company during the early 1920s. A supervisor in a central control room observed the temperature records of the heat treatment furnaces and passed commands to the furnace operator to increase or decrease the heat in the furnace by placing pegs in a switchboard and thus illuminating different coloured lights placed by the furnace. The furnace operator adjusted the heat according to the colour of the lights: red meant reduce the heat; green meant increase the heat, and white meant no change. It was not long before instruments were introduced first to operate the lights directly and then to operate the valves which controlled the supply of heat to the furnace, thus making the furnace operator redundant.

2.3 Introduction of automatic controllers

The early automatic controllers were, with a few exceptions, of three types: electrical relay with a solenoid operated valve which gave on-off action, electrical relay with motor operated valve which gave floating control action, and pneumatic relay with a diaphragm valve which initially gave on-off action but was soon modified to give the so-called narrow-band proportional action (typical controllers had proportional bands of between 1% and 5%).

The pneumatic relay was either a single stage device using a poppet valve operated directly from the measuring device, or a two stage relay with a flapper-nozzle amplifier preceding the poppet valve. Figure 2.1 illustrates the principle of the latter arrangement. Organ manufacturers working at the beginning of the nineteenth century had used pneumatically operated relays, and a patent for a temperature regulator granted to W.S. Johnson in 1885 shows a flapper-nozzle arrangement.[15] Their use in industrial instruments stems from the work of E.H. Bristol of the Foxboro Company during the winter of 1913–14 and a patent application for the device was filed during 1914.[16] The original prototype is shown in Figure 2.2. Figure 2.3 shows the general arrangement of the flapper-nozzle amplifier based controller in the form in which it was introduced in 1915. Extensive improvements to this device were made during the period 1920–1930. In the early controllers a single bellows was used and in 1922 the design was modified with a new control head (the bellows and pilot valve are referred to as the 'Control Head' in Foxboro literature). The main change was to replace the single diaphragm with a double bellows arrangement (F in Figure 2.4). The general form of construction is shown more clearly in Figure 2.5 which shows the 1929 version of the controller. The temperature bulb is connected to a standard Foxboro helical tube transducer by a capillary tube. Attached to the outside of

32 *Process control: technology and theory*

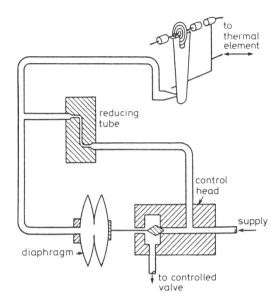

Figure 2.1 *Schematic diagram of an on-off controller with bellows operated pilot valve*

Reproduced (with partial redrawing) by permission of T.J. Rhodes, from Industrial instruments for measurement and control (McGraw-Hill, New York, 1941), p.485

Figure 2.2 *Prototype of Foxboro air-operated controller circa 1914*
Smithsonian Institute, Photo No. 89-267

Process control: technology and theory   33

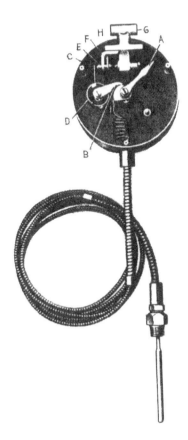

Figure 2.3   *Foxboro temperature controller circa 1915*
Reproduced from Foxboro Bulletin 96 (1915), p.20

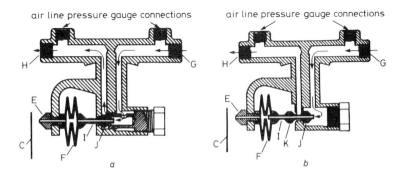

Figure 2.4   **Cross section of control heads**
Reproduced (with partial redrawing) from Foxboro Bulletin 112-1 (1924), p. 5

34 *Process control: technology and theory*

Figure 2.5 *Interior view showing the construction of an automatic temperature recorder of 1929*

Reproduced from Foxboro Bulletin 112-2 (1929), p. 16

the spiral is the flapper arm. A bellows, fed from the back pressure from the nozzle, operates the pilot valve (see J in Figure 2.4).

The key to the success of the early Foxboro controllers was the helical tube element (see Figure 2.6) that was used to convert the pressure changes generated in the temperature bulb or other transducer into mechanical movement to operate either a recorder or controller. From the time of William Bristol's modification of the standard Bourdon tube element in 1890 instrument manufacturers invested in research and development to find improved materials and construction techniques for such elements.[17]

2.4 Problems of on-off control

From the initial on-off (relay) systems there was a gradual change to proportional, continuous control. The problems of hunting caused by on-off systems were well known: wear in the controls, sudden changes causing damage to plant, and low accuracy. The advantages were equally well known: familiarity, simplicity, and the fact that the continuous oscillation (dither) prevented or ameliorated friction in valves and other components. J.T. Hawkins, in 1887, warning against the use

Process control: technology and theory 35

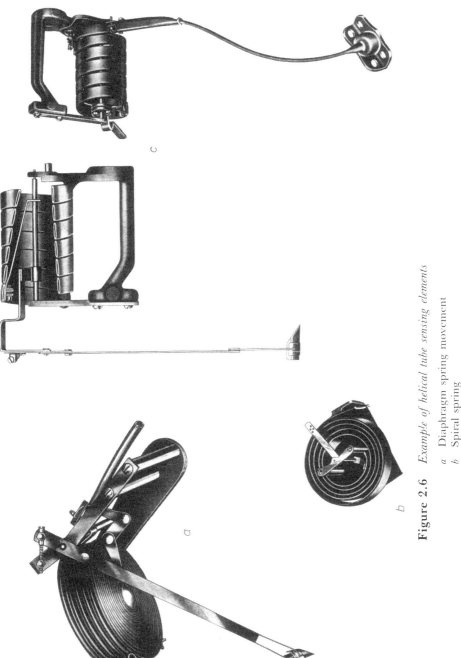

Figure 2.6 *Example of helical tube sensing elements*
  a Diaphragm spring movement
  b Spiral spring
  c Helical spring movement
  Reproduced from various Foxboro company publications

of thermostatically operated electrical contacts for automatic control of heating said, 'they are extreme in their operation, requiring a certain range of temperature to be submitted to in an apartment before a change will occur...the dampers are either wholly closed or wholly opened...the result is a continual succession of maximum and minimum temperatures in the apartment, instead of an equably maintained one...[the] fluctuations are quite beyond the limits of comfort'.[18] The problem was how to avoid oscillation but still maintain good control.

With close setting of the limits on the thermostat, and careful positioning in the room, Hawkins' concern about comfort could be answered. However, on-off control systems, even when providing acceptable control of the measured (controlled) variable, can cause other problems. In 1925, the Carrick Engineering Company explained that the constant fluctuations of the damper in response to the on-off control of boiler steam pressure 'subject the fuel bed first to excess air then to insufficient air.... Furnace efficiency and fuel economy are...sacrificed for pressure regulation.... The furnace brickwork [is damaged] because of sudden changes in furnace temperature.... Wear and tear on equipment is excessive because of the constant speeding up and slowing down of stokers, blowers and other equipment'.[19] Their solution was sound in principle — the use of proportional action — but in implementation it owed more to the nineteenth century than to contemporary technology. Steam pressure was measured by a large, mercury filled U-tube connected to the steam outlet pipe from the boiler. A float in the tube sensed the level and moved a lever on a hydraulic cylinder operating the dampers on the boiler. With proportional action and correct adjustment of the gains, hunting is easily prevented; however, if the load on the boiler (steam flow out) changes, another type of error, offset, occurs. (With a change in load, offset error is unavoidable because a proportional control system can give the correct damper position for only one value of the load.)[20]

By-pass valves were commonly used to mitigate the adverse effects of oscillating controller action. Only part of the controlled medium, steam for instance, passes through the control valve while the rest passes through a manually set by-pass valve. Rapid fluctuations of the control actuator thus produce less disturbance to the system. Alternative methods used included reducing the speed at which the control valve moves (slowing down the control valve movement reduces the gain of the system and can produce 'floating' control action), and increasing the hysteresis in the on-off action of relays used in electrical systems.

It is not difficult to get acceptable performances from on-off controllers applied to processes with large capacities and hence long time constants. However, during the 1920s, manufacturers were changing from batch to continuous processing methods. For example, the adoption of 'flash' pasteurisation of milk changed a batch process, in which a large quantity of milk was placed in a tank, heated, held at a constant temperature for 30 minutes and finally flowed out over a cooler, to a 'flow' process in which a continuous stream of milk passed through two sets of coils, where it was heated in the first set and cooled in the second. The consequence was to change a process which had a large time constant (arising from the size of the tank) to one with a small time constant (since the quantity of milk in the heating section of the pasteuriser at any one time is small).[21]

These changes in process techniques caused problems. The loss of the smoothing effect of large capacity, and hence the sudden changes induced by an on-off controller, became more apparent in the output and, more importantly, were fed back to the controller. And other time constants in the closed loop system became more significant. In processes with large capacity the dominant time constant is typically that of the process itself, while the time constants of the measuring system are insignificant. However, if the time constant of the main process is reduced then other time constants in the system, especially the time constant of the measuring device, may begin to have significant effects on behaviour. There are two effects: the actual inherent time constant of the transducer, and the time delay between a change in process condition and that change reaching the transducer. This time delay is sometimes referred to as a transport lag and is typical of flow processes where it is often difficult to locate a measuring device close to the point where the process action is occurring.

The response to these problems was diverse: the addition of damping devices to various parts of the control loop; increases in size of by-pass valves thus reducing the effect of the control valve; and reduction in the sensitivity of the controller, initially by the provision of two sensitivity settings — high and low — quickly followed by the development of the narrow-band proportional response controller.[22] The Foxboro Company modified their flapper mechanism and nozzle design thus converting their on-off controller to a narrow band proportional controller.[23] The Smoot Engineering Corporation introduced a proportional system for steam boiler control; and the Neilan Company developed a system with negative feedback round the pilot valve and with four settings of the proportional band.[24] Widening the proportional band reduced the likelihood of oscillation but there was a penalty to be paid. As a Foxboro bulletin of 1929 explained, 'close limits of control must be sacrificed, if throttling [proportional] action is desired when the process is out of balance'. In other words, if the proportional band is made too wide then a change in load results in substantial steady state error in the controlled variable.

The Foxboro Company later claimed (in 1933) that in developing their adjustable flapper arm they had 'recognised this inherent weakness in wide throttling range': in the development of their adjustable flapper movement they provided for a maximum range of 5% and employed auxiliary means of retarding the valve reactions, and 'in this manner did not completely sacrifice accuracy of control'.[25]

The Foxboro Company like others, was reticent about explaining how the valve was retarded and in the earlier description (1929) of the adjustable flapper arm there is no mention of the adjustment being designed to limit the band to 5%. Retardation of the valve can be produced by inserting a variable orifice bleed valve in the air line between the pilot valve thus providing floating action for the main valve. The bleed slows down the movement of the main valve. Both the Foxboro Company and The Bristol Company offered this as an option from 1929 to about 1936. The length of time during which this type of controller was offered for sale suggests it provided acceptable control.[26]

Electrically operated valves were widely used as alternatives to the air operated diaphragm valve. If they were operated by a solenoid then the overall control action obtained was on-off but if an electric motor was used to move the valve stem, a different form of control action, one in which the valve stem was said to float, was obtained. The rate of valve movement is set by the motor speed. By choosing a slow rate of movement relative to the natural frequency of the system then even if the controller is oscillating between its upper and lower limits the valve will oscillate only a small amount about a mean position. With a change in demand the controller spends a greater proportion of its time against one of the limits rather than equal times at each, and hence the valve moves slowly to a new mean position to reflect the change in load.

For processes with some inherent self-regulation, floating control action can give acceptable performance and avoid the oscillation of the control valve which occurs with on-off action. However, it results in a slow response to changes in either demand or set point. Increasing the control valve speed in an attempt to give a faster response results at best in an oscillatory response and at worst, because of the inevitable time lags and delays, instability.

The evidence shows that the instrument companies were successful in getting manufacturers to buy instruments, but how effective were such instruments, particularly controllers? Major E. Behar, editor of the journal *Instruments*, had no doubts about the efficacy of automatic control instruments and he enthusiastically supported their use,[27] but others were more cautious.

The inadequacy of the controllers being marketed led some process companies to develop their own systems.[28] For example, the Dow Chemical Company designed many controllers for their own use between 1924 and 1929.[29] They investigated various modifications to floating action controllers. In 1924 they introduced a two speed motor such that when the error was large the valve movement was fast, but once the error came within certain limits, a change was made to a slower speed. They later modified the system to hold the valve stationary when the error was reducing, thus producing a behaviour approximating to reset (integral) action.

In 1929 they produced a control block which could be attached between a standard potentiometric recorder and the control valve motor. The control block caused:

> ...the control motor to operate in the desired direction for a period of time which is determined by (1) the amount by which the reading is off, (2) the rate at which the reading is moving away from the right value, thus eliminating the effect of inventory lag, and (3) the time during which the control has been operating, in the direction it is operating at that moment, within a period of time measured by the time lag.[30]

Care is required in interpreting the above passage: it has to be remembered that what is being controlled is the length of time for which the motor runs in each two-second interval (the sampling time of the potentiometric recorder). The valve position is thus determined by the integral of the motor speed taken over each time step and thus in terms of the valve position, (1) above gives reset or integral of error action and (2) gives action proportional to error, while (3) gives an additional term which acts to speed up movement of the valve. J.J. Grebe, the director of Dow's Research Laboratory, claimed that when they replaced an on-off controller with this new controller they were able to reduce the speed of the actuating valve from 20 minutes for full stroke to 20 seconds. The slow speed necessary to prevent hunting was no longer required.[31]

The Dow controller belongs to a class of controller referred to as definite correction controllers. Other controllers of this type include an instrument produced by the Bailey Meter Company during the 1920s and the Leeds & Northrup controller introduced in 1921 in which the time interval during which the motor ran each cycle was made proportional to the error.

The measuring units in the Bailey system operated electrical contacts that connected electric motors, run at a constant speed, to the control actuators (valves, dampers, variable speed motors) so as to either increase (more) or decrease (less) the value of the manipulated variable. The system was sold mainly for steam boiler control operations. It is 'definite correction' in that every 10 seconds the measurement pointer is pushed against a stepped wedge. The wedge slides back to make an electrical contact which operates the appropriate motor to give 'more' or 'less'. The larger the error the longer the contact time hence the control action is proportional, albeit narrow band, as the wedge has a limited number of steps. The contact does not move back with the return of the wedge but is slowly returned by the 'anticipator' (moving through its full range in about one hour). Therefore, when the error begins to decrease, no further movement of the control valve takes place until the error changes sign. The overall control action is approximately proportional plus reset.

During the period 1925 to 1940 there are many references to anticipatory control. This was normally used with on-off control and valves operated by electric motors. The control signal included a term related to the rate of change of the error and the effect was to advance the switching point of the on-off controller, thus if the error was decreasing the controller switched before the error reached zero. Using anticipatory control meant that higher motor speeds could be used and hence a faster response obtained. The Leeds & Northrup Company introduced an *Anticipating* controller in 1930, and the Bristol Company its *Degree-Splitting Anticipatory* controller in 1931.[32]

Other attempts to modify in some way the signal applied to the on-off control unit include the Foxboro Company's *Deoscillator* unit (1934), offered as an addition to its potentiometric based controller; George Kent's *Gradient-Analyzing* controller; and the Brown Instrument Company's *Trend Analyzer* (1931).[33] The Foxboro *Deoscillator* unit generated a signal that was a function of both the current control action and the thermocouple reading potential, and this signal was added to the thermocouple potential.[34] The *Gradient Analyzing* and *Trend-Analyzer* controllers were based on detecting the change in the gradient of the error signal. In the Brown controller the recorder follow-up motor and the electric motor, used to operate the control valve, were connected in parallel. This arrangement forced the control valve to follow the movement of the recorder contact carriage and provided proportional action. An additional circuit provided reset action. If there was an error between the controlled variable and the set point the contact carriage of the recorder was re-positioned without changing the position of the control valve.[35]

2.5 Wide band proportional control

The primary cause of many of the problems involved with the control of low capacity processes was the high gain inherent in on-off and narrow band controllers and the solution lay in providing controllers having a wide, and adjustable, proportional band. The first of such controllers was the Foxboro Company's *Stabilog* introduced in 1931. The *Stabilog* was based on the work of Clesson E. 'Doc' Mason and W.W. Frymoyer.

Mason had observed the problems of on-off controllers while operating a small petroleum plant in the early 1920s. He joined the Tulsa office of the Foxboro Company in 1925, moving to Foxboro, Massachusetts, in 1929 and was director of control research from 1930 to 1941. W.W. Frymoyer, the factory superintendent at Foxboro and later also a director of the company, apparently independently, investigated ways of avoiding some of the problems of the narrow-band controller.[36] On August 14th 1928, two separate patents for pneumatic process controllers were filed, one by Mason and the other by Frymoyer. Both patents were eventually granted and assigned to the Foxboro Company.[37] A test system based on the Mason patent was constructed as a separate unit and was attached to the air output line of the flapper-nozzle amplifier of an existing controller. The unit was tried out at an oil refinery and was claimed to have worked well. Unfortunately the diaphragm units kept rupturing because of the repeated flexing and the unit had to be removed.[38]

Mason worked actively on the problem of producing a Wide Band recorder during 1930. He experimented with a method based on using two steam-valves placed in parallel. One was operated directly by the air-relay pressure output; for the other the air-relay pressure output was first passed through a capillary into a large 'tank' before it operated the valve — in other words the pressure output was processed by a resistance-capacitance network. A patent for this system was filed in May 1930.[39] Work on the problem continued and a few months later patents for two further ideas were filed. The first of these, filed jointly in the name of Mason and A.M. Dixon,[40] described a device which used two separate air relays to provide a coarse and a fine adjustment of a valve. The valve had specially shaped ports that provided a linear displacement-flow relationship over part of the valve movement and a much higher displacement-flow ratio when the valve was nearly fully open.

The second patent, in Mason's name alone, formed the basis for the development of the Foxboro *Stabilog*. The Model 10 *Stabilog* was announced in September 1931. Figure 2.7 shows its general structure and the principle of operation is illustrated in Figure 2.8. The importance of the *Stabilog* was that it provided a combination of Wide Band proportional action and reset or integral action to remove the steady state offset that occurs with simple proportional action. The mechanism that provided the combined action was called a differential pressure motor by the Foxboro Company (see Figure 2.7). It is formed from units 39, 53 and 55 in Figure 2.8. In the patent specification Mason shows that the pressure P_1 in chamber 39 is given by

$$\frac{dP_1}{dt} = \frac{d}{dt}(I/K) + (T - T_c^d)\frac{K_1}{K} \qquad (2.1) \text{ [eqn. 7 in patent]}$$

where K and K_1 are constants; T is measured temperature, T_c is the desired temperature and I would appear to be a misprint for T. It is easily seen since pressure P_1 is applied to the control valve that the controller provides proportional plus integral action. Later in the patent, Mason states that the change in pressure in 53 is 'proportional to the integral of the deviation in temperature relative to time' [p. 5 lines 37–39].

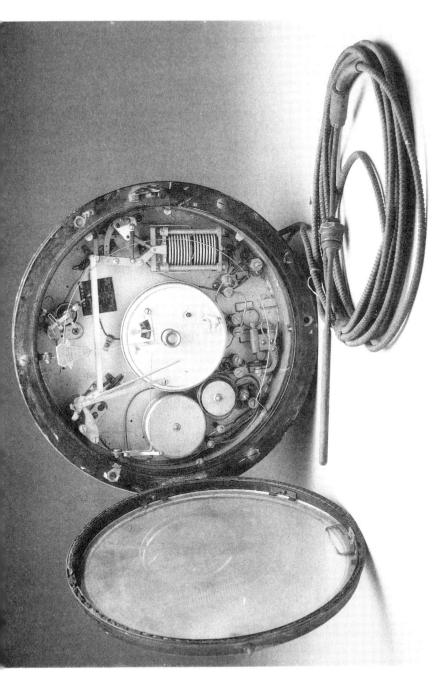

Figure 2.7 *Foxboro Stabilog controller*
Smithsonian Institution, Photo No. 86-600-11

42 *Process control: technology and theory*

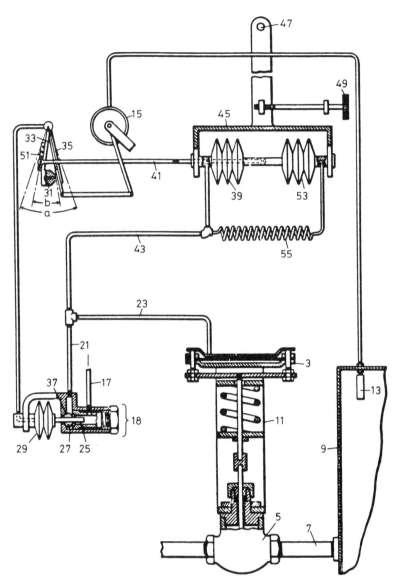

Figure 2.8 *Basic mechanism of the Stabilog diagram*
Mason patent 1897135 (1933)

The key to the practical success of the *Stabilog* was the use of the recently developed *Hydron* all-steel, welded bellows. These were used for the chambers marked 39 and 53 (Figure 2.8) replacing the fragile and unreliable diaphragms used in the test system. As in the previous range of Foxboro controllers a worm drive mechanism turned by a clock key was used to adjust the set point. The throttling range could be changed to one of four preset values by moving a link on the bell crank lever.[41]

In publicising the *Stabilog* Foxboro focused on the fact that it provided reset control action. However, Mason had also solved the problem of providing a linear amplifier: he had invented the pneumatic equivalent of the electronic negative feedback amplifier, invented by Black in 1927 (the patent application filed in 1928 was, in 1930, still under consideration). The simple amplifier is obtained if the reset bellows are replaced by a spring. Units without the reset bellows were sold by Foxboro as *Reactor Controllers*.[42]

The path by which Mason and his colleagues arrived at negative feedback was similar to that followed by Black. They saw clearly that the amplifier had to be linear. They sought initially to achieve this by trying to make each individual element behave linearly; the angled flapper was an attempt to linearise the pressure-displacement relationship of the flapper-nozzle combination, as were modifications made to the pilot valve in the control head. They also realised that the components must not change their behaviour over time — something that was difficult to achieve. A difference, however, was that Black in 1934 clearly explained how, with negative feedback, linearity and stability of the active components (in the electronic amplifier the tubes, in the pneumatic the flapper-nozzle combination and the pilot valve) was not necessary. He showed that linearity and stability were obtained if the passive components in the feedback loop were linear and stable.[43]

It is not clear if Mason was aware of the full significance of the use of negative feedback. In an undated paper, R.A. Rockwell a Foxboro engineer, explained that in the pneumatic amplifier 'all throttling characteristics are dependent only upon the spring characteristic [of the] throttling bellows. [It] is ruggedly designed and has ample power so that its calibration will remain fixed throughout the life of the Controller. The prime requisite is a [control head] relay made very sensitive purposely'. The advantages of the new pneumatic amplifier, he claimed, were that 'the pneumatic throttling range is uniform and does not have the dead zones which always exist in mechanical throttling ranges as a result of lost motion or friction' and it 'is independent of variations in air supply pressure. Changes in supply will not affect the operating air prerssure, provided the supply pressure is sufficient to meet the maximum operating air pressure requirements'.[44]

Although Foxboro was not alone in marketing a controller with proportional plus reset action — controllers using electric motors as described above also provided proportional plus reset action — the *Stabilog* gave Foxboro a competitive advantage in terms of air-operated controllers for difficult applications. Foxboro's lead did not last long for in 1933 Taylor offered its *Dubl-Response* unit.[45] This was a separate unit which fitted onto the Taylor *Evenaction* control valve. The unit used feedback of the valve position and hence in addition to providing reset action also (as the brochure pointed out) eliminated 'the effect of friction of the control valve stem and hysteresis of the valve motor'. Taylor engineers were also aware of the effect of feedback on the stiffness of the valve positioning unit, pointing out that it 'has 500 times greater power for correctly positioning the valve disc than one without it'. The recognition that the valve positioning part of the unit — without the reset feature provided by the damping unit — could provide an improvement in performance of some systems led Taylor to develop and market a control valve positioner which overcame many of the well known practical problems caused by sticking valves.

The advertising literature called the *Dubl-Response* unit the 'most amazing development in the history of process control'. Unfortunately process control engineers were not quite ready for the unit: there was as yet no established technique for determining the settings for optimum response and so trial-and-error was used which, according to Ziegler, 'could be very time consuming and earned for the double response unit the unflattering name of "doubtful response" unit'.[46] Although the *Dubl-Response* unit was announced in 1933, there appears to have been no serious attempt to market it until 1935 when it was re-launched as a new controller.

At the same time the Taylor Company redesigned its basic controller, which was introduced in 1934 under the name *Fulscope*. This controller provided field adjustable, infinitely variable throttling settings between 1% and 150%. The Taylor Company was quick to point out to potential customers the benefits of infinitely variable settings 'research and field experience prove that the "unit sensitivity"...of a controller must be matched with the process time lag of the apparatus under control.... Too-low unit sensitivity causes an unnecessary deviation from the control point when there is a change in load.... The *infinite* number of operating conditions...demand a controller with an *infinite* number of unit sensitivities'.[47] The challenge to Foxboro was even greater in that the *Dubl-Response* unit which could be attached to the *Fulscope* also had provision for changing the setting of the reset action time.

A further challenge to Foxboro came from Tagliabue with its *Damplifier*, a proportional-plus-reset controller which had negative feedback round the pilot valve but not round the flapper-nozzle amplifier. It thus had high gain but the inner control loop on the pilot valve provided damping of the system.[48]

Foxboro extended the range of applications of the *Stabilog* system with the introduction in 1933 of the *Potentiometric Stabilog* which combined the *Stabilog* with the Wilson-Maeulen potentiometric controller.[49]

The very limited provision for modifying the *Stabilog*'s performance — a link could be changed to give low, medium or high proportional band but there was no means of changing the reset-action time — had been a serious drawback. Foxboro engineers, using experience and knowledge of the process to be controlled, chose appropriate values; but the reset action time was difficult to change because it was determined by a resistance capacitance network formed from lengths of capillary tubing joined to capacity chambers. The only feasible method was to change the length of the capillary tube which was not easily done. The Foxboro controllers continued to have limited means of adjustment until the introduction of the Model 40 *Stabilog* in 1946.[50] Another proportional-plus-reset controller was introduced into the market in 1936 when the Bristol Company began to sell its *Ampliset Free-Vane* controller. This controller provided a fully adjustable proportional range but it was not a true proportional-plus-reset system since it used floating response techniques with no feedback around the flapper-nozzle amplifier. It had, however, an easy method of adjusting the floating action: a bleed valve was connected between the pilot valve and the main control valve. Bristol did not introduce a true proportion-plus-reset system until 1938.[51]

Having invented the *Stabilog* and having put it into production, how was the company to sell it? Potential users were often unaware of the reasons for their control problems (some were not even aware that they had a problem), and hence the company had to educate its customers: 'we had to deliver a short course in

automatic control with every Model 10 we sold.'[52] To do so they produced a
brochure that explained in simple terms the potential of the *Stabilog*.[53] It began
with an explanation of the difference between the control needs for batch and
continuous processes, 'the principle difference is that in the batch process the effect
of any disturbance develops slowly, because of the relatively great capacity of the
process. On a continuous process the effect develops rapidly, due to the small
capacities.' It continued with the argument that in a continuous process 'the flow
of the controlling medium must be continuous, and any change made by the
instrument to compensate for process change must be as fast as, but no faster
than it can be absorbed by the process'.[54] The writers claimed that previous
attempts to modify intermittently acting controls for use with continuous
processes — presumably including the previous range of Foxboro modulating
controllers — were not successful and such controllers had produced hunting. The
new controller 'functions to stabilise the process and maintain it in a state of
equilibrium from which it cannot deviate', and hence its name *Stabilog*.[55]

The behaviour of the *Stabilog* was explained with aid of the diagram (reproduced
as Figure 2.9). The crucial element is the 'Differential Pressure Motor' which moves
the nozzle part of the flapper-nozzle amplifier and thus provides the reset action.

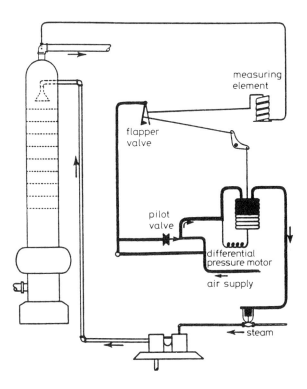

Figure 2.9 *Principle of operation of the Stabilog*

Reproduced (with partial redrawing) from the Foxboro Bulletin No. 175 (1932), p. 15

Its operation is described as follows:

> A Differential Pressure Motor constantly readjusts the 'pilot valve' so that the flapper valve is always just tangent to the air nozzle. If the temperature tends to fall, this constant readjustment builds up a pressure in the Differential Pressure Motor. Since this motor is in the 'valve air system', the control valve is operated throughout its range without a change in control-point. In addition, the design is such that the rate of the readjustment depends on the velocity of the adjustment. In this way the amount of change of the control is a function of the rate at which the temperature changes. This brings the temperature back on a curve that is *tangent* to the control point instead of cutting sharply across the line of control.[56]

2.6 Derivative control action

The final stage in the development of the standard pneumatic controller during the 1930s was the addition of derivative action to the controller. Derivative action control implies that the manipulated variable — the position of the control valve, say — is a function of the rate of change of the deviation of the controlled variable from its desired value. In the literature of the late 1920s and early 1930s there are frequent references to control action being proportional to the deviation and its rate of change but on inspection the control action produces a rate of change for the control valve and consequently the overall control effect is that of proportional plus integral action. Controllers of this type were dealt with above. The Mason-Neilan Company introduced its Compensated Temperature Controller about 1932 and Figure 2.10 shows it schematically. As with the earlier Neilan controller there was feedback round the pilot valve but now pressure feedback to the flapper arm through the control valve compensator, and feedback (steam pressure) from the steam flow line supplying heat to the kettle is added. As the flow rate of the steam is proportional to the pressure the latter provides the equivalent of velocity feedback, a technique which was to be widely used after 1940 for the stabilisation of servomechanisms. The provision of velocity feedback allows the use of higher gains and thus a faster response without the penalty of system oscillation and instability.[57]

The mainstream development of controllers with derivative action came from the difficulties experienced in controlling the shredding of the cellulose crumb used in manufacturing viscose rayon. The mechanical work of shredding causes the temperature to rise and cooling is used to maintain the temperature at 24°C. Measuring the temperature is difficult since the temperature bulb has to be strong enough to withstand considerable force when the cellulose is in chunks but not so well shielded as to have long time lags when the chunks are reduced to a fluffy mass (in which condition the cellulose is an excellent insulator). Ziegler recalls thinking that 'control of this awful piece of equipment seemed impossible' for 'proportional control sensitivity had to be set so low that intolerable temperature offsets developed as the batch progressed. Automatic reset was tried but only made results worse due to the reset wind-up and attendant temperature over-peak at the start of the batch.'[58]

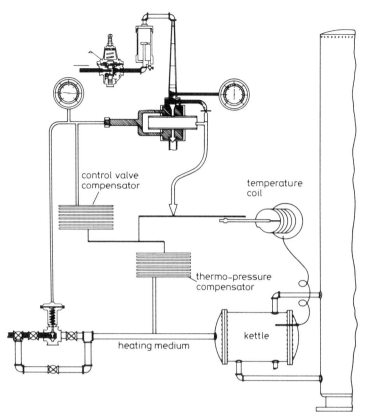

Figure 2.10 *Mason Neilan temperature controller with thermo-pressure compensator*

Reproduced (with partial redrawing) from a Mason Neilan Bulletin 2000-B brochure circa 1932

During 1931, Ralph Clarridge, working in the Research Department at Taylor Instruments, conducted a series of experiments on various controller configurations which could be achieved with pneumatic devices. He noticed that when a restriction was introduced in the feedback line of a proportional response controller the controller gave an unexpected response, in the form of a 'kick', when the set point was suddenly changed. The controller thus anticipated the change in the controlled variable. Figure 2.11 is a schematic diagram of the experimental system. The chambers marked B and C and the moveable frame marked J formed the circuit that gives rise to the anticipating response.

When consulted about the rayon production process Clarridge remembered this 'kick' response and suggested that it might help. As a result a 'unit was made up consisting of a piece of capillary tubing, soldered into a small tank, and was installed in the proportional response controller'. The plant was first run with the unit installed on March 20, 1935, and it was found that 'the controller sensitivity could be increased, the period of oscillation decreased to a fraction of its previous value, and the temperature came to its set point and held practically constant'. The Taylor engineers named the effect pre-act.[59]

48 Process control: technology and theory

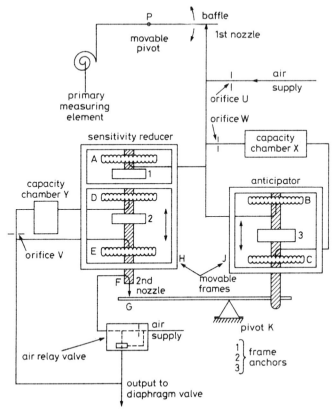

Figure 2.11 *Schematic diagram of an anticipator regulator*
                Reproduced (with partial redrawing) from *Taylor Technology*, 1951, **4**(1), p. 16

The pre-act obtained through the restriction in the proportional feedback line provides a controller output which is proportional to the derivative of the error signal (the difference between the actual temperature and the desired temperature). Thus the overall control action consists of the sum of three terms:

- proportional to the error,
- proportional to the integral of the error (reset), and
- proportional to the derivative of the error, thus giving PID or three term control.

The value of such a controller had been shown theoretically in 1922 by Minorsky in his work on the automatic steering of ships. Its use was also suggested by J. Grebe (1933), A. Ivanoff (1934) and S.D. Mitereff (1935), and the same combination of terms had been used in governor designs from the 1890s.[60] Subsequently, it has been shown repeatedly that in the absence of any knowledge (in terms of a dynamical model) of the process to be controlled, the PID controller is the best form of controller.

Taylor Instruments subsequently offered controllers which could be fitted at the factory with capillary units to give a fixed derivative action time in the range 0.2 to 8 minutes but, except for units used on rayon shredders, few were sold. In 1939, they introduced a completely redesigned *Fulscope* which offered pre-act as an option. In this instrument a needle valve which allowed for continuous adjustment of the derivative action time replaced the capillary units. In the same year, 1939, Foxboro introduced its *Stabilog* 30 with *Hyper-Reset* which provided continuously variable setting of the proportional band but had only four fixed settings of the derivative-plus-reset terms.[61] As there was no recognised way of choosing appropriate controller settings at that time, this was perhaps a sensible decision because it relieved the user of attempting to find the right balance. Unfortunately, though, according to Ziegler, the combinations chosen by Foxboro did not give an optimum response. And by this time many engineers were beginning to see the need for controllers which could be adjusted in the field.[62] The first Foxboro controllers to be fully field adjustable were in the Model 40 *Stabilog* series introduced in about 1946. Ed Smith, in 1936, drew attention to the importance of user adjustable parameters when he wrote:

> There is generally considerable advantage to a user in being able to readily adjust the constants of the individual regulator to suit the particular installation on which it is then used. It is seldom that the pertinent data affecting the performance of a regulator are available at the time of ordering it, or even when putting it in service. Consequently, the regulator's flexibility and usefulness is in general increased if it is provided with conveniently accessible adjusting means. It is particularly important that such adjustments can be made without interfering with the action of the regulator in any way.[63]

2.7 Process control theory

C.E. Mason was blunt in his criticism of those who had a naive view of automatic control when, in 1933, he wrote, 'the greatest breach of sound engineering is in the assumption that automatic control will reproduce the results of perfect manual control simply because it is automatic'. This comment appeared in a series of articles written for *World Petroleum*. These articles, and those of Major Behar in *Instruments* during the early 1930s, and the paper by J.J. Grebe, R.H. Boundy and R.W. Cermak of the Dow Chemical Company,[64] although perceptive and authoritative, were largely descriptive in approach. The *Stabilog*, *Fulscope*, and other controllers were developed and used without the benefit of an extensive theory of automatic control.

In the mid-1930s, A. Ivanoff and S.D. Mitereff attempted to go beyond the purely descriptive and they tried to develop a theoretical basis that would support analysis and synthesis of controllers. 'In spite of the wide and ever-increasing application of automatic supervision in engineering', wrote A. Ivanoff in 1933, 'the science of the automatic regulation of temperature is at present in the anomalous position of having erected a vast practical edifice on negligible theoretical foundations'.[65] A few months later his views were echoed by S.D. Mitereff, an engineer employed in the chemical industry who, in referring to the great need for a fundamental analysis of automatic regulators, wrote, 'automatic-control problems are solved at present by purely empirical methods' and after installation 'the usual cut-and-try method of adjustment is very tedious and unreliable.'[66]

Absence of published information on a 'rational' approach to the design of controllers does not necessarily mean that such an approach was lacking. The work of Clarridge at Taylor Instruments referred to above is an example of a rational approach; C.E. Mason's patent application of 1930 contains a mathematical analysis of the controller behaviour; and in the discussion of Mitereff's paper R.L. Goetzenberger of the Industrial Regulator Division of the Minneapolis-Honeywell Regulator Company defended the industry stating:

> The fact that little may have been published on the subject of automatic regulators does not imply that manufacturers of this apparatus have not explored scientific grounds and are not familiar with the fundamental equations expressing characteristics.[67]

Although engineers and scientists working for the instrument companies were familiar with the fundamental equations that could be used to model their regulators, they were less familiar with and less knowledgeable about the dynamic behaviour of the overall system comprising the plant and controller. On-off controllers were still used in most applications and, for processes that were difficult to control, such controllers could not compete with a good operator who, 'in regulating a process from a curve drawn by a recorder, perhaps unwittingly brings into play some highly mathematical considerations.' As I.M. Stein the Director of Development for Leeds & Northrup, went on to explain, the operator 'automatically observes not only the momentary condition and the direction of change of that condition, but observes also the rate of change of that condition with respect to time (the first derivative) and the rate of change of the rate of change (the second derivative). These observations are very essential to close regulation, particularly in processes involving appreciable time lag.'[68]

Ivanoff approached the problem by comparing the response of an ideal process — one with no time lag and hence no phase shift, and with gain independent of the frequency of the applied signal — with that of the real process when both were subjected to a sinusoidal input and with different control devices (on-off, proportional, floating and proportional-plus-floating). He pointed out that a real process distorts the applied signal:

> The sine wave of the potential temperature, as it passes through the plant to the chart of the pyrometer, can suffer changes summarised as those in
> (a) the form of the wave,
> (b) the amplitude of the wave, and
> (c) the displacement of the wave along the time axis.[69]

Ivanoff used temperature control as his example and by 'potential temperature' he means the response of the ideal process.

Designers of radio oscillator circuits used the criteria that for sustained oscillation the difference in phase between the output wave and the input wave must be 180°.

Ivanoff adapted this criteria in his consideration of closed loop stability of the controlled system, arguing that if the amplitude of the oscillation were not to grow continuously, the loop gain must be less than or equal to unity when the phase shift was 180°. He was not, however, able to take this any further and his next step undermined the credibility of his analysis. He used empirically determined values to represent gain and phase of the temperature wave passing through the plant (he assumed that both gain and phase were functions of the square root of frequency and were equal in value), and showed that for stability the permissible maximum controller gain is 23.1 and that it is independent of the plant characteristics. He called this result 'surprising'![70]

The assumption that the functions representing gain and phase were equal in value was criticised by T. Barratt and O.A. Saunders, and they also questioned the validity of the empirical results.[71] Other discussants suggested that most plants could be represented by a simple lag giving a maximum time shift (phase lag) of $\pi/2$; but, as Ziegler observed many years later, Ivanoff's plant model approximated to 'a dead period lag, L, as major lag element plus a time constant of about $4L$, and a minor element with time constant about $0.1L$', which was much closer to many industrial processes than the simple lag model.[72] Attention was drawn to the work of T. Stein in Germany,[73] and concern was expressed about the mathematical complexity of the paper. It was, however, generally welcomed and thought to have set out the way forward for a better understanding of automatic control.[74]

Mitereff's treatment was in some ways more limited than that of Ivanoff: he did not attempt to analyse the overall control problem, or seek to examine the stability of the system. However, in providing mathematical models of typical plant units, and in classifying controllers in terms of differential equations representing their behaviour, he went beyond Ivanoff and opened the way to further analysis. In all he listed twelve classes of controller, six which he said were in common use and six which he claimed were novel. All the novel controllers include terms involving the derivatives of the error signal (his classification list is shown in Figure 2.12). In discussing the characteristics of the various control laws he commented that class IV was an example of a controller based on 'the misconception that a regulator could be stabilised by means of a dash pot or a similar retarding device... Since hunting is the result of the time lag inherent in the system, the addition of an artificial time lag cannot possibly correct the situation'. He admitted, however, that the addition of damping could make a hunting system appear more stable and argued that 'the only rational method of combatting hunting is by counteracting the time lag', which can be done by adding derivative and second derivative of error characteristics to the controller. He noted that 'elimination of hunting by the rational method involves advancing the response of the regulator.'[75]

Mitereff also gave careful consideration to the causes of time lag in the plant. Several discussants took issue with him, advancing arguments about the relative importance of application lag, controller lag, and metering reaction time. Mitereff in reply commented that the 'effect of time lag is substantially the same irrespective of the place of its introduction'.[76] Most discussants welcomed Mitereff's paper; adverse criticism was largely directed at the simplifying assumption made (inevitable when attempting to develop an analysis of well established engineering systems), the lack of definition of terms used, and the importance of considering

I $F = k_1 \int P dT + (C)$ .... [1]

II $F = k_1 P$ .... [2]

III $F = k_1 \int P dT + k_2 P + (C)$ .... [3]

IV $F + k_2 \dfrac{dF}{dT} = k_1 P$ .... [4]

V $k_3 \int F dT + F = k_1 \int P dT + k_2 P$ .... [5]

VI $F + k_3 \dfrac{dF}{dT} = k_1 \int P dT + k_2 P + (C)$ .... [6]

The following classes of automatic-control apparatus are more or less new to the art:

VII $F = k_1 P + k_2 \dfrac{dP}{dT}$ .... [7]

VIII $F = k_1 \int P dT + k_2 P + k_3 \dfrac{dP}{dT} + (C)$ .... [8]

IX $F = k_1 P + k_2 \dfrac{dP}{dT} + k_3 \dfrac{d^2 P}{dT^2}$ .... [9]

X $F = k_1 \int P dT + k_2 P + k_3 \dfrac{dP}{dT} + k_4 \dfrac{d^2 P}{dT^2} + (C)$ .... [10]

XI $k_5 \int F dT + F + k_4 \dfrac{dF}{dT} = k_1 \int P dT + k_2 P + k_3 \dfrac{dP}{dT}$ .... [11]

XII $k_7 \int F dT + F + k_6 \dfrac{dF}{dT} + k_5 \dfrac{d^2 F}{dT^2} =$
$k_1 \int P dT + k_2 P + k_3 \dfrac{dP}{dT} + k_4 \dfrac{d^2 P}{dT^2}$ .... [12]

Figure 2.12 *Mitereff's classification of process controllers*

Reproduced by permission of S.G. Mitereff, from *Trans. Amer. Soc. Mech. Eng.*, 1935, 57, p. 161

the action of the actuator. Several contributors commented on the practical importance of friction and other non-linearities in the actuator.

The discussion revealed differences in understanding, in terms of application of mathematics, of the different instrument manufacturing companies. P.S. Dickey of The Bailey Meter Company implied that much more detailed and comprehensive analyses of control systems than Mitereff's were being carried out; M.J. Zuchrow of The Republic Flow Meters Company said that 'a truly rational solution of a control problem cannot divorce the controller from the application', adding that 'such mathematical analyses have been made'; whereas H.A. Rolnick of the Brown Instrument Company said, 'the rational solution of automatic-control problems has been scarcely attempted, or if it has been attempted, the published results have been meager'.

Mitereff referenced German work by Tolle and Wunsch, and the influential book, *The Manual of Instrumentation*, written by Behar; he was also aware of the patent literature but made no reference to Ivanoff's paper.[77] A number of discussants also made reference to the German work.[78] Zuchrow pointed out that an important question was the determination of the stability of the system and he quoted the Routh rules for second, third and fourth order systems.[79] Ed Smith drew attention to the work of Ivanoff and Hodgson in England. He also proposed holding a symposium at which other viewpoints could be presented, the symposium to be followed by the formation of a research committee for the purposes of agreeing on definitions and basic relationships. A.E. Sperry and C.E. Mason supported his proposals and they all urged the American Society of Mechanical Engineers to support and encourage research in the subject.[80] Smith then contributed a paper for discussion to encourage the submission of papers by others.[81] The paper was a revised and expanded version — the main addition was a mathematical analysis of the behaviour the controllers — of one published in *Instruments* in 1933.[82] In an attempt to establish a common terminology and to avoid unnecessary contention he circulated the draft of his paper widely during January of 1936, and the revised version was published in *Transactions* of the American Society of Mechanical Engineers, in June of that year.[83]

The major part of Smith's paper was taken up with the classification of different controllers, a description of their behaviour, and suggestions for determining how well they performed. It indicates the number of extant controller configurations. He also attempted to define very carefully the terms used within the industry. John J. Grebe of the Dow Chemical Company, discussing this paper commented on the 'endless confusion of terms and explanations that accompany the discussions of what various instruments do' complainted that 'each manufacturer uses a different terminology, and often deliberately chooses different phrases to avoid the appearance that a given control effect might be similar to that of another manufacturer'. He gave strong support to Smith, saying that 'there is no one individual item of standardisation...in which the ASME can become more useful and valuable than in this particular development.'[84]

The worst confusion was in the use of the terms rate control, anticipatory control, and follow-up. The problem was compounded by concentration on the way the controller behaved and failure to treat all the components, including the process as a system. Thus controllers were described as rate controllers — which was strictly true in that the output of the controller was proportional to the rate of change of the deviation — but such controllers were typically used with motorised valves, and because of the integral action in the actuator the actual affect on the controlled variable was that it was proportional to the deviation. J.C. Downing of the Niagara Hudson Power Corporation, Buffalo, NY, emphasised the need to deal with the whole system; 'each part...relies upon the other; remove any one part and you destroy the system' and he proposed that we should talk about 'classes of control systems' instead of 'classes of regulators.'[85]

Smith and his colleagues succeeded in persuading ASME to support and encourage work in control for in 1936 the Process Industries Division of ASME agreed to the formation of an Industrial Instruments and Regulators Committee, thus becoming the first major professional engineering body to form a specific section dealing with control.[86] This committee, led by Smith, was active, particularly in attempting to standardise the nomenclature. One of its first acts was to form a nomenclature sub-committee which reported in 1940, 1944, 1945, and 1946: the recommendations produced in 1946 received wide publicity through the articles of Donald Eckman published in *Chemical Industry* and *Instrumentation* 1946.[87]

Progress towards understanding the nature of typical process plant, the development of simplified plant models, and the classification of commonly used controllers was made in papers by C.E. Mason (1938), A.F. Spitzglass (1938, 1940), E.S. Bristol and J.C. Peters (1938), G.A. Philbrick (Mason and Philbrick, 1940), all published in the *Transactions* of the American Society of Mechanical Engineers.[88] Mason and Spitzglass writing in 1938 both stressed the need for methods of process analysis that would provide models expressible in simplified mathematical equations.[89] Mason observed that the various parts have the ability to absorb or store energy thus giving rise to two types of lag: 'capacity lag' characterised by a first order differential equation; and 'transfer' lag which involves at least a second order differential equation. He stressed that these differed from a third type of lag which he termed 'distance-velocity' lag, that is, transport delay. Spitzglass emphasised the need for linear models and for the use of a 'standard disturbance'. Following Stein he advocated the use of a 'sudden supply disturbance', that is a step input. The need for a clear, precise, agreed terminology was stressed by Bristol and Peters.

By 1940 Spitzglass, and Mason and Philbrick were turning attention towards the performance of controllers; both how to evaluate performance — ratio of amplitudes of successive oscillations, maximum deviation — and how to choose settings to give the desired performance. In the paper by Mason and Philbrick we also find the beginnings of a block diagram approach as shown in Figure 2.13.

Restricting mathematical analysis to linear operation, as in all the papers mentioned above, raised some doubt as to its practical use: 'real' controllers were known to be non-linear, they possessed friction, backlash, and dead-space. Engineers within the industry, with long experience of using 'cut-and-try' methods for designing and setting controllers, were reluctant to accept these 'rational' methods. The old hands such as Behar and Fairchild argued that there were no such things as standard disturbances, and they considered the mathematics being introduced to be unnecessarily complicated. In a paper published in 1940 Fairchild described a simple approach which produced formula showing that high gain gave small steady state errors. Beyond this, however, his analysis was unhelpful as it led to the conclusion that a controller with the high gain necessary to give small steady state errors also required high damping. Such a controller has a very sluggish response. From this conclusion he postulated what he termed the 'fifth principle' namely 'that high sensitivity made possible for the continuous process by damping the controller is incompatible *per se* with quick response of the controller to a disturbance'. Although correct in the strict sense that a controller with high

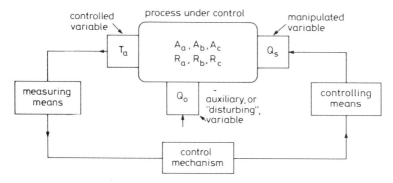

Figure 2.13 *Diagrammatic control circuit*
Reproduced (with partial redrawing) by permission of C.E. Mason and G.A. Philbrick, from *Trans. Amer. Soc. Mech. Eng.*, 1940, **62**, p. 297

damping cannot respond quickly, his principle ignores alternative formulations for the controller—for example, the use of velocity feedback to provide damping as in the Mason-Neilan controller; compensation for low sensitivity through the use of reset; speeding-up the response to a disturbance by means of pre-act.[90]

In some ways Behar's and Fairchild's caution about the use of mathematical analysis was sensible, practical, and useful, even if their arguments in detail were incorrect. The performance obtained from a well tuned two or three term controller could be improved on only if detailed and accurate dynamic models of each specific process were available. For many industrial processes such models are extremely difficult to construct, if they can be constructed at all. The assistance the plant engineers required was not a method of analysing the stability of the system, or a method of designing a controller with certain specific characteristics but a method for choosing the best parameters for the controller.

2.8 Process simulators

The work of George Philbrick and Clesson E. Mason during the period 1938–1941 did much to convince people of the value of the theory being developed. As a means of testing the validity of their ideas Philbrick built an electronic analogue of the controllers and a process. This was a forerunner of the electronic analogue computer.[91] It was not the first use of a computing mechanism to analyse a control system: Douglas Hartree and Arthur Porter of the University of Manchester, UK, used a differential analyser to solve a problem posed by A. Callender who worked in the Research Department of ICI at Northwich.[92]

The system studied, based on a temperature control mechanism, was presented in a generalised form of a linear continuous control system with three term control (PID) and it included what the authors termed a time-lag. The time-lag was in fact a pure time delay. In the first paper they assume a control law of the form:

$$-\dot{C}(t+T) = n_1\theta(t) + n_2\dot{\theta}(t) + n_3\ddot{\theta}(t) \tag{2.2}$$

where $\theta(t)$ = error signal (or measured variable in a regulator system with constant input); $C(t)$ = the control signal; T = time lag; and n_1, n_2, and n_3 are constants. In the second paper they assume that

$$-\dot{C}(t+T) = n_1 z(t) + n_1 \dot{z}(t) + n_3 \ddot{z}(t) \tag{2.3}$$

where $z(t)$ is an auxiliary variable related to $\theta(\tau)$ by

$$-\dot{z}(t) + B_1 z(t) = B_2 \theta(t) + B_1 \dot{\theta}(t) \tag{2.4}$$

The overall control law in this second case is the equivalent of passing the error signal through a lead network prior to the PID control (since equations 2.2 and 2.3 are equivalent to PID control). The authors note that under particular circumstances the use of the auxiliary form of control has certain advantages: they do not say what they are but the form of the equations suggest that they would have been improved response for systems with large pure time delay. In both cases the plant is modelled as a simple lag with variable time delay T in the application of the control signal. Figure 2.14 shows the system in schematic diagram form. The system is treated as a regulator and the effects of a disturbance $D(t+T)$ are investigated. Figure 2.15 shows the Differential Analyser set-up for carrying out the analysis; it should be noticed that the feedback control loop on the analyser is completed by an operator tracing the output of the controller with the tgracer point offset by a distance equivalent to the time delay T.

Their basic approach was to find the normal modes of the system (that is the location of the roots of the characteristic equation), and to investigate how these varied with the parameters of the system. The location of the roots gives an indication of the stability of the system but they were interested in more than just stability, 'we shall be concerned to determine the suitable ranges of these

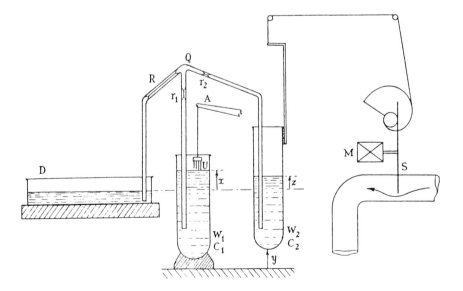

Figure 2.14 *Schematic diagram of the Callender, Hartree and Porter system*
Reproduced from *Philosophical Trans. Roy. Soc. London*, 1935-6, **235**, p. 440

Process control: technology and theory 57

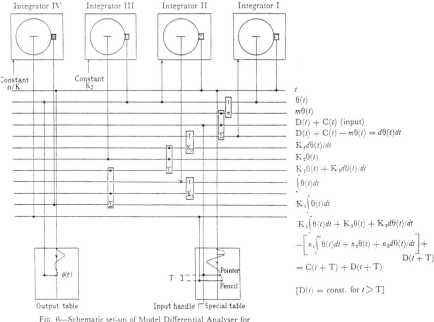

Fig. 6—Schematic set-up of Model Differential Analyser for

$$\int d\theta(t)/dt = D(t) + C(t) - m\,\theta(t)$$
$$-C(t+T) = n_1 \int^t \theta(t)\,dt + n_2\theta(t) + n_3 d\theta(t)/dt$$

The values of K_1, K_2, and K_3 are chosen so that $n_1/K_1 = n_2/K_2 = n_3/K_3$, and so that K_1 and K_3 can be obtained by combination of available gear wheels; the multiplication by the ratio n/K is carried out on Integrator IV. The 't' shaft is driven by the independent-variable motor.

Figure 2.15 *Schematic of a differential analyzer set-up*

Produced (with partial redrawing) by permission of Callender, Hartree and Porter, from *Philosophical Trans. Roy. Soc. London*, 1935-6, **235**, p. 434

constants for practical control rather than the boundary of the region for which control is stable' [p. 418]. The constants referred to are, in the notation used in the paper, u_1, u_2, u_3 being integral, proportional and derivative gains respectively. They argued, qualitatively, that 'a law of control involving u_1 and u_2 alone will give satisfactory control with suitable values of control constants' and that the addition of derivative action (u_3) 'by giving... still earlier indication of the incipient deviations of $\theta(\tau)$, greatly hastens the checking of these deviations, and, provided v_3 is not made too large, values of v_1 and u_2 can be found so that the control is still satisfactory from the point of view of stability.' [p. 420]

They assumed that although the presence of a time-delay in the system leads to an infinite number of roots, the practical behaviour of the system will be determined by the fundamental mode or first harmonic given by the complex roots with the smallest value for the imaginary part. That is, they assumed a solution of the form

$$\theta = \alpha e^{-\alpha t} \sin(\beta\tau + \phi) \qquad (2.5)$$

where α and β are defined by $\sigma = \alpha \pm ib$.

Using a chart of the form shown in Figure 2.16 which shows lines of constant parameter values, it is possible to determine the required parameters for a specified performance. Callender and Stevenson gave a simplified version of the theory in a paper presented at a two day conference on automatic control organised by the Chemical Engineering Group of the Society of the Chemical Industry in London in 1936.[93] They illustrated the procedure with the example: 'if a decrement of 10 and a period of 8 time-lags be taken as a requirement of an automatic two term control, then

$e^{-8a} = 0.1$, or $\alpha = 0.3$

and $K = 2\pi/8 = 0.78$

From Figure 2 [Figure 2.16] it is seen that when $k_1 = 0.25$ $k_2 = 0.65$ these conditions are fulfilled' [k_1 and k_2 are the equivalent of b_1 and b_2 in the notation used in the Royal Society papers, and K is the equivalent of β]

The method was not practical for general use as a different chart was 'required for each pair of values (b_1, b_2)' that is for each setting of gain and integral action time. They produced several such charts, and they also simulated the behaviour

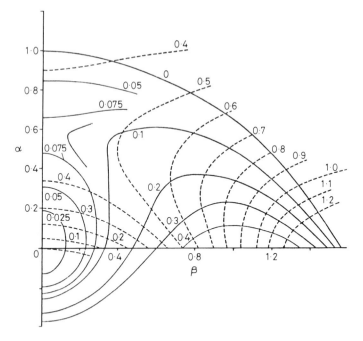

Figure 2.16 *Chart for determining controller parameters*

Diagram shows v_1 and v_2 contours for $v_3 = 0$, $u = 0$
——— contours of constant v_1
- - - contours of constant v_2
Reproduced (with partial redrawing) by permission of Callender, Hartree and Porter, from *Philosophical Trans. Roy. Soc. London*, 1935-6, **235**, p. 424

of the control system with different settings. The simulation was carried out using a step disturbance as the standard input on the grounds that 'by the superposition principle, the effect of a general disturbance can be analysed into the effects of disturbances of this particular class, multiplied by suitable factors and beginning at different times, so that by studying the effects of typical disturbances of this kind we shall get a survey of the general behaviour of the system.' [p. 421]

The paper presented at the automatic control conference concentrated mainly on the apparatus developed to demonstrate the controller (see Figure 2.17). The discussion of the paper was largely an argument between the authors and Ivanoff on the value of the process reaction curve method used to estimate the time lag

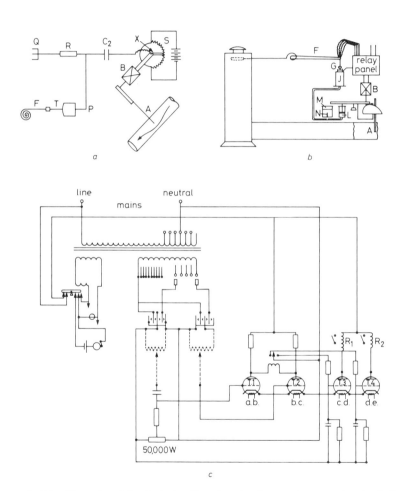

Figure 2.17 *Examples of electric, hydraulic and ac electric control systems given by Callender and Stevenson, 1936*

    a   Electric control
    b   Hydraulic control
    c   Connection diagram of all-electric ac controller

in the plant. In using the process reaction curve the authors anticipated the work of Ziegler and Nichols.

The work of Callender, Hartree, Porter and Stevenson epitomises academic-industrial collaboration, not only through the ICI-University of Manchester collaboration, but also because it was only through industrial sponsorship that the University was able to construct the differential analyser.[94] It further illustrates the importance of personal links and academic exchanges for rapid dissemination of ideas. Douglas Hartree was a frequent visitor to MIT and as such had taken a close interest in the development of the differential analyser: as a consequence he was privileged to receive a copy of the drawings in 1931. Hartree was also familiar with Hazen's work (particularly the 1934 papers) and his use of the Heaviside operator technique follows Hazen's approach. The general discussions of the control laws also suggests a familiarity with the work of Minorsky to whom Hazen made reference.

However, despite all that Hartree, Callender, Porter and Stevenson's work contributed to the understanding of the control problem, engineers were provided with little practical help. Without access to a differential analyser the construction of the appropriate design charts and the simulation of the control system were not practical. Some other form of chart or technique for relating the parameters of the system to the location of the roots of the characteristic equation in the complex plane was required. This was to come in 1948 with the Root Locus technique of Walter Evans.

2.9 Tuning of three term controllers

The average plant engineer — particularly if he or she had been attempting to understand the theoretical papers — must have been relieved, encouraged, and even felt a little self-righteous, to read in a paper published in 1942 that

> A purely mathematical approach to the study of automatic control is certainly the most desirable course from a stand-point of accuracy and brevity. Unfortunately, however, the mathematics of control involves such a bewildering assortment of exponential and trigonometric functions that the average engineer cannot afford the time necessary to plow through them to a solution of his current problem.[95]

This was the opening paragraph of a paper by J.G. Ziegler and N.B. Nichols in which they expounded their now famous method for determining the appropriate settings for three term controllers. Their method is still in widespread use.[96] In the first of two papers, Ziegler and Nichols described the oscillatory method of determining the settings: that is, the method in which the controller gain is increased until the system oscillates steadily. If the gain at which this occurs is K_u and the period of oscillation is P_u then for a controller $K[1 + T_d s + 1/T_i s]$ the settings for P control are $K = 0.5\ K_u$, for PI control $K = 0.45\ K_u$, $T_i = 0.83\ P_u$ and for PID control are $K = 0.6\ K_u$, $T_i = 0.5\ P_u$ and $T_d = 0.125\ P_u$.[97] An alternative method of finding the optimum settings, the so called 'process reaction curve' method, was described in a second paper published in 1943.[98]

In this second paper Ziegler and Nichols expressed some strong views about the design process:

> The chronology in process design is evidently wrong. Nowadays an engineer first designs his equipment so that it will be capable of performing its intended function at the normal throughput rate... The control engineer... is then told to put on a controller capable of maintaining static equilibrium for which the apparatus was designed... When the plant is started, however, it may be belatedly discovered that...the control results are not within the desired tolerance. A long expensive process of 'cut-and-try' is then begun in order to make the equipment work...[then it is realised that] some factor in the equipment design was neglected... The missing characteristic can be called 'controllability', the ability of the process to achieve and maintain the desired equilibrium value.

The words are still relevant today, for the design process should involve 'not steady-state but transient characteristics of the process and controller'. Behar many years earlier (1930) had made a similar plea for plant builders to 'make "controllability" part of design routine'.[99]

2.10 Summary

In developing a theoretical base the process engineers faced immense difficulties for not only were the processes difficult to model but so were the controllers. The pneumatic and hydraulic devices gave approximate integral and derivative action (differentiation in the real world of noisy signals is not a nice operation) with interaction between the two components. The electrical and electro-pneumatic systems presented even greater difficulties since they operated intermittently — they were what we now term sample-data systems.

The development of process control concepts, theory and devices, during the late thirties was largely advanced by engineers working for instrument companies. A. Ivanoff worked for the George Kent Company in England, E.S. Bristol and J.C. Peters worked for the Leeds & Northrup Company, C.E. Mason and G.A. Philbrick were employed by the Foxboro Company and J.G. Ziegler and N.B. Nichols by the Taylor Instrument Companies, A.F. Spitzglass worked for the Republic Flowmeter Company before becoming a consultant, and Ed S. Smith worked for Builders Iron Foundry and then for the C.J. Tagliabue Company. In disseminating information about control devices and their application the sales and application engineers of the industrial instrument manufacturing companies played an important role, in particular in encouraging and persuading small companies to follow the example of the larger companies. Frequently application engineers played a dual role of consultant and sales person.[100]

In that there was virtually no academic involvement, the development followed a similar pattern to that of the electronic negative feedback amplifier which will be described in the next chapter. However, no single instrument company could provide comparable resources to those available within the AT&T and the Bell Laboratories to support a sustained effort to develop a theoretical understanding.

Also, with several competing companies, there was a reluctance to share information or to publish more than was necessary to stimulate sales.

Why the companies chose to permit publication of technical papers by their employees in the late 1930s is an open question. Was it because they felt they had to respond to Mitereff's paper and demonstrate that they had similar technical ability and knowledge? Or was it, as Smith's comments about the low cost of industrial controllers leaving little margin for research would suggest, that the problems were too great for the companies to solve individually and cooperation was required? The Industrial Instruments and Regulators Committee, as part of the Process Industries Division, provided a forum for the exchange of theoretical ideas and for general education.[101] If support could not be provided internally as it was in large organisations (for example, research support leading to the negative feedback amplifier was provided by AT&T's research group, the Bell Laboratories), then it had to be provided externally.[102] This is a good example of the role which the various professional societies have played in the development of technology. One of its contributions was the standardisation of nomenclature: the provision of a common language that contributed to the recognition of the commonality of the problems,[103] a major achievement when the physical manifestations range from a governor on a turbo-alternator to the control of temperature during the pasteurisation of milk. The impact of the ASME group was such that much of terminology that they established still persists in process applications despite attempts to replace it with the terminology that emerged from the study of servomechanisms and communication systems.

The journal *Instruments*, largely a trade journal whose publisher, Richard Rimbach, and editor, Major Behar, gave tireless support for the adoption of instruments and controllers, was an effective transmitter of information. It continued to be the main source for the dissemination of information until the formation, in 1946, of the Instrument Society of America as an organisation for instrument technicians, plant operators, and others interested in control applications but who could not fulfil the membership requirements for the major engineering societies. A similar society, the Society of Instrument Technologists (now the Institute of Measurement and Control) was formed in England in 1945.

Recognition of the great importance of instrumentation in scientific and industrial work came in 1942 when the American Association for the Advancement of Science chose the subject of instrumentation for one of its Gibson Island conferences. Attendance at the Gibson Island conferences of the AAAS was by invitation only and no proceedings were published—everything that was said was supposedly 'off the record'. Lecturers at the 1942 conference included M.F. Behar, W.G. Brombacher of the National Bureau of Standards, C.S. Draper and T.S. Gray of MIT, C.O. Fairchild of C.J. Tagliabue Manufacturing Co., J.J. Grebe of Dow Chemical Co., C.E. Mason of the Mason-Neilan Regulator Co., N.B. Nichols of MIT (later to return to the Taylor Instrument Cos.), Bradford Noyes of Taylor Instruments, J.C. Peters of Leeds & Northrup, H. Ziebolz of Askania Regulator Co., and J.G. Zigler of Taylor Instrument Cos. Other Gibson Island conferences on instrumentation were held during the war at which Philbrick demonstrated his Polythemus, Ziegler and Nichols explained their methods for tuning controllers and process engineers were introduced to the frequency domain techniques developed by the communications engineers.[104]

By the time the USA entered the Second World War there is little doubt that the major expertise in the wide scale application of automatic control devices was to be found among engineers employed by the instrument manufacturing companies. We can only speculate what the course of automatic control history might have been without the massive investment in military application, both during the war and immediately post-war. After the war, process control, although of great economic importance and although it continued to present difficult and challenging problems, was overshadowed by developments in other areas.

2.11 Notes and references

1. Process control devices before the twentieth century are dealt with by Mayr, O.: *The Origins of Feedback Control* (MIT Press, Cambridge, MA., 1970) and he has also briefly dealt with more recent process controllers *Feedback Mechanisms in the historical collections of the National Museum of History and Technology* (Smithsonian Institution Press, Washington DC, 1971). The development of direct-acting regulators invented during the 19th century is described in Ramsey, A.R.J.: 'The thermostat or heat governor: an outline of its history', *Trans. Newcomen Society,* **25,** 1945-7, pp. 53-72
2. Perazich, G., Schimmel, H. and Rosenberg, B.: *Industrial Instruments and Changing Technology,* Works Progress Administration, National Research Project on Reemployment Opportunities and Recent Changes in Industrial Techniques. Report No. M-1, October 1938, reprinted in *Research and Technology,* ed. I. Bernard Cohen (Arno Press, 1980), for sales of instruments see Figure 8, p. 32 and Table 4, p. 33; estimate of total value of sales p. 71
3. *Ibid.* For information on controllers see Figure 10, p. 40. The global figures have to be used with caution since the penetration of controllers was not evenly spread across various sections of the industry, eg. by 1932 over 90% of one type of heat treating furnace were fitted with automatic controls, pp. 41-42. (The Leeds & Northrup Company manufactured and sold heat treatment furnaces which were offered with automatic control as a standard feature). The ratio of instrument sales to total machinery sales may represent transient behaviour since it is known that during this period instruments were commonly fitted to existing machinery. The estimate of over 75,000 controllers is based on using the data from Table G-2, p. 116, and an average price per instrument of $200. Price lists of the period from The Taylor Instrument Companies, Leeds & Northrup and The Foxboro Company suggest that there was very little change in the dollar price during the period 1925-1935. Typical prices range from $140 to $280, in choosing $200 I have assumed that sales were biased towards the lower end of the price range
4. Among the leading companies were The Brown Instrument Company of Philadelphia which was formed in 1860 and merged with the Minneapolis-Honeywell Company in 1935. The Bristol Company originated in 1889 when William H. Bristol with his father and one brother — Franklin B. — formed a company to manufacture the patent Bristol Recording Pressure Gauge and Bristol's Patent Steel Belt Lacing. The first company was the Bristol Manufacturing Company, it was incorporated as the Bristol Company in 1892. William H. Bristol formed the William H. Bristol Electric Pyrometer Company in 1906 to manufacture and market the base metal thermocouple which he had invented. The company was merged with the Bristol Company in 1908. The Foxboro Company was formed when two younger Bristol brothers Bennet B. and Edgar H. left the Bristol Company in 1908 after a disagreement on policy and formed their own company, originally called the Industrial Instrument Company. It used the name Foxboro as a trade mark from 1912 and the company formally became the Foxboro Company on 1st January 1914. In 1899 Morris E. Leeds who had previously worked for the Philadelphia instrument company James W. Queen & Company bought an instrument company owned by Elmer Willyoung and renamed it the Morris E. Leeds Company. He took into partnership a theoretical physicist Dr Edwin F. Northrup in 1903 and the company became Leeds & Northrup Company. Leeds bought out Northrup in 1910 and Northrup left the company. The C.J. Tagliabue Company of Brooklyn, New York,

claimed in its advertising and in its catalogues to have been manufacturing instruments since 1796. The company was actually founded in 1874 but Charles J. Tagliabue was a descendant of Caeser Tagliabue who began making thermometers in England in 1796. The precursor of the Taylor Instrument Companies was formed in 1851 by David Kendall and George Taylor and the company grew through mergers with the Watertown Thermometer Company, Hohman and Maurer Mfg. Co., Davis and Roesch, it also established agreements with the Cambridge Scientific Instrument Company and with Short & Mason in London who acted as the distributor and agent for Taylor Instruments in Great Britain

5 Noble, D.: *Forces of Production: A Social History of Industrial Automation* (Alfred A. Knopf, New York, 1984), pp. 59–61

6 The mechanism was invented in 1851 by Bourdon to measure differences in steam pressure. It was used in 1880 as part of a self-recording barometer the Jules Ricard barograph, see Sydenham, P.H.: *Measuring Instruments: Tools of Knowledge and Control* (Stevenage, Peter Peregrinus, 1979), pp. 364–5. Another early example of a recording device is the self-recording thermometer manufactured by the Draper Mfg. Co. of New York in 1887 an example of which is in NMAH Accession No. 1984.0386.01; the Draper Mfg. Co. also made a recording barometer, samples of charts for which are in NMAH Accession No. 1984.0539.01

7 Some non-contact techniques were also used, for example, radiation pyrometers but they tended to require higher skill levels to operate and could not be connected to recording apparatus as easily as the thermocouple or resistance thermometer based devices

8 William H. Bristol's pressure recorder of 1890 was direct acting. He claimed in response to questions at the ASME meeting at which the recorder was described that the mechanism was resistant to vibrational disturbances (a problem if the free end of the tube was not supported in some way) and that the pen friction was negligible. Bristol, W.H.: 'A new recording pressure gauge', *ASME*, **11** (1890), pp. 225–234. He also observed that the recorder could also be used to record temperatures and in 1893 the company was marketing both pressure and temperature recorders. In 1900, with E.H. Bristol, he produced an improved recorder with a Bourdon tube constructed in the form of a helix with two complete turns. Bristol, W.H.: 'A new recording air pyrometer', *ASME*, **22** (1900), pp. 143–151

9 The conversion of galvanometric instruments from laboratory to industrial use was not an easy task. Edwin F. Northrup of the Leeds & Northrup Company pointed out 'a galvanometer needle, steady and smoothly moving in the manufacturer's laboratory of reinforced concrete, is often an uncertain and dancing imp when located beside an annealing oven with a trip hammer near by' Northrup, Edwin F.: 'Modern electrical resistance pyrometry', *Trans. American Institute of Chemical Engineers*, **1** (1908), p. 120

10 The company claimed that in 1906 118 recorders were in use in 88 organisations, see P.H. Sydenham, *op. cit.* (n.6), p. 438

11 Bristol, William H.: 'Pyrometers', *ASME*, **27** (1906), pp. 552–589, suggests that a recorder with a depressor bar mechanism could be used with a galvanometric system. The Bristol Company displayed a circular-chart, recording pyrometer at the Panama-Pacific Exhibition in San Francisco in 1915 (Bristol Company leaflet, 1915); Tycos Book 4014 (Taylor Instrument Companies, 1917), contains a reference to pyrometer chart recorder based on the deflection principle (Tycos was the Taylor trade name)

12 Vannevar Bush in the work leading up to the development of the differential analyser was faced with a similar problem. See Bennett, S.: 'Harold Hazen and the theory and design of servomechanisms', *Int. J. Control*, **42** (1985), pp. 989–1012 and Owens, L.: 'Vannevar Bush and the differential analyser: the text and context of an early computer', *Technology and Culture*, **27** (1986), pp. 63–95

13 A brief account of the Callendar recorder is given in Sydenham, P.H., *op. cit.* (n.6 above), pp. 115–9, 433, 437–9. For accounts of the Leeds and Northrup recorders see Vogel, W.P.: *Precision, People and Progress* (Leeds & Northrup Company, Philadelphia, 1949); and also Williams, A.J.: 'Bits of recorder history', *ASME, J. of Dynamic Systems, Measurement, and Control*, **1** (1973), pp. 6–16; Williams worked in the Research Department of Leeds & Northrup and was the designer of the Speedomax recorder. He was the recipient of the 1972 Rufus Oldenburger Medal of ASME. For a

contemporary survey of recorders see Borden, P.A. and Behar, M.F.: 'Recording electrical instruments', *Instruments*, **8** (1935), pp. 7-14, 34-44
14 *Instruments*, **8** (1935) pp. 34-44; Catalogue No. 1101C, (C.J. Tagliabue Co., 1937)
15 A detailed account of the development of pneumatic controllers is to be found in Stock, J.T.: 'Pneumatic process controllers: the early history of some basic components', *Trans. Newcomen Society*, **56** (1984-5), pp. 169-77, and Stock, J.T., 'Pneumatic Process Controllers: the ancestry of the proportional-integral-derivative controller', *Trans. Newcomen Society*, **59** (1987-88), pp. 15-29. The Johnson patent is Johnson, W.S.: 'Thermo-pneumatic temperature regulator', US Patent 314,027, 1885
16 Bristol, E.H.: 'Control System', US Patent 1,405,181, 1922
17 Bristol, William H.: 'A new recording pressure gauge', *Trans. American Society of Mechanical Engineers*, **11** (1890), pp. 225-234. For a history of the development of the Bourdon tube see Exline, P.G.: 'Pressure-responsive elements', *Trans. American Society of Mechanical Engineers*, **60** (1938), pp. 625-632
18 Hawkins, J.T.: 'Automatic Regulators for Heating Apparatus', *Trans. American Society of Mechanical Engineers*, **9** (1887), p. 432
19 Carrick Engineering Company, Catalogue 99, 1925, MHT tlbk 6732; similar comments about the adverse effects of stop-start action and rapid changes in control valve position were made by Grebe, J.J., Boundy, R.H., Cermak, R.W.: 'The control of chemical processes', *Trans. American Institute of Chemical Engineers*, **29** (1933), p. 228
20 This is not strictly true, it applies only to the so called Class 0 systems that is systems in which the process transfer function does not contain any pure integral terms. Such systems have a stable open-loop response and adequately model the majority of industrial processes
21 H. Dow founder of the Dow Chemical Company stressed the economic advantages of 'continuous' processes and actively sought to use such processes, see Dow, H.H.: 'Economic trend in the chemical industry' *Industrial and Engineering Chemistry*, **22** (1930), pp. 113-116. Rolnick drew attention to the increased accuracy demanded and to a demand for greater flexibility in controllers by which I assume he meant the ability to adjust or 'tune' controllers on site, *ASME* **58** 1936, p. 58 contribution to discussion of paper by Mitereff, 1935
22 Ziegler, J.G.: 'Development of industrial control responses' unpublished memoir, 1968, copy in MAH, E & I
23 Stock, (1984-5) *op. cit.* (n.15), p.176
24 See Smoot Engineering Corporation Bulletins 25 (1926), 31 (1926) and 33 (1927); see also Neilan Company Bulletin 2000-B n.d. circa 1932 (Columbia Collection)
25 Bulletin 177-1 (Foxboro Company, 1933), p 5. Care is required in interpreting texts from the 1930s as the terms throttling and floating are sometimes used as synonyms. Throttling is the broader, older, term which was in general use and includes proportional action. However, during the latter part of the period throttling began to be used to signify proportional action only for example in the usage 'throttling plus reset action'
26 Other companies were also aware of the importance of the behaviour of the actuator: 'particular attention is required in the application of the elements that vary the supply of heat in response to the controller impulses... The controller... is adapted to operate the valve, rheostat or damper, controlling the supply of heat, in accordance with the individual characteristics of that unit'. See Catalogue No. 84 (Leeds & Northrup, 1927), p. 5
27 In a series of editorials from 1932 through to 1940 Behar attacked any suggestion that instruments and automation were responsible for unemployment. In 1938 he gave as a 'fundamental doctrine' the following rules: '(1) Eliminate the *variables* by reducing them to *measurables*; (2) whatever is to be controlled should first be measured; (3) if you can measure it you can control its effects; (4) if you can control it by hand you should control automatically.' *Instruments*, **11** (1938), p. 55
28 The Dow Chemical Company developed a range of controllers during the period 1924-29 because they were unable to purchase suitable devices, see Grebe, J.J., Boundy, R.H., Cermak, R.W., 1933, *op. cit.* (n.19), p. 211; the United Electric Light and Power Company of New York in the early 1930s developed their own proportional plus derivative action controller, see Keppler, P.W. and Salo, E.A., contribution to discussion of paper by Mitereff, ASME, **58**, (1936), pp. 61-2

29 H.H. Dow, *op. cit.* (n.21), pp. 113–116 and Grebe, J.J., Boundy, R.H. and Cermak, R.W., *op. cit.* (n.10), p. 211
30 Grebe, J.J., *op. cit.* (n.10), p. 226.
31 Grebe, J.J.: 'Elements of automatic control', *Industrial and Engineering Chemistry*, **29** (1937), p. 1228
32 The Leeds & Northrup controller was announced in 1930, it was, however, produced only to special order, it was a modification of the 1921 'proportional-floating' controller. The Bristol Controller was available from 1931, Catalogue 2050 (Bristol Company, 1931)
33 I have not been able to establish the date when the George Kent controller was introduced but it is described in an article 'Kent recording and controlling apparatus', *Engineering*, **132** (1931), pp. 407–408, and in a brochure Kent Automatic Boiler Control, Publ. No. 791/532 (George Kent, n.d.)
34 Bulletin 186 (Foxboro Company, n.d., circa 1934). No details are given about the actual control function but it is claimed that by the use of 'anticipating action' oscillation is reduced
35 Catalogue 8008, 'Automatic Control for Temperature—Pressure—Flow' (Brown Instrument Company, 1931); Behar, M.F.: 'Automatic Temperature Control', *Instruments*, **3** (1930), pp. 457–8
36 Much of the information about Mason is drawn from *A Little of Ourselves*, an anonymous memoir 'A more or less technical history of the Foxboro Company and its research department', copy in MAH, E & I; and from Catheron, A.R.: 'Clesson E. Mason, Oldenburger Medallist, 1973', *Trans. ASME, J. of Dynamic Systems, Measurement, and Control*, **96**, (1974), p. 12. Mason joined the Mason-Neilan Company as Director of Engineering in 1941 and in 1945 became Technical Director of the Bristol Company. Control research under Mason was the only research performed by The Foxboro Company until the Research Department was formed in 1937
37 Frymoyer, W.W.: 'Control Mechanism', US Patent 1,799,131, 1931, filed 14th Aug. 1928; Mason, C.E.: 'Control Mechanism', US Patent 1,950,989, 1934, filed 14th Aug. 1928. Stock has pointed out the similarity in the patent drawings even down to the same numbering for the parts (Stock (1987–88), *op. cit* (n.15), p.19). I have found no evidence relating to possible collaboration (Mason was in Tulsa, Oklahoma, and Frymoyer in Foxboro, Mass) it may be simply that the patent applications were prepared at the same time by one patent agent
38 A photograph of the system is given on page 40 of *A Little of Ourselves*, *op. cit.* (n.36)
39 Mason, C.E.: 'Control Mechanism', US Patent 1,881,798—1932, filed 12th May 1930
40 Mason, C.E. and Dixon, A.M.: 'Controller', US Patent 1,899,705—1933, filed 15th Sept. 1930
41 It was advertised in the September issue of *Instruments*, **4**, (1931)
42 Rockwell, R.A.: *'Pneumatic versus mechanical throttling range'*, Foxboro Company, n.d. circa 1940
43 See Black, H.S.: 'Stabilised feedback amplifiers', *Bell System Technical J.*, **13**, (1934), pp. 1–18; Black, H.S.: 'Inventing the negative feedback amplifier', *IEEE Spectrum*, **14**, (1977), p. 54
44 Rockwell *op. cit.* (n.42 above)
45 Taylor Dubl Response unit
46 Ziegler, *op. cit.* (n.22) p. 3
47 The Taylor Instrument Companies, Catalogue 56R April 1936 p. 7.
48 *Instruments*, **7**,(1934), p. 14.
49 The Foxboro Company took over the Wilson-Maeulen Company at the beginning of 1932.
50 The advertising campaign run by Foxboro during the latter part of the 1930s with its claim that 65% of all applications could be handled adequately with on-off control suggest that they faced strong competitive pressure from companies able to offer easily adjustable throttling control who saw no need for the added complications and difficulty of adjustment of the *Stabilog*.
51 Bulletin 507 (Bristol Company, 1938).
52 A Little of Ourselves op. cit. (n.36), p. 41.
53 *Stabilog*, Bulletin 175 (Foxboro Company, 1935)
54 *Ibid.* p. 3.
55 *Ibid.* p. 5.

56 *Ibid.* p. 14.
57 Mason-Neilan Bulletin 3000-C 1933 Compensated Temperature Controllers.
58 Ziegler, *op. cit.* (n.22), p. 4.
59 Ziegler, *op. cit.* (n.22), see also, 'History of the Pre-Act response', *Taylor Technology,* **4**(1) Summer 1951, pp. 16–20; this reads like a 'cleaned up' account of how the development actually took place. Ziegler's account is probably nearer the truth.
60 See Bennett, S.: 'Nicolas Minorsky and the automatic steering of ships', *IEEE Control Systems Magazine,* (November 1984), pp. 10–15.
61 'Hyper-Reset' was the invention of George A. Philbrick. While working for the Foxboro Company he also developed in 1937–8 an electronic simulator. This was a hard-wired analogue computer which could be used to simulate particular process loops – process and controller. The process could contain up to four time lags and the controller could be three-term. The simulator allowed the parameters of the controller to be adjusted independently. See Per Holst, A.: 'George A. Philbrick and POLYPHEMUS, the first electronic training simulator' (undated report – after 1980) in NMAH E & I Foxboro file.
62 The importance of user adjustable parameters had been emphasised by Ed. Smith who, in 1936, argued that, 'it is seldom that the pertinent data affecting the performance of a regulator are available at the time of ordering it, or even when putting it in service. Consequently, the regulator's flexibility and usefulness is in general increased if it is provided with conveniently accessible adjusting means. It is particularly important that such adjustments can be made without interfering with the action of the regulator in any way.' Smith, E.S.: 'Automatic regulators, their theory and application', *ASME,* **58** (1936), p. 302.
63 Smith, E.S.: 'Automatic regulators, their theory and application', *Trans. American Society of Mechanical Engineers,* **58** (1936), p. 302. In attempting to modify the characteristics of controllers – both relay and continuous – many stratagems were adopted. The patent literature of the 1930s is full of claims for mechanical linkages, pneumatic bellows, restrictors, balances and other units. A mathematical analysis of over 100 such devices was made by Ziebolz of the Askania Co. and presented as a catalogue of relay devices to solve mathematical equations. See Ziebolz, H.W.: *Relay Devices and their Application to the Solution of Mathematical Equations* (Askania Regulator Co. Chicago, 1940), 100 copies of the first edition were produced, it was reprinted in 1942 and again in 1944.
64 Mason, C.E.: 'Science of automatic control in refining', *World Petroleum* (June–September, 1933) quotation is from p. 188; Behar, M.F.: *The Manual of Instrumentation* (Instrument Publishing Co., Pittsburgh, 1932), this was orginally published in sections in *Instruments* during 1930–31; Grebe, J.J., Boundy, R.H., Cermak, R. W. (1933) *op cit.* (n. 19).
65 Ivanoff, A.: 'Theoretical foundations of the automatic regulation of temperature', *J. Institute of Fuel,* **7,** pp. 117–30, 1934 (the paper was presented at a meeting on 13th December, 1933). Ivanoff, a physicist with a PhD, was employed by the George Kent Company.
66 Mitereff, S.D.: 'Principles underlying the rational solution of automatic-control problems', *Trans. American Society of Mechanical Engineers,* **57** (1935), pp. 159–63; the paper was presented at the Semi-Annual meeting in Ohio, June 1935; quotation is from p. 159.
67 Discussion of paper by Mitereff (1935) in *Trans. American Society of Mechanical Engineers,* **58** (1936), p. 62.
68 Stein, I.M.: 'Precision industrial recorders and controllers', *J. Franklin Institute,* **209** (1930), p. 210. Nicolas Minorsky, in his work on the automatic steering of ships, also observed that the operator, in this case the helmsman, took into account more than just the actual heading of the ship. Minorsky expressed in mathematical terms the actions which the controller needed to perform in order to match the behaviour of the helmsman. See Minorsky, N.: 'Directional stability of automatically steered bodies', *J. American Society of Naval Engineers,* **34** (1992), pp. 280–309; and Bennett, S.: 'Nicolas Minorsky and the automatic steering of ships', *IEEE Control Systems Magazine* (November, 1984), pp. 10–15.
69 Ivanoff, *op. cit.* (n.65), p. 117.
70 *Ibid.* p. 123.
71 Comments by T. Barratt p. 131 and O.A. Saunders p. 132 on Ivanoff, *op. cit.* (n.65).
72 Ziegler, J.G.: 'Those magnificent men and their controlling machines', *Trans. ASME, J. of Dynamic Systems, Measurement, and Control, (September 1975), p. 279–80.*
73 Stein, T.: *Regelung und Ausgleich in Dampfanlagen* (Springer, Berlin, 1926), 'Selbstregelung

ein neues Gesetz der Regeltechnik', *Zeitschrift des Vereins deutscher Ingenieure,* **72,** (1928); theoretical work on controllers was also carried out by Wunsch, G.: *Regler für Druck und Menge* (Oldenbourg, Munich, 1930).
74 Six years later Ivanoff was still being criticised for complicating what was a simple application of Fourier's heat conduction equation — see (n.90 below)
75 Mitereff, *op. cit.* (n.66), p. 163.
76 *Trans. American Society of Mechanical Engineers,* **58** (1936), p. 64.
77 Tolle, M.: *Regelung der Kraftmaschinen* (Springer, Berlin, 1921). Wunsch, G., *op. cit.* (n.73); Behar, M.F., *The Manual of Instrumentation* (Instrument Publishing Co., Pittsburgh, 1932). In the reply to the discussion Mitereff refers to five patients which he held (US Patents 1,955,680; 2,015, 861; 2,015,862; 2,020,847; 2,022, 818) all relating to 'rate-of-change' responsive regulators; and to Minorsky patent US Patent No. 1,436,280 relating to the steering of ships.
78 Hort, W.: *Technische Schwingungslehre* (Springer, Berlin, 1910). Stein, T.: *Regelung und austgleich in Dampfanlagen* (Springer, Berlin, 1926). Stodola, A.: *Dampf — und Gasturbinen* (Springer, Berlin, 1922). Tolle, M.: *Die Regelung der Kraftmaschinen* (Springer, Berlin, 1909), and Wunsch, G.: (1930) *Regler für Druck und Menge* (Oldenbourg, Munich, 1930).
79 Discussion *ASME,* **58** (1936), p. 57.
80 *ASME,* **58** (1936), p. 57; Catheron, A.R.: 'Clesson E. Mason, Oldenburger Medallist, 1973', *Trans. ASME, J. of Dynamic Systems, Measurement, and Control,* **96** (1974), p. 12.
81 Smith, E.S.: 'Automatic regulators, their theory and application', *Trans. American Society of Mechanical Engineers,* **58** (1936), pp. 291-303.
82 Smith, E.S.: 'Analysis of fluid rate control systems', *Instruments,* **6** (1933), p. 97.
83 The companies and individuals circulated were listed in the paper as: Detroit Edison Co.; Philadelphia Electric Company; Standard Oil Company of New York; M.F. Behar and R. Rimbach, editors of *instruments*; the Barber-Colman Co.; the Brown Instrument Co.; Foxboro Co.; Leeds and Northrup Co.; Smoot Engineering Company (Division of Republic Flow Meters Co.; and C.J. Tagliabue Manufacturing Co. He also acknowledged the assistance of E.R. Loud and A.E. Mignone of Builders Iron Foundry, D.J. Stewart of Barber-Colman; C.O. Fairchild of C.J. Tagliabue and A.F. Spitzglass of Smoot Engineering.
84 Grebe, J.J.: contribution to discussion *ASME,* **59** (1937), p. 128.
85 Downing, J.C.: contribution to discussion, *ASME,* **59** (1937), pp. 136-7.
86 For details of development of support by the professional institutions see Bennett, S.: 'The emergence of a discipline: automatic control 1940-1960', *Automatica,* **12** (1976), pp. 113-21.
87 Eckman, D.P.: 'Automatic control terms', *Chemical Industry,* **58** (1946), p. 832, p. 1020, **59,** p. 528, p. 710; 'Automatic Control Terminology', *Instrumentation,* 2 (1946), pp. 9-14.
88 Mason, C.E.: 'Quantitative analysis of process lags', *Trans. American Society of Mechanical Engineers,* **60** (1938), pp. 327-?; Mason, C.E., and Philbrick, G.A.: 'Automatic control in the presence of process lags', *Trans. American Society of Mechanical Engineers,* **62** (1940), pp. 295-308; Brisol, E.S. and Peters, J.C.: 'Some fundamental considerations in the application of automatic control to continuous processes', *Trans. American Society of Mechanical Engineers,* **60** (1938), pp. 641-50; Spitzglass, A.F.; 'Quantitative analysis of single-capacity processes', *Trans. American Society of Mechanical Engineers,* **60** (1938), pp. 665-74; Spitzglass, A.F.: 'Quantitative analysis of single-capacity processes', *Trans. American Society of Mechanical Engineers,* **62** (1940), pp. 51-62.
89 Mason, *op. cit.,* 1938, (n.88), p. 327.
90 C.O. Fairchild, 1940, p. 338. Fairchild's paper had been presented at a symposium 'Temperature, its measurement and control in science and industry' organised by the American Institute of Physics and held in New York, 2nd-4th November 1939. It was the only paper dealing with automatic control and must have given the impression that industrial process control was less advanced than it actually was. The Symposium Proceedings were published as Fairchild, C.O., Hardy, J.D., Sosman, R.B., Wensel, H.T.: *Temperature — Its Measurement and Control in Science and Industry,* papers presented at the 1939 Temperature Symposium of the American Institute of Physics (Reinhold, New York, 1941).
91 The original Philbrick electronic analogue computer for process control analysis is in the Museum of American History, Smithsonian Institution, Washington DC, see Mayr, O.: *Feedback Mechanisms in the historical collections of the National Museum of History*

  and Technology (Smithsonian Institution Press, 1971), p. 125; see also Paynter, H.M..: 'In memorium, George A. Philbrick (1913-1974)', *Trans. ASME, J. of Dynamic Systems, Measurement, and Control,* **97** (1975), pp. 213-5; a detailed account of Philbrick's work on the simulator is given in Per A. Holst, n.d. 'George A. Philbrick and POLYPHEMUS, the first electronic training simulator', copy in MAH, Division of E & I.
92 Callender, A. Hartree, D.R., Porter, A.: 'Time-lag in a control system', *Philosophical Trans. Royal Society of London,* **235** (1935-6), pp. 415-444 (the paper was first submitted on 15th July 1935, read on 2nd February 1936 and published on 21st July 1936); Hartree, D.R., Porter, A., Callender, A., Stevenson, A.B.: 'Time-lag in a control system — II', *Proc. Royal Society of London,* **161** (1937, series A), pp. 460-476; Callender, A., Stevenson, A.B.: 'The application of automatic control to a typical problem in chemical industry', *Society of Chemical Industry, Proc. Chemical Engineering Group,* **18** (1936), pp. 108-116.
93 The conference took place on 15 and 16 October and the following papers were presented: Lauchlan, A.D.E.: 'Electical control of chemical processes'; Griffiths, E.: 'Simple forms of automatic regulators'; Harrison, D.W.: 'The installation factor in automatic controls'; Callender and Stevenson, *op. cit.*; Wingfield, B.T.: 'Applications and limitations of self-operating temperature regulators'; Copeland, W.J.A.: 'Instruments for automatic control of temperature, pressure, and flow in chemical industry'; Clark, W.J.: 'The automatic control of chemical processes'; Ivanoff, A.: 'The influence of the characteristics of a plant on the performance of an automatic regulator'; Lambert, L.B., Walton, H.R.: 'The means of control as a factor of plant design'; Lowe, E.I., Frisken, J.: 'Some experimences of the use of instruments as aids to plant control'.
94 Hartree had been using a differential analyser build out of meccano until he persuaded MacDougal (Macdougal's flour) to provide £5000 to finance the building of a fully engineered differential analyser by Metropolitan Vickers (this machine is now in the Science Muscum, South Kensington) which was brought into use early in 1935 — Porter, A., Interview with author, 1975.
95 Ziegler, J.G., Nichols, N.B.: 'Optimum settings for automatic controllers', *Trans. American Society of Mechanical Engineers,* **64** (1942), pp. 759-768; in a simplified version of this paper Ziegler laid out a step-by-step method of setting controllers, see Ziegler, J.G.: ' "On-the-job" adjustments of air operated recorder-controllers', *Instruments,* **16** (1941), pp. 394-7, 594-6, 635.
96 The increasing use of microprocessor-based three-term controllers with automatic tuning is making the method obsolete.
97 N.B. Nichols recalls that on running tests on a model set-up on the differential analyser they found that for proportional action the ideal response was obtained by setting $K = K_u/1.57$ but to be safe they chose $K = K_u/2.0$. Interview with author, September 1982.
98 Ziegler, J.G., Nichols, N.B.: 'Process lags in automatic control circuits', *Trans. American Society of Mechanical Engineers,* **65** (1943), pp. 433-444.
99 Behar, M.F., *Instruments,* **3** (July 1930), p. 427.
100 P.H. Sydenham has observed that through until the 1970s process instrument companies were expected to provide consultancy as part of the contract to supply instruments — a practice that still largely continues. See Sydenham, P.H. *op. cit.* (n.6), pp. 118.
101 Bennett, S., *op. cit.* (n.86), 1976, pp. 116-7.
102 For a detailed analysis of the role played by the large industrial research laboratories see L.S. Reich,: *The Making of American Industrial Research: Science and Business at GE and Bell, 1987-1926* (Cambridge University Press, Cambridge, 1985).
103 Zuchrow discussion of Mitereff paper, *op. cit.* (n.66), p. 56.
104 See *Instruments,* **16** (1943), p. 337; Ziegler, J.G.: 'Those magnificent men and their controlling machines', *Trans. American Society of Mechanical Engineers, J. Dynamic Systems, Measurement, and Control,* **97** (September 1975), p. 279; Ziebolz, H.W.: 'The Oldenburger Award Lecture', *Trans. American Society of Mechanical Engineers, J. Dynamic Systems, Measurement, and Control,* **97** (March 1975), pp. 8.

Chapter 3
The electronic negative feedback amplifier

> Telephone history is full of dreams come true. Few rosier dreams could be dreamed than that of an amplifier whose overall performance is perfectly constant, and in whose output distortion constitutes only one-hundred millionth of the total energy, although the component parts may be far from linear in their response and their gain may vary over a considerable range. But the dreamer who awakes in amazement to find that such an amplifier can be built has additional surprises in store for him. These benefits can be obtained by simply throwing away some gain, and by utilising 'feedback action'.
>
> H.S. Black, 1933[1]

The principles of the negative feedback amplifier were described by H.S. Black in a paper presented at the Winter Convention of the American Institution of Electrical Engineers in New York in January 1934. The paper subsequently appeared in *Electrical Engineering* and in the *Bell System Technical Journal*.[2] The invention was already seven years old when information about it was made public and repeater amplifiers using the principle had already been developed and tested in the field. Thus the paper was not a tentative account of a new invention, but a succinct and confident description of a maturing device that had passed through the difficult years of development and come to practical fruition. However, to many people the idea was so revolutionary that they found it difficult to accept: the American patent took nine years to obtain and the British Patent Office treated the invention as they treated applications for patents for perpetual motion machines: they demanded a working model. Even within the Bell Laboratories there were doubts, for when Black first proposed using the amplifier for the Morristown trials (see below) the then director of the laboratories forbad its use and instructed Black to design push–pull amplifiers. This Black did but he also continued development work on the feedback amplifier and the director eventually relented.[3]

The invention of the negative feedback amplifier exemplifies applied research carried out in industrial research laboratories. The AT&T company formed an industrial research laboratory as part of its strategy of controlling all American telecommunications, summarised by its then President, Theodore Vail, as 'One

Policy, One System, Universal Service.' To implement the strategy the company needed to control the rate and direction of technical change by obtaining, or preventing others from obtaining, key patents; and it also needed to avoid being broken up under the Sherman Antitrust Act. The research laboratories played a major part in ensuring that the company kept control of the technology and patent rights; Vail's pragmatic, flexible approach, his publicly pronounced willingness to accept regulation, and his eventual 'voluntary' divestment of Western Union, protected the company from an antitrust case. The monopoly was lost but the integrated network was maintained. Also, and perhaps more importantly, acceptance of government regulation was seen as legitimising the Bell System's dominance of American telecommunications.[4]

3.1 The repeater amplifier

The embryo of a research laboratory was formed in 1907 when the research and development effort was consolidated under the lead of John J. Carty.[5] Two years later in April 1909, Carty wrote a memorandum to the directors:

> Whoever can supply and control the necessary telephone repeater will exert a dominating influence in the art of wireless telephony and the number of able people at work upon the art create a situation which may result in some of these outsiders developing a telephone repeater before we have obtained one ourselves, unless we adopt vigorous measures from now on. A successful telephone repeater, therefore, would not only react most favorably upon our service where wires are used, but might put us in a position of control with respect to the art of wireless telephony should it turn out to be a factor of importance.[6]

The board accepted Carty's view and long term research and development work on the repeater and on wireless systems began.

For trans-continental telephone links, amplification of the electrical signal representing the speech pattern was vital. The first amplifiers used were electro-mechanical devices; the electrical signal representing the speech pattern was converted into sound using a loud-speaker and the sound reconverted to an electrical signal using a microphone. The amplifiers were used in pairs (each pair was referred to as a repeater), one amplifier for each direction of transmission. The practical limit of electro-mechanical repeaters combined with heavy-gauge, loaded, open-wire lines was reached in 1911 when the east coast of the USA was linked to Denver, Colorado. Conversation was just possible!

H.D. Arnold, as part of his research on wireless systems, investigated the potential applications of the vacuum tube. He recognised its importance both for wireless and telephone applications and on his advice AT&T bought the rights to the de Forest audion tube in 1913.[7] They began work on making the as yet little understood audion into a reliable amplifying device (the so called triode) that could be used as an alternative to the electro-mechanical repeater. By 1915, the improved vacuum tube went into service for the first trans-continental line (New York to San Fransisco). Three electronic (vacuum-tube) repeaters were used, with three electro-mechanical repeaters provided for stand-by purposes. A year later, three further electronic repeaters were added, and two more in 1918, giving a total of eight.

The trans-continental service was provided over an open wire transmission system with four wires and, by using the so called phantom circuit, provided three separate voice circuits. The original open wire system used heavy inductance loading to reduce transmission losses. In 1920 the line was unloaded and additional repeaters added; the change doubled the bandwidth, reduced the overall loss, and increased the propagation velocity.

Open wire lines required a large amount of space, were unsightly, and caused problems at river crossings, on intercity routes and in high density areas in towns. They were also susceptible to damage in severe weather conditions—a storm caused the loss of all lines out of Washington DC during the inauguration of President Taft in 1909.[8] Cables which collected together several circuits in one outer sheath were intensively investigated during the early 1920s. Unfortunately, cables brought their own problems: the increased capacitance due to the shielding necessary to prevent cross-talk increased signal losses and thus cable systems required more repeater amplifiers for a given distance than open wire systems. As the number of repeaters increased so did the overall signal distortion.

Growth in demand led to attempts to find methods of carrying multiple conversations over a single pair of wires, called the carrier system. In a carrier system each of the conversations is allocated to a particular carrier frequency and the sound input is used to amplitude modulate the carrier. The carriers for the different conversations are separated by means of wave filters. Work on such systems began in about 1910. A low pass filter at input of a telephone repeater was used for the first time in 1912, and shortly afterwards bandpass filters were being studied by E.H. Colpitts and others in an attempt to separate individual channels in a carrier system. The major advances came through the work of G.A. Campbell (who originally proposed the use of such filters in 1909), and J.R. Carson, J.J. Carty, E.H. Colitts, K.W. Wagner and O.J. Zobel who provided techniques for the analysis and design of electric wave filters.[9]

Carrier systems were introduced into the Bell System in 1918 (Type A) and in 1920 (Type B). Based on the experience gained from these two systems the Type C system, which provided three voice channels in the frequency band 5 to 30 kHz, was introduced in 1924. Type C systems continued to be installed until the 1940s and the last were removed from service in the 1950s.[10] The use of carrier techniques, however, exacerbated the repeater amplifier problem since carrier systems need higher bandwidths, and for a given wire size, the losses in the wires increase with the signal frequency.

The designers of carrier systems encountered problems caused by the nonlinear characteristics of the triode vacuum tube. Owing to the nonlinear behaviour, amplification of a complex waveform, such as speech, results in an output containing harmonic components and intermodulation products. For a single voice channel with a small amplitude signal, the linear portion of the tube characteristic could be used, but with multiple channels, as in the carrier system, the intermodulation problem becomes more severe and, for speech to remain intelligible, the distortion introduced by each amplifier needs to be very low.

Soon after joining the Western Electric Company in 1921, Black produced a report in which he evaluated the requirements for transmitting thousands of channels over a transcontinental link. This was an ambitious and audacious proposal given that at the time the company engineers were struggling to make three channel systems with 10 to 12 repeaters work, and the Type C three channel carrier system had yet to be introduced. In the report he considered a system with 1000 amplifiers in a series. Assuming that the distortion products for each successive amplifier are not phase related, he argued that the overall distortion would be given by the root mean square of the individual distortion products; thus with 1000 amplifiers each could be allowed to contribute 1/30 of the total allowable distortion. R.V.L. Hartley quickly pointed out that the products arising from the third order distortion are phase related and thus to get an acceptable reduction in third order distortion each amplifier must not contribute more than 1/1000 of the total distortion. The amplifiers then in use were barely capable of coping with systems with 10 to 20 repeater stations and designers were struggling in an attempt to reduce the allowable distortion in each amplifier to 1/100 of the total.[11]

3.2 The invention of the negative feedback amplifier

Black requested permission to work on amplifier design which was granted on the condition that it did not interfere with his other work. He first used the conventional, well tried approach of attempting to produce vacuum tubes with linear characteristics. Others in the Bell Laboratories and elsewhere were attempting to do the same. An effort was also being made to improve the production techniques for vacuum tubes. Studies were carried out relating to quality control, and to the sensitivity of designs to quality variations, with the intent of reducing variations in characteristics between vacuum tubes.[12] Both approaches produced substantial improvements to vacuum tubes during the 1920s. The improvements achieved were not sufficient to solve the problems. Black was close to giving up. 'This might have been the end of it', he explained many years later, but having attended a lecture by Charles Proteus Steinmetz on 16 March, 1923, he was impressed by the fundamental way in which Steinmetz formulated problems, and on his return home at 2 am he restated the amplifier problem as being the need to 'remove all distortion products from the amplifier output'. Putting the problem in this way enabled him to accept that the amplifier might be imperfect and that 'its output [was] composed of what was wanted plus what was not wanted'. The problem was transformed into how to manipulate electrical signals. Distortion was not the only problem; there was also a problem with gain variations in the amplifiers which made it difficult to keep constant the overall net equivalent loss over the line.[13]

Very quickly, Black formulated a possible solution. He reasoned that if the amplitude of the output was reduced to that of the input and the two signals subtracted, the resulting signal would consist of the distortion products. By amplifying the distortion products in a separate amplifier the resulting signal could be subtracted from the original amplifier output, thus giving a distortionless final output. A repeater based on this idea was built, tried in the laboratory in March 1923, and found to work as expected. The repeater, however, was not suitable for general application, as it required the two amplifiers to have precisely balanced gains which had to be maintained over a wide frequency range, and over long periods of time.

For example, every hour on the hour — 24 hours a day — somebody had to adjust the filament current to its correct value. In doing this, they were permitting plus or minus 1/2-to-1dB variation in amplifier gain, whereas, for my purpose, the gain had to be absolutely perfect. In addition, every six hours it became necessary to adjust the B battery voltage, because the amplifier gain would be out of hand. There were other complications too, but these were enough.[14]

For several years Black wrestled with the problem. The solution came to him one Saturday morning (2 August, 1927) while on the ferry crossing the Hudson river on his way to work. He realised that by adding part of the output signal to the input signal but *in reverse phase* the distortion could be reduced to any desired extent at the expense of sacrificing some overall amplifier gain. Black sketched the basic circuit and equations on a copy of the *New York Times* — the only writing material he had to hand — and had the invention witnessed when he reached the office.

The basic idea was simple: if the amplifier has a gain of μ and the network in the feedback circuit a loss of β then

$$\frac{\text{output}}{\text{input}} = \frac{A}{F} = \frac{\mu}{1-\mu\beta} = \frac{1}{-\beta}\left[1 - \frac{1}{1-\mu\beta}\right] \qquad (3.1)$$

The gain with feedback is thus $\mu 5(1 - \mu\beta)$ abd if $\mu\beta$ is much greater than 1 then the amplification factor reduces to $1/-\beta$. Under this condition the gain is dependent solely on the feedback circuit network, usually a passive network; changes in the behaviour of the active components in the forward path, for example because of the effect of variations in power supply voltage or ageing of the tube, have almost no effect on the output. The feedback network can also be used to modify the overall frequency characteristics of the system. As Black was to explain in his 1934 paper, the noise and distortion components are also reduced by the factor $1/(1 - \mu\beta)$.

Black was not alone in investigating the effects of 'feedback' circuits on amplifiers and oscillators: one of the earliest studies was carried out by Friis and Jensen (in connection with AT&T's work on wireless systems).[15] In 1924, J.W. Horton in experiments with feedback circuits on vacuum tube oscillators noticed that negative feedback reduced the influence of component variations in the circuit. He wrote:

> There is another advantage in keeping the feed-back resistance high. In making it the major element in the network shunted across the resonant circuit, the effect of any variations in the output impedance of the tube or in the load impedance is reduced.[16]

However, most investigations were concerned with means of avoiding feedback since it was seen as undesirable and a cause of 'singing'.[17] Part of the reason for this was that positive feedback had been used in early radio sytems as a means of increasing the amplification, but by the mid-1920s, as E.W. Kellogg explained, 'it is so easy to get all the amplifications you need in the regular way that it hardly seems worthwhile to go to the difficulties that you encounter with feedback'.[18]

Within a few months Black, with the assistance of colleagues was able to demonstrate a negative feedback amplifier which reduced the distortion by 50 dB. They also showed that the feedback system was self correcting in that any change in the amplifier gain μ was automatically compensated by a change in the feedback signal. Black was aware that the applications of his invention went far beyond telephone amplifiers: 'the invention is applicable to any kind of wave transmission such as electrical, mechanical or acoustical', he claimed in the patent application. In total, he made 126 separate claims in the patent, including circuits for carrier amplifiers and for radio broadcast receivers.

The first patent application was submitted in 1928 and supplementary applications were filed in 1930 and 1932. After long drawn out arguments with the US patent office, it was finally granted in December 1937.[19] A reason for the delay in granting the patent, and for the treatment it received from the British Patent Office, was the claim that an amplifier with a loop gain greater than one could be stable: this was contrary to current belief. Many years later, Nyquist explained the problem as it was seen at the time in the following way:

> For best results, the feedback factor, the quantity usually known as $\mu\beta$, had to be numerically much larger than unity. The possibility of stability with a feedback factor greater than unity was puzzling. Granted that the factor is negative it was not obvious how that would help. If the factor was minus 10, the effect of one round trip around the feedback loop is to change the magnitude of an original current from, say, 1 to -10. After a second round trip around the loop the current becomes 100, and so forth. The totality looks much like a divergent series and it was not clear how such a succession of ever-increasing components could add to something finite and so stable as experience had shown.[20]

In the discussion of the 'prior art' in the patent application, Black clearly distinguishes between positive and negative feedback, arguing that the negative feedback used in radio frequency amplifiers was solely to oppose the positive feedback inherent in such amplifiers, whereas he proposed using negative feedback for a very different purpose. As well as drawing attention to the general improvements obtained (lower distortion, increased linearity, noise reduction and improved stability with respect to component variations), he also specifically mentions application to multiplex carrier systems, and the possibility of modifying the feedback path components to modify the attenuation-frequency characteristics of a transmission line.

Black's work marked a departure from the pulse response techniques for design then in use by telegraph engineers. Their concern was to maximise the pulse transmission rate for any real system but they did not have a method of determining what that rate might be, or of relating the pulse rate and the frequency response. Nyquist, in 1924, provided an answer to some of the designers' problems when he showed there was an upper limit to the pulse transmission rate and that this rate was related to the bandwidth of the circuit (in numerical terms he showed that the maximum pulse rate is twice the bandwidth). This work, he noted, provided a steady state approach to the design of telegraph transmission circuits.[21] John Carson in the early 1920s provided a similar steady state approach for systems with randomly varying signals (telephone and radio) as opposed to pulse signals (telegraph and radio-telegraph) when he showed that a randomly varying signal (for example the speech signal in a telephone system) could be represented by a set of steady state sinusoids, and that by using the convolution theorem the frequency and pulse responses of a circuit could be easily related. The importance of Carson's work was stressed by Hendrik Bode in his address made when receiving the Oldenburger Medal in 1975:

> It is worth noting, however, that before one can have the complex variable approach at all we must have converted the original randomly varying signals to an equivalent bundle of steady state sinusoids, using Fourier — later Laplace — transforms. This is almost a philosophical matter involving a basic question of what we mean by signalling anyway.[22]

To what extent Black was aware and influenced by Carson's work we do not know. Black was certainly working in an environment in which these ideas would have been discussed and there was a growing body of technical literature on the subject. For example, L.C. Pocock argued in 1924 that although the vowels could be treated as sustained signals and expressed in terms of the Fourier series, the consonants could not be so treated. In the discussion of his paper, however, he admitted that, 'it is also possible to deduce from the work of J.R. Carson that in a low-pass filter the response to an impulsive impressed force becomes less sluggish as the cut-off frequency is raised'. Pocock also examined the relationship between the gain and phase of a signal, concluding:

> It is seen, therefore, that the plot of the modulus of the mutual factor [mutual admittance or gain] of any physical system defines also the phase angle of the mutual factor, provided that the modulus is expressible as a fraction in integral powers of p [the Heaviside operator p].[23]

For many years — until Bode showed that there was a minimum phase shift associated with a gain characteristic — engineers searched in vain for circuits that combined particularly desirable gain characteristics with small phase shifts.

3.3 The practical carrier amplifier

During the period 1925–1929 the growth in telephone traffic was such that it threatened to outgrow the cable capacity. AT&T engineers began to investigate the possibility of introducing the carrier system on to the cable lines and in May 1928 Nyquist and others conferred with Black about the properties of his amplifier and its potential for use on a carrier system applied to cable lines. In November 1928 a specific proposal was made for a field trial of a cable carrier system. Morristown, New Jersey, was chosen as the site for the trial. The total length of the circuit was 850 miles and this was achieved by installing twenty-five miles of cable in existing ducts. The cable was specially made and comprised 68 pairs of 16 gauge wire. By using two pairs 34 four wire circuits each of 25 miles could be obtained and these were connected in series at the Morristown end to form one 850 mile circuit.

Installed at the Morristown end were the repeater amplifiers, 34 in total, providing a repeater every 25 miles. The circuit carried nine channels in the band 4 to 40 kHz. In addition to the amplification problem there were major problems with cross-talk and with the variation in cable loss caused by temperature changes. The cross-talk was reduced to acceptable levels by shielding in the cable and the use of balancing capacitors at the Morristown end. The loss variation was compensated for by using a feedback system to provide automatic adjustment of regulating networks placed in the line. Figure 3.1 shows the system used. It is based on the use of a self-balancing Wheatstone bridge and is similar to the widely used systems in recorders and in speed regulators for sectional drives (see Chapter 1). It is interesting to note that in the type K carrier system developed during the 1930s and brought into service in 1937, much of the contactor and relay mechanisms for controlling the adjustment to resistors had been replaced by synchro transmitter-receiver mechanisms — a similar pattern of development to the industrial sectional drive.

Moving from laboratory demonstration to practical implementation did not prove easy. The high gain necesssary in order to have gain to sacrifice could not be achieved with a single stage amplifier but there were many difficulties with multi-stage amplifiers. H.T. Friis and A.G. Jensen, in 1924, had investigated multi-stage amplifiers for radio work, and found that the large grid plate (anode) capacitance introduced unwanted positive feedback at radio frequencies which caused the amplifier to sing (oscillate). They also observed that in some circumstances the coupling could provide negative feedback and in this case the amplification was decreased. A further problem was that each stage of the amplifier introduced a large phase shift.[24]

In 1928 amplifier designers did not know how to handle these problems since they did not understand fully the conditions under which an amplifier with feedback would be stable: yet this knowledge was vital since the benefits of negative feedback as outlined by Black depended on the amplifier being stable. They knew they needed a large negative value of $\mu\beta$ over the useful bandwidth; they knew the gain decreased and the phase lag increased as the frequency increased; they knew that in multi-stage amplifiers the phase lag could quickly reach 180°; they also knew that if the gain $\mu\beta$ was greater than 1 when the phase lag reached 180° the amplifier would be unstable. But how could stability be determined? The only known method was the Routh test.[25] But in order to use the Routh test the characteristics of the amplifier system had to be expressed in the form of a set of differential equations. This posed little difficulty for the passive network components, but the equations describing the vacuum tube characteristics were non-linear, and a complex multi-stage amplifier required approximately 50 differential equations to describe its behaviour.

By the time Black's paper was published in 1934, many of these problems had been solved and the field trials had been successfully completed. Development of cable carrier systems was delayed because one of the effects of the Great Depression was the slowing down of traffic growth.[26]

In the introduction to the paper, after referring briefly to the difficulties engendered by component variations and the extraneous frequencies generated in carrier systems, Black wrote:

78 *The electronic negative feedback amplifier*

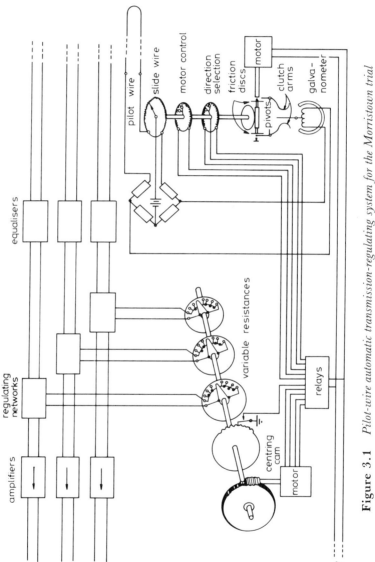

Figure 3.1 *Pilot-wire automatic transmission-regulating system for the Morristown trial*

Reproduced (with partial redrawing) by permission of E. F. O'Neill, from *A history of engineering and science in the Bell System: transmission technology (1925–1980)* (AT&T Bell Laboratories, 1985), p. 78

However, by building an amplifier whose gain is deliberately made, say 40 decibels higher than necessary and then feeding the output back on the input in such a way as to throw away the excess gain, it had been found possible to effect extraordinary improvement in constancy of amplification and freedom from non-linearity. By employing this feedback principle, amplifiers have been built and used whose gain varied less than 0.01 db with a change in plate voltage from 240 to 260 volts and whose modulation products were 75 db below the signal output at full load.

Stabilized feedback possesses other advantages including reduced delay and delay distortion, reduced noise disturbance from the power supply circuits and various other features best appreciated by practical designers of amplifiers.

It is far from a simple proposition to employ feedback in this way because of the very special control required of phase shifts in the amplifier and feedback circuits, not only throughout the useful frequency band but also for a wide range of frequencies above and below this band.[27]

In a simple, succinct and clear way Black described the benefits to be achieved from negative feedback — benefits which were being sought by engineers in many different fields — and the problem that existed in obtaining the benefits.

Figure 3.2 shows a reproduction of the schematic diagram used by Black to illustrate the amplifier. He derived the following equation;

$$E + N + D = \frac{\mu e}{1 - \mu \beta} + \frac{n}{1 - \mu \beta} + \frac{d(E)}{1 - \mu \beta} \qquad (3.2)$$

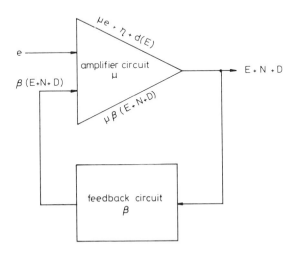

Figure 3.2 *Amplifier system with feedback*

Reproduced (with partial redrawing) by permission of H.S. Black, from *Bell System Technical Journal*, 1934, **13**, p. 3

where E, N, D represent the signal, noise, and distortion outputs, respectively, from the amplifier; hence the sum is the total output signal; μ is the gain of the amplifier (Black was careful to draw attention to the fact that this was 'complex ratio of the output to the input voltage of the amplifier circuit' not as 'it is sometimes used, namely, to denote the amplification constant of a particular tube'); and β is the gain of the feedback circuit.

From the equation Black deduced that when $|\mu\beta| \gg 1$ then $E \simeq -e/\beta$ and commented:

> Under this condition the amplification is independent of μ but does depend on β. Consequently the over-all characteristic will be controlled by the feedback circuit which may include equalizers or other corrective networks. [p.3]

He further noted that the noise and distortion are reduced by the loop gain and are given by $N \simeq -n/\mu\beta$ and $D \simeq d(E)/\mu\beta$.

Following on from the general description of the properties of the amplifier Black considers how changes in the feedback β modify the gain, the stability, and modulation (distortion and noise) characteristics of the amplifier. He uses a sensitivity analysis approach in that he considers the changes in gain, stability etc. caused by a change in β. For design purposes he proposes the use of polar and rectangular plots of the vector field $\mu\beta = |\mu\beta| < \Phi$ on which are plotted contours of constant change of gain. Using these contours he attempts to identify areas of the plot in which the amplifier will be stable. Although the plots are similar to Nyquist plots it would appear from the paper that Black had devised the charts before consulting with Nyquist about the conditions of stability, because he wrote:

> one noticeable feature about the field of $\mu\beta$ is that it implies that even though the phase shift is zero and the absolute value of $\mu\beta$ exceeds unity, self-oscillations or singing will not result. This may or may not be true. When the author first thought about this matter he suspected that owing to practical non-linearity, singing would result whenever the gain around the closed loop equalled or exceeded the loss and simultaneously the phase shift was zero, that is, $\mu\beta = |\mu\beta| + j0 \geq 1$. Results of experiments, however, seemed to indicate something more was involved and these matters were described to Mr. H. Nyquist, who developed a more general criterion for freedom from instability. [p.11–12].

Many years later, when describing how he had invented the amplifier, Black said that the reason why he knew he 'could avoid self-oscillations over very wide frequency bands when many people doubted such circuits would be stable' was because of the confidence he gained from work he had done earlier on oscillator circuits. There he had shown that for oscillations to occur, 'the loop transfer factor must be real, positive and greater than unity at some frequency. Consequently, I knew that in order to avoid self-oscillation in a feedback amplifier it would be sufficient that at no frequency from zero to infinity should $\mu\beta$ be real, positive and greater than unity.'[28] Arnold Tustin reached the same conclusion in 1930 by using a similar argument, although he had no knowledge of Nyquist's or Black's work.[29]

In the rest of the 1934 paper, Black presents experimental results based on the amplifiers used in the Morristown field trials. For example, Figure 3.3 shows the improvement in gain stability over the frequency range. Other figures are used to show the effect of feedback on the modulation products: the gain variation with changes in output load and changes in plate voltage. The paper also hints at the problem which was to occupy the Bell Telephone Laboratories for the rest of the 1930s: equalisation.[30]

Although the 1934 paper was published under Black's name alone — in keeping with his status as the originator of the idea — the idea was brought to practical fruition through team work.[31] It has been pointed out by Leonard Reich that one of the benefits of industrial research laboratories was (and is) their ability to bring together teams of experimentalists, theoreticians, technicians and clerical staff who together are more productive than they would be as individuals.[32] Members of such teams do not have to be individually brilliant to contribute to the success of the work. The practical negative feedback amplifier is a product of such team work.

3.4 Nyquist and the stability criterion

An outcome of Black's request for assistance in understanding the conditions under which the feedback amplifier is stable was Harry Nyquist's paper 'Regeneration

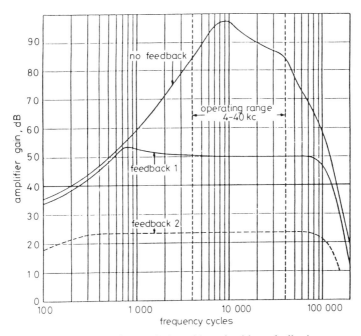

Figure 3.3 *Gain frequency characteristics with and without feedback*

Reproduced (with partial redrawing) by permission of H.S. Black, from *Bell System Technical Journal*, 1934, **13**, p. 12

82   The electronic negative feedback amplifier

Theory', published in 1932.[33] Underlying the whole paper, although not explicitly referred to, is the understanding at which he had arrived in 1924, that the behaviour of a system can be analysed in terms of its frequency characteristics and that all impressed signals can be described in terms of their Fourier components.[34] The paper commences with a definition of stability:

> The circuit will be said to be stable when an impressed small disturbance, which itself dies out, results in a response which dies out. It will be said to be unstable when such a disturbance results in a response which goes on indefinitely, either staying at a relatively small value or increasing until it is limited by the non-linearity of the amplifier. [p.126]

In starting with this definition Nyquist was trying to avoid confusion with oscillator circuits which relied on the non-linearity characteristics to limit the amplitude of oscillation and which were termed stable if the frequency of this oscillation remained constant. Further to avoid confusion he assumed 'the use of a strictly linear amplifier of unlimited power carrying capacity' and considered whether with such an amplifier 'an initial impulse dies out or results in a runaway condition'. He noted that if a runaway condition developed in the assumed, linear, amplifier it would also occur in a similar non-linear, real, amplifier.

Following on from this definition Nyquist argues that previous attempts to determine the conditions for stability, based on expressing successive passages of the signal round the loop in the form of an infinite series, failed because they neglected transient effects. His approach assumes that a signal—a 'disturbing wave'—is introduced somewhere in the loop. The signal or wave is represented by its Fourier transform, as is the loop gain $AJ(i\omega)$. He then uses the convolution property to argue that after n traverses of the circuit the voltage is given by

$$s(t) = \sum_{k=0}^{n} f_k(t) = \lim_{n \to \infty} \frac{1}{2\pi i} \int_{S'} S_n(z) e^{zt} \, dz \quad [18] \qquad (3.3)$$

where

$$S_n = F + Fw + Fw^2 + \ldots Fw^n = \frac{F(1 - w^{n+1})}{1 - w} \quad [19] \qquad (3.4)$$

(Numbers in square brackets are the equation numbers given in Nyquist's paper.)

He shows that $s_n(t)$ converges absolutely as n increases. His next question is what happens to $s(t)$ as t increases, 'what properties of $w(z)$ and further what properties of $AJ(i\omega)$ determine whether $s(t)$ converges to zero or diverges as t increases indefinitely?' Using complex variable theory he argues that in the limit $n \to \infty$ equation 3.3 reduces to

$$s(t) = \frac{1}{2\pi i} \int_{s^+} [F/(1-w)] e^{zt} \, dz \quad [33] \qquad (3.5)$$

and that the values of the integral along the imaginary axis approaches zero at very large values of t, thus it suffices to evaluate the integral round the semicircle

$$\int_C [F/(1-w)]e^{zt}\,dz \quad [35] \tag{3.6}$$

By assuming that $1-w$ does not have a root on the imaginary axis and that the disturbing function $F(z)$ has the special value $w'(z)$ allows him to write the integral in the form

$$\frac{1}{2\pi i}\int_C [w'/(1-w)]e^{zt}\,dz \quad [36] \tag{3.7}$$

Changing the variable gives

$$\frac{1}{2\pi i}\int_D [1/(1-w)]e^{zt}\,dw$$

where D is the curve in the w-plane which corresponds to the curve C in the z-plane. The imaginary axis of the z-plane transforms into the locus $x = 0$ in the w-plane and since he had defined $w(z)$ such that in the limit $z \to 0$

$$w(z) = w(iy) = AJ(iy) \tag{3.8}$$

then the locus corresponding to $x = 0$ is the locus of $AJ(i\omega)$.

Using the method of residues, Nyquist then argues that the circuit will be stable providing that the point $\omega = 1$ lies wholly outside the locus $x = 0$ and unstable if the point is within the locus. Based on this, he gives the following rule:

> Rule: Plot plus and minus the imaginary part of $AJ(i\omega)$ against the real part for all frequencies from 0 to ∞. If the point $1 + i0$ lies completely outside this curve the system is stable; if not it is unstable [p.136].

The formulation of the stability problem in terms of the open-loop gain and phase, quantities which can be directly measured, was of enormous practical significance. As we have already observed, the Routh-Hurwitz criterion depends on the availability of a differential equation based system model whereas experimentally determined data is all that is required to apply Nyquist's criterion. It has a further advantage over the Routh-Hurwitz criterion in that it provides a measure of the degree of stability of the system, and shows the frequency range to which attention should be given in order to improve the stability.[35]

Nyquist made reference to a possible feedback amplifier circuit which would have the properties that:

> For low values of A [the amplifier gain] the system is in a stable condition. Then as the gain increased gradually, the system becomes unstable. Then as the gain is increased gradually still further, the system again becomes stable. As the gain is still further increased the system may again become unstable [p.137].

E. Peterson, J.G. Kreer and I.A. Ware experimentally investigated Nyquist's criterion and also this 'striking conclusion'; their results, reported in 1934, confirmed Nyquist's analysis and showed that an amplifier which behaved in the way outlined above could be constructed.[36]

The mathematics underlying Nyquist's analysis of the problem are derived from the work of Cauchy.[37] Nyquist formulated the problem such that Cauchy's Principle of the Argument could have been used but then proceeded by way of a simple argument based on the calculation of residues. It was L.A. MacColl, another mathematician who worked for the Bell Laboratories, who some years later gave a proof for a restricted class of functions which could be specified as rational functions of a complex variable which was based on the Principle of the Argument and this became the standard method of proof used in many textbooks.[38]

Nyquist specifically excluded from consideration systems which had poles in the closed right-half plane. Many practical servomechanisms have a pure integration term in the open loop which give rise to poles at the origin and both MacColl, and A.C. Hall, provided appropriate extensions to the Nyquist criterion to deal with such systems.[39] In 1945, Frey extended the criterion to deal with systems which were open loop unstable.[40]

Nyquist was not alone in investigating stability problems. In Germany, Barkhausen, who was studying oscillator circuits, gave the formula for self excitation as

$$KC(j\omega) = 1$$

where K is the amplifier gain and $F(j\omega)$ the frequency dependent characteristic in the feedback loop. As we have seen this was the same criterion ($\mu\beta = 1$) which was being used by the Bell engineers and others in the USA. The criterion was widely used in Germany for determining the stability of both positive and negative feedback amplifiers. K. Küpfmuller made an important contribution to the development of frequency response methods in papers dealing with the relationships between frequency transmission characteristics and transient behaviour, and with closed-loop stability. His approach to closed loop stability was to represent the dynamical behaviour of the system in the form of an integral equation and he developed an approximate criterion for closed loop stability in terms of time-response quantities measured and calculated from the system transient response. F. Strecker, in his book published in 1950, claims to have presented a frequency response approach and criterion equivalent to Nyquist's at a colloquium at the Central Laboratory of Siemens and Halske in 1930, although publication in the technical literature did not occur until after the Second World War.[41]

A.V. Mikhailov in 1938 suggested a frequency response method that has superficial similarity to Nyquist's method. His technique, however, involves obtaining the characteristic polynomial $p(s)$ from the differential equations of the system and plotting the locus of $p(j\omega)$ in the p-plane. The crucial feature of the Nyquist method is that knowledge of the differential equations of the system is not required. Cremer (1947) and Leonhard (1940) proposed methods similar to Mikhalov's.

3.5. Hendrik W. Bode

Although the work of Black and Nyquist provided the Bell Laboratories' engineers with some of the tools to support amplifier design, there were still uncharted territories. Engineers spent many fruitless and expensive years in attempts to build amplifiers with both a fast cut-off in gain and a small phase shift.[42] The problem was eventually solved by Hendrik Bode who observed that, 'it is apparent that no entirely definite and universal relation between the attenuation and the phase shift of a physical structure can exist.'[43] Thus a designer has the freedom to trade attenuation against phase with a resulting ambiguity in the gain-phase relationship but, he continued,

> While no unique relation between attenuation and phase can be stated for a general circuit, a unique relation does exist between any given loss characteristic and the *minimum* phase shift which must be associated with it. [p.424]

The search for fast cut-off and small phase shift had been futile.[44]

Bode's involvement in feedback circuits began in 1934 when he was asked to design a variable equaliser to compensate for the effect of temperature variations in a coaxial line transmission system that was being developed. The problem was difficult because the loss varied both with temperature and also with frequency. The proposal was to compensate for — or equalise — the loss characteristic by modifying the feedback (β) path of an already designed amplifier. Unable to obtain the required characteristics and stability by changing the feedback components, Bode had to redesign the complete feedback loop. The difficulties encountered in this design led Bode to study the extensive literature on networks dealing with the gain and phase relationships. Extending the work of Y.W. Lee and Norbert Wiener,[45] he obtained the relationship between phase and gain under the minimum phase condition which he expressed in two different ways:

$$\int_{-\infty}^{\infty} B \, du = \frac{\pi}{2}[A_\infty - A_0] \tag{3.9}$$

$$B(f_c) = \frac{1}{\pi} \int_{-\infty}^{\infty} \frac{dA}{du} \log \coth \frac{|u|}{2} \, du \tag{3.10}$$

where $u = \log f/f_c$, (f_c is an arbitrary reference frequency); B = phase shift in radians; and A_∞, A_0 are the attenuation (in nepers) at infinite and zero frequencies respectively.

Bode explained that the relationships state 'that the total area under the phase characteristic plotted on a logarithmic frequency scale depends only upon the difference between the attenuations at zero and infinite frequency, and not upon the course of the attenuation between these limits'. He continued by remarking that

> [The] significance of the phase area relation for feedback amplifier design can be understood by supposing that the practical transmission range of the amplifier extends from zero to some given finite frequency. The quantity $A_0 - A_\infty$ can be identified with the change in gain around the feedback loop required to secure cut-off. Associated with it must be a certain definite phase area. If we suppose that the maximum phase shift at any frequency is limited to some rather low value the total area must be spread out over a proportionately broad interval of the frequency scale.' [p.425]

In other words, if the phase shift at some specified frequency is to be kept small, then the cut-off rate must be kept small so that the necesssary total area under the phase curve can be obtained. Using the second of the above equations he showed that if the attenuation was $6k$ dB per octave, then the phae shift is $k\pi/2$ radians. Using this feature, and other attenuation characteristics, he developed the asymptotic construction techniques for drawing the so-called Bode diagrams used in feedback control design.

Bode related his work to that of Nyquist. He is responsible for rotating the Nyquist diagram through 180°, thus making the critical point − 1,0 rather than + 1,0. He argued that the rotation of the diagram allowed the phase reversal in the amplifier tubes to be ignored and showed the phase shifts with values that were of direct interest for the design.[46] Of more importance, however, were the concepts of a phase margin and gain margin which he introduced. He argued that the theoretical condition for stability — that the phase shift must not exceed 180° until the loop gain is reduced to 1 or less — was not adequate, because in practice there will be inaccuracies in the design and construction of the amplifier. The limiting phase angle must therefore be less than 180° by some definite amount. He called this amount the phase margin. Similarly, he argued that in practice it is physically impossible to restrict the phase shift to less than the maximum specified once the frequency increases beyond the required bandwidth. Hence for a stable system the gain must be reduced: the amount by which the gain is less than 1 (0 dB) at a phase shift of 180° is the gain margin.

In the final section of the paper, Bode examined the behaviour of the feedback system if the frequency band did not extend to zero. He concluded that 'the feedback which is obtainable in an amplifier of given general configuration and with given parasitic elements and given margins depends only upon the breadth of the ban in cycles and is independent of the location of the band in the frequency spectrum'.[47]

The value of Bode's 1940 paper and of the book published in 1945[48] was the way in which they linked the theoretical ideas with practical examples, and the way in which he explained the limitations of the theory.

3.6 Wireless communication systems

The development of radio — wireless communication — was seen by AT&T as a potential threat to its telephone business, and the company invested in research in the field. It was from work on wireless receivers that the word 'feedback' emerged.[49] And we have already seen from Black's discussion of the prior art in his patent application that both negative and positive feedback were commonly used in radio receivers by the 1920s.

During the 1920s and 1930s a further control problem emerged in radio, that of maintaining accurately the frequency of transmission, and there was a major effort to develop frequency stable oscillators. The journals of the period abound with papers on the stability of oscillators, with circuit designs to improve stability and statements about the valve characteristics required. Few of the papers make any attempt to distinguish, in a fundamental sense, between changes to circuit parameters to reduce frequency drift, modifications to circuits involving internal feedback, and systems which involved the monitoring of the output frequency and comparison with some reference value.[50]

Engineers found difficulty in dealing with the problem largely because it was perceived as a steady state problem: the need to maintain steady oscillations. Because the problem was seen in this way, the solutions developed focused on the inherent regulation characteristics of components. And engineers sought, just as Black did initially, components whose characteristics did not change over time, or with changes in environmental conditions.

In many of the early systems the oscillator circuit also provided power for the transmission antenna as Figure 3.4 illustrates. Designers quickly realised the improvement in the regulation of frequency achievable if the oscillator circuit was separated from the circuit driving the antenna and was limited to low power output. Moving to the circuit shown in Figure 3.5 was a much bigger step for there was an assumption that frequency variations caused by changes in the power amplifier characteristics were second order effects. Introducing negative feedback round the system changed the problem to one of obtaining a stable master oscillator. The provision of a stable reference for the set point of a system is a problem which has plagued many regulatory control systems. C.W. Siemens met the problem in the 1840s when he used a differential gear system to compare the engine speed with a constant reference speed in an engine governor. All he achieved was to transform the control problem into the design of a reference system running at constant speed — a different control problem.

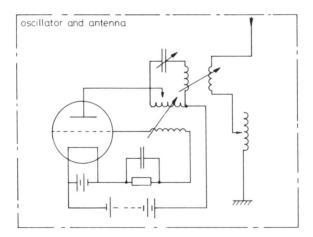

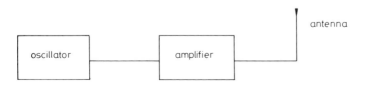

Figure 3.4 *Combined oscillator and transmitter*

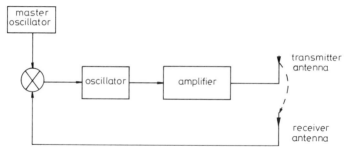

Figure 3.5 *Radio transmitter system with feedback*

Large numbers of schemes were proposed, all based on using a low-power control oscillator followed by an amplifier. Many devices were investigated as a basis for the control oscillator but by 1930 there was widespread use of the piezo-electric effect (using quartz crystals) for certain types of oscillator[51] and a realisation that accurate frequency control depended on maintaining the crystal at a constant temperature.[52]

Elaborate schemes for constant temperature enclosures were investigated and as the stability of oscillations obtained from the piezo-electric crystal itself improved, variations caused by the rest of the components in the circuit became more significant. V.E. Heaton noted that the temperature of all the components should be kept constant.[53] Alternative methods of producing reference or master oscillators ran into the same temperature regulation problem. J.W. Conklin, J.L. Finch and C.W. Hansell[54] claimed success in 1931 with a method using long transmission lines; however, five years later they reported that compensation for temperature variations was necessary.[55]

Alternative methods using feedback to correct for errors due to variations in component characteristics began to be used. Figure 3.6 illustrates a typical arrangement (an accurate control oscillator was still required to provide the reference frequency). In these schemes there was still considerable difficulty in escaping from two established ideas: one was the idea that the connection between the measurement and the actuating system must be some type of electromechanical device; the other that the use of electronic valves for amplification involved the use of positive feedback.[56] Typical of the former is a system described by Y. Kusonose and S. Ishikawa in 1932.[57] Figure 3.7 shows the principle of the system. A low powered control oscillator, assumed to have good frequency stability,

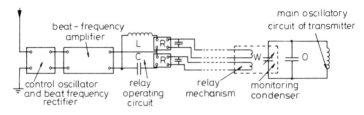

Figure 3.6 *General circuit arrangement of an automatic monitor*

Reproduced (with partial redrawing) by permission of H.A. Thomas, from *J. IEE*, 1936, **78**, p. 18

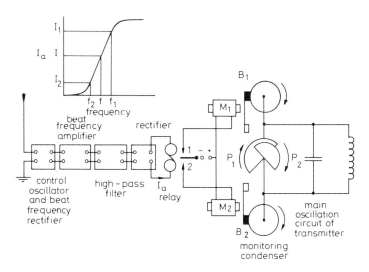

Figure 3.7 *Schematic diagram of Kusonose mechanical monitor system*
Reproduced (with partial redrawing) by permission of H.A. Thomas, from *Theory and design of valve oscillators* (Chapman & Hall, 1944), 3rd edn., p. 246

provides the frequency set point. Mixing the output of the transmitter (obtained from a receiving aerial) with the output of the control oscillator gives a beat frequency signal which is amplified and rectified. The rectified signal is fed to the operating coil of a relay controlling the current to two electromagnets $M1$ and $M2$. If the frequencies are equal then the relay is in its null position and both electromagnetics are off. In this condition the two brakes $B1$ and $B2$ are on and the motors driving the vernier condensers $P1$ and $P2$ cannot rotate. If the frequencies are not equal, the increase or decrease in current through the rectifier activates one or other of the electromagnets, thus releasing one of the brakes. The resulting rotation of the vernier capacitance plates placed across the main capacitor changes the frequency of the transmitter. The system obviously has a deadband and hence there would be some variation in frequency. The size of the deadband was controlled by the gain of the beat frequency amplifier.

The problem of inertia in such electromechanical systems was becoming well known and an improved, lower, inertia system was devised by H.A. Thomas in 1936.[58] Reeves (1931) proposed an all electronic system, as did Y. Kusonose and S. Ishikawa in 1932, and gradually there was a move to such systems.

The importance of oscillators both for receivers and transmitters led to attempts to model such systems and to develop a body of theory. Analysis is difficult because an oscillator is a nonlinear device. Major contributions to theoretical understanding were made by B. Van der Pol (1920, 1926 and 1934), E.V. Appleton (1921, 1922, and 1923), Lienard (1928), Llewellyn (1931) and Le Corbeiller (1931 and 1936).

3.7 Summary

Comparing developments in telephone and then wireless communication systems illustrates the great significance of Black's recognition that negative feedback has benefits other than simply providing stability. The period marks a change in emphasis in automatic control from the problem of regulation, for which an open loop solution involving stable components with inherent regulation can be satisfactory in many cases, to the problem of automatic following of signals, the resolution of which requires closed loop feedback systems.

In examining the way in which 'feedback' has been used in radio applications, Black clarified understanding and through his realisation that the behaviour of the complete system could be modified by changing the passive components in the feedback network, he pointed the way forward. His concept of treating the problem as one of manipulating signals opened the way to the modern treatment which represents components by their transfer function and manipulates systems in terms of blocks. The approach was made more powerful by being able, through the use of the Nyquist diagram, to handle both blocks whose characteristics were known in terms of a transfer function (that is a mathematical model) and blocks described solely in terms of an empirically determined frequency response.

Recognition of the importance of Nyquist's work spread slowly. It has been suggested that the paper was designed to conceal and confuse as much as to reveal.[59] A more likely explanation of the slowness with which it spread was the lack of a common nomenclature across the various fields of application, and the lack of recognition of the commonality of many of the problems. Few companies had gone as far as the Bell Laboratories and the General Electric Company in improving components, so that for many systems greatly improved performance could be achieved by reducing the variability of components and their immunity to disturbance and drift. Where feedback was being used the problems were still at the stage of asking the basic question, 'is the system stable'? not 'how well does it perform and how can we improve its performance?'.

Whether earlier publication of information about Black's negative feedback amplifier — say in 1928 or 1929 — would have resulted in the Nyquist criterion being more widely and quickly utilised is impossible to say with certainty. There was certainly considerable work during the 1920s and early 1930s and many other researchers who were on the edge of a breakthrough with feedback control could have benefited from the ideas. Earlier publication may have avoided the entrenchment of the $|\mu\beta| \leq 1$ criterion. The issue does raise questions about the publication policies of industrial research laboratories, which obviously have to balance several conflicting aims: patent rights, commercial potential, company standing and staff aspirations. It is regrettable that the otherwise excellent histories of the research work of the Bell Laboratories do not address this question.

John Taplin of MIT adapted the Nyquist tools for use in general feedback systems in about 1937 when he applied the techniques to a servomechanism. The major step in the integration of the approach in general feedback design was taken by Herbert Harris in a report produced for the NDRC in 1942 in which he introduced the general idea of a transfer function and showed how it could be applied to process plants and mechanical servomechanisms as well as to the negative feedback amplifier.

The use of frequency response techniques for investigating system performance led to attempts to relate frequency and transient response data. For example, L.H. Bedford and O.S. Puckle attempted to relate the frequency characteristics of an amplifier — which they expressed in the form of separate amplitude and phase plots against logarithm of frequency — to the transient response of the system obtained by the use of the Heaviside operational calculus. They commented:

> Although the amplitude and phase characteristics contain implicitly the whole information as the behaviour of the amplifier, it is not entirely an easy matter to deduce from the response of the amplifier to a transient.[60]

The dominant position occupied by frequency domain techniques for control system design during the 1940s and 1950s has led to an overemphasis of the work carried out by Harold Black, Harry Nyquist and Hendrik Bode during the 1930s. The quality, originality and significance of their work is beyond doubt. The content, style, and assurance of their papers published during this period demonstrates a much deeper understanding of feedback concepts than that of their contemporaries; it is perhaps only Haraold Hazen and Nicolas Minorsky who come close to this level of understanding. Their success is partially attributable to the management and organisation of the Bell Laboratories and exemplifies the importance of industrial research laboratories in carrying out a sustained, large scale, research and development program. The Bell Laboratories recruited the best qualified staff they could find and encouraged their staff to be open minded, to publish and to further their education.[61] The Laboratories also had a clear aim and commitment to maintain control of the technology in order to sustain a monopoly position in the market.

In process control there was almost the opposite position: there were many companies competing and a reluctance to share information or to be open about developments. Many of the companies had emerged through the 'inventor' route so although there was often a small number of people within the company with a sound understanding of basic physics and mechanics there was little support for long term basic research. The instrument manufacturing companies gradually established research departments; the Leeds & Northrup Company was probably the first to do so, but the number of professionally qualified staff employed was small. The instrument makers used the patent system but were too small to control the market by such means and they relied more on the 'trade secret' to protect production techniques, in particular.[62] However, drawing from hydraulic servo-mechanism practice, engineers began to develop negative feedback amplifiers for process controllers and to recognise the importance of negative feedback in improving linearity and in reducing the effects of external and internal disturbances. The proportional plus reset controller of Clesson E. Mason, patented in 1930, introduced negative feedback around the flapper-nozzle amplifier, thus creating the pneumatic equivalent of Black's electronic negative feedback amplifier.

The idea that linear operation and improved rejection of noise and disturbances could be obtained through the use of negative feedback was difficult to grasp; engineers still sought after devices, components, mechanisms, and systems exhibiting inherent self-regulation. The search of the communication engineers for methods of achieving stable frequency oscillators epitomises this approach.

In the colleges, apart from some work on engine governors, control was not taught, and neither was much attention given to dynamics of systems , until the 1930s. 'Engineering graduates from American colleges...usually did not have sufficient training along these [mathematics and fundamentals] lines', complained A.R. Stevenson and A. Howard of the General Electric Company in 1934. 'Most of them had no mathematical training beyond the calculus, and those who had studied differential equations had not learned how to make use of them in the analysis of physical problems'. The General Electric Company was recruiting 'many excellent, ingenious designers, but relatively few who were thoroughly grounded in methods of analysis' and the problem was blamed on the colleges whose courses were thought to be 'too practical; men were being taught routine rule of thumb methods of design, and there was insufficient emphasis on thinking problems through by the use of fundamental principles'.[63] The General Electric Company remedied the problem by running its so-called 'Advanced Course', which, according to Fritz Johnson, in 1932, concentrated on fundamentals such 'as proving the operational formulas of Heaviside by means of the Bromwich integral'. It was not until later years that the Laplace and Fourier transform methods were introduced. Other companies were also concerned about graduate training for their engineers. The Westinghouse Company, in conjunction with the University of Pittsburgh, set up a training programme for its engineers. This was, however, considerably more practically oriented than General Electric's Advanced Course.[64]

The colleges were not entirely to blame. Operational techniques were taught as part of the electrical engineering courses at MIT in the late 1920s and such courses were taken by 'co-operative' students supported by General Electric. The company, however, failed to recruit many of the students who graduated from the course.[65]

Some educationalists were aware of the need for engineers to be fluent in the handling of differential equations: 'this subject probably has a greater practical utility to engineers than does calculus', wrote P.W. Ott in 1934. 'It is an extremely essential tool for researchers and experimenters, and all those engaged in engineering requiring much analytical thinking. In modern electricity it is almost essential'.[66] Changes were slow, according to Arnold Tustin many engineers in the late 1930s still tended to think in terms of static design charts and, even if they had studied differential equations during their education, they found no use for them in their every day work.[67]

3.8 Notes and references

1 'Feedback Amplifiers' in *Bell Laboratories Record,* **12** (1933–34), p. 290
2 Black, H.S.: 'Stabilised feedback amplifiers', *Bell System Technical J.,* **13** (1934), pp. 1–18; this paper also appeared in *Electrical Engineering,* (January, 1934) and a shortened and simplified version appeared under the title 'Feedback Amplifiers' in *Bell Laboratories Record,* **12** (1933–34),pp. 290-296
3 Black, H.S.: 'Inventing the negative feedback amplifier', *IEEE Spectrum,* **14** (1977), pp. 55-60. The major sources for this section are: O'Neill, E.F.: *A History of Engineering and Science in the Bell System: Transmission Technology (1925–1975)* (AT&T Bell Laboratories, 1985); Reich, L.S.: *The Making of American Industrial Research: Science and Business at GE and Bell, 1987–1926,* (Cambridge University Press, Cambridge, 1985); and Shaw, T.: 'The conquest of distance by wire telephony', *Bell System Technical J.,* **23** (October 1944), pp. 337–421
4 For a detailed discussion of the Vail's attempts to maintain a monopoly of the telecommunications system see Reich *op. cit.* (n.3 above) chapters 6, 7 and 8. The Company also used every form of publicity available in an attempt to create a favourable image

with the public, even to the extent of trying to persuade a film director to cut out a sketch showing hysteria in a central exchange: see Fischer, C.S.: 'Touch somebody: the telephone industry discovers sociability', *Technology & Culture* **29** (1988), p. 37

5 The research laboratories originally formed part of the Engineering Department of Western Electric in New York but were formed into a separate entity — the Bell Telephone Laboratories — in 1925

6 Quoted from Reich, *op. cit.* (n.3 above), p. 158; the memorandum is reproduced in full in an appendix to the paper by Shaw, *op. cit.* (n.3 above)

7 For a detailed account of the work done by Arnold and others see *A History of Engineering and Science in the Bell System*, **Vol. 1** *The Early Years* and **Vol. 4** *Physical Sciences (1925-80)*, (AT&T Bell Laboratories, 1985)

8 Shaw cites this as being one of the deciding factors in the move towards the use of cables — as a result of the failure Vail took the decision to establish an underground system using cable along the Atlantic seaboard from Washington to Boston, Shaw, *op. cit.* (n.3 above), pp. 354-355

9 Atherton, W.A.: *From Compass to Computer*, (Macmillan, 1984), pp. 228-229; Colpitts, E.H.: *Collected papers of George Ashley Campbell*, (AT&T, New York, 1937), pp. 1-2; see also Campbell (1922): Carson (1922)

10 O'Neill, *op. cit.* (n. 3 above), pp. 73-74

11 *Ibid.* p. 63; Black recalls being determined to find out how Western electric worked and going in on a Sunday to read through the memorandum files which were kept as a means of recording the work of research engineers; he discovered that repeater amplifiers were a major source of trouble in the telephone system, Black, 1977 *op. cit.* (n.3 above), p. 55

12 See, for example, Pocock, L.C.: 'Distortion in thermionic tube circuits', *Electrician*, **86** (1921): p. 246; Bethenod, J.F.J.: 'Distortion free telephone receivers', *Proc. Institute of Radio Engineers*, **11** (1923); p. 163; Pocock, L.C.: 'Faithful reproduction in radiotelephony', *J. of the Institution of Electical Engineers*, **62** (1924), pp. 781-815; Crisson, G.E.: 'Irregularities in loaded telephone lines', *Bell System Technical J.*, **4** (1925); Kellogg, E.W.: 'Design of non-distorting power amplifiers', *Trans. American Institute of Electrical Engineers*, **44** (1925), pp. 302-17. A survey of early work is given in Hoyt, R.S.: 'Probability theory and telephone transmission engineering', *Bell System Technical J.*, **12** (1933), pp. 35-75

13 Black, 1977 *op. cit.* (n.3 above), p. 58; see Clark, A.B.: 'Telephone transmission over long cable circuits', *Trans. American Institute of Electrical Engineers*, **42** (1923), pp. 86-97.

14 Black, 1977 *op. cit.* (n.3 above), p. 58

15 Friss, H.T., Jensen, A.G.: 'High frequency amplifiers', *Bell System Technical J.*, **3** (1924), pp. 181-205

16 Horton, J.W.: 'Vacuum tube oscillators', *Bell System Technical J.*, **3** (1924), pp. 508-24

17 See, for example, Fletcher, H.: 'The theory and the operation of the howling telephone with experimental confirmation', *Bell System Technical J.*, **5** (1926); investigations continued while Black was developing the negative feedback amplifier, see, for example, Crisson, G.: 'Negative impedance and the twin 21-type repeater', *Bell System Technical J.*, **10** (1931), pp. 485-513

18 Kellogg, E.W.: 'Design of non-distorting power amplifiers', *Trans. American Institute of Electrical Engineers*, **XLIV** (1925), p. 315

19 Black, H.S.: 'Wave Translation System', US Patent No. 2,102,671 issued 21st December, 1937: the first application was filed 8th August, 1928, with supplementary applications filed on 26th March, 1930 and 22nd April, 1932

20 Nyquist, H.: Luncheon address presented at the Frequency Response Symposium, held at the Annual Meeting, New York, NY, 29th November-4th December, 1953, of the American Society of Mechanical Engineers, published in Oldenburger, R. (ed.): *Frequency Response*, (Macmillan, New York, 1956); reprinted in *Trans. ASME, J. of Dynamic Systems, Measurement, and Control*, (June, 1976), pp. 128-9

21 Nyquist, H.: 'Certain factors affecting telegraph speed', *Bell System Techical J.*, **3** (1924), pp. 324-346; the paper was also published in *Trans. American Institute of Electrical Engineers*, (1924); see also Nyquist, H.: 'Certain topics in telegraph transmission theory', *Trans. American Institute of Electrical Engineers*, **47** (1928), pp. 617-44; and Hartley, R.V.L.: 'Theory of information', *Bell System Technical J.*, **7** (1928)

22 See Bode, H.W.: *Trans. American Society of Mechanical Engineers, J. of Dynamic Systems, Measurement, and Control* (June 1976); p. 126
23 Pocock, L.C.: 'Faithful reproduction in radio-telephony', *J. Institution of Electrical Engineers*, **62** (1924), pp. 791-815
24 Friis and Jensen, *op. cit.* (n.15 above)
25 Telephone engineers were aware of Routh by this time. See, for example, Foster, *Bell System Technical J.*, **3** (1924); and Peterson, E., Kreer, J.G., Ware, L.A.: *Bell System Technical J.*, **13** (1934)
26 See O'Neill, *op. cit.* (n.3 above), p. 80
27 Black, *Bell System Technical J.*, 1934 (n.2 above), pp. 1-2
28 Black, 1977 (n.3 above), p. 59. The work to which he was referring is Black, H.S.: *Trans. American Institute of Electrical Engineers*, **48** (1929), pp. 117-140
29 Arnold Tustin in interview with author
30 Simply put the equalisation problem is that of obtaining flat frequency response over a wide bandwidth for the amplifier-cable combination and maintaining that response in the face of temperature variations along the cable
31 The Morristown trials were described by Clark, A.B., and Kendall, B.W.: 'Carrier in cable', *Bell System Technical J.*, **12** (1933), pp. 251-263; Kendall had done early work on repeaters and had been in charge of development work on the toll transmission system since 1919 and Clark was a toll transmission system development engineer; methods of recording frequency characteristics of transmission lines were developed, see, for example, Best, E.H.: 'A recording transmission measuring system for telephone circuit testing', *Bell System Technical J.*, **12** (1933), pp. 22-34, and Curtis, A.M.: 'An oscillograph for ten thousand cycles', *Bell System Technical J.*, **12** (1933), pp. 76-90, and of course there was the theoretical contribution of Nyquist
32 Reich, *op. cit.* (n.3 above), pp. 7-8
33 Nyquist, H.: 'Regeneration theory', *Bell System Technical J.*, **11** (1932), pp. 126-147
34 Nyquist, H.: 'Certain factors affecting telegraph speed', 1924, *op. cit.* (n.21 above)
35 A neat method of estimating the real part of a pair of complex roots associated with the dominant mode for a Nyquist locus which passed close to the critical point was given by Ludwig (1940)
36 Peterson, E., Kreer, J.G., Ware, L.A.: 'Regeneration theory and experiment', *Bell System Technical J.*, **13** (1934), pp. 680-700; the paper also appeared in *Proc. Institute of Radio Engineers*, **22** (1934), pp. 1191-1210
37 This account is drawn from MacFarlane, A.G.J.: 'The development of frequency response methods in automatic control', *IEEE Trans. on Automatic Control*, **AC-24** (1979), pp. 250-265; this paper forms an introduction to a collection of papers dealing with frequency response technique *Frequency-response methods in control systems*, (IEEE Press, New York, 1979). See also Rörentrop, K.: *Entwicklung der Modernen Regelunkstechnik*, (Oldenbourg, Munich, 1971). Nyquist himself makes no reference to Cauchy's theorems: he refers to Osgood, W.F.: *Lehrbuch der Funktionentheorie*, 5th ed., Hobson, E.W.: *Functions of a Real Variable*, **1** 3rd ed., and to Thomson, W. and Tait, P.G.: *Natural Philosophy*, **1**, no edition given; as well as to the work of Carson, J.
38 MacColl, L.A.: *Fundamental Theory of Servomechanisms*, Van Nostrand, 1945; reprinted with new Preface by R.W. Hamming, Dover Books, 1968; the same method of proof was used by Bode, H.: *Network Analysis and Feedback Amplifier Design* (Van Nostrand, 1945) and by James, H.J., Nichols, N.B., Phillips, R.S.: *Theory of Servomechanisms*, Radiation Laboratory Series, **25**, (McGraw-Hill, 1946) both of which were influential books
39 MacColl, L.A.: 'The analysis and synthesis of linear servomechanisms', *NRDC Rep. Sec. D-2 Fire Control*, n.d. circa 1941; Hall, A.C., (1946)
40 Frey, 1945
41 See MacFarlane, *op. cit.* (n.37)
42 Bell Laboratories had been aware of the problems of distortion introduced by phase shifts and by transmission delays for many years, the problems of phase shift were exacerbated by the use of amplifiers in series in cable circuits and in the developing use of telephone lines to relay music for broadcast to radio transmitters.
43 Bode, H.W.: 'Relations between attenuation and phase in feedback amplifier design', *Bell System Technical J.*, **19** (1940), pp. 421-54
44 A story is in circulation that during a period in the mid 1930s one of Nyquist's duties

was to advise the directors of AT&T on the merits of technical proposals originating elsewhere in the company. The majority of these came from the staff at Bell Labs and Nyquist gained the reputation among such staff of being outspoken and critical. At one meeting a team from Bell Labs made a presentation of a new equalisation circuit which they claimed gave a perfectly flat response over a very wide bandwidth. Nyquist listened to the presentation without saying a word; at the end of the meeting he rose and made to leave the room, the leader of the Bell group, possibly feeling that at last they had got something he could not criticise, stopped him and said 'Have you nothing to say, Dr. Nyquist'? The reply was brief: 'Have you tried speaking through it?' The group had not — the phase shift was so great that the circuit was useless for speech transmission. Story told to me by N.B. Nichols
46 Bode *op. cit.* (n.43 above), p. 431 footnote
47 *Ibid*, pp. 448-9
48 Bode, H.W.: *Network Analysis and Feedback Amplifier Design*, (Van Nostrand, Princeton, NJ, 1945)
49 See Bennett, S.: *A History of Control Engineering 1800-1930* (Peter Peregrinus, Stevenage,
49 See Bennett, S.: *A History of Control Engineering 1800-1930* (Peter Peregrinns, Stevenage, 1979). For the history of the general development of radio and radio technology see Aitken, H.G.J.: *The Continuous Wave: Technology and American Radio, 1900-1932*, (Princeton University Press, Princeton, 1985)
50 See, for example, Thomas, H.A.: *Theory and Design of Valve Oscillators*, (Chapman & Hall, London, 1944) 1st ed. 1939
51 See, for example, von Handel, P., Kruger, K., Plendl, H.: 'Quartz control for frequency stabilisation in short-wave receivers', *Proc. Institute of Radio Engineers*, **18**(2) (1930), pp. 307-320; Watanabe, Y.: 'The piezo-electric resonator in high frequency oscillation circuits', *Proc. Institute of Radio Engineers*, **1**(5) (1930), pp. 862-93
52 See Morrison, W.A.: 'Thermostat design for frequency standards', *Proc. Institute of Radio Engineers*, **16**(7) (1928), p. 976; Clapp, J.K.: 'Temperature Control for frequency standards', *Proc. Institute of Radio Engineers*, **18**(2) (1930), pp. 2003-2010, and Heaton, V.E., Brattain, W.H.: 'Design of a portable temperature controlled piezo-electric oscillator', *Proc. Institute of Radio Engineers*, **18**(7) (1930), pp. 1239-46. Clapp noted problem of ripple through use of on-off control and recommended that a thick wall be placed between the heating element and the constant temperature enclosure to smooth
53 out the ripple.
piezo-oscillators', *Proc. Institute of Radio Engineers*, **20**(2) (1932), pp. 261-71
54 Conklin, K.W., Finch, J.L. and Hansell, C.W.: 'New methods of frequency control employing long lines', *Proc. Institute of Radio Engineers*, **19** (1931), pp. 1918-30
55 Hansell, C.W., Carter, P.S.: 'Frequency control by low power factor line circuits', *Proc. Institute of Radio Engineers*, **24**(4) (1936), pp. 597-619
56 A typical example of the type of circuit being proposed is given in Carwile, P.B., Scott, F.A.: 'Automatic neutralisation of the variable grid bias in a direct current feed-back amplifier', *Rev. Sci. Int.*, **1** (1930), p. 203, the assumption in the title and in the body of the paper is that feed-back is positive
57 Kusonose, Y., Ishikawa, S.: 'Frequency stabilisation of radio transmitters', *Proc. Inst. Rad. Eng.*, **20** (1932), pp. 310-339
58 Thomas, H.A.: 'A method of stabilising the frequency of a radio transmitter by means of an automatic monitor, *J. Institution of Electrical Engineers*, **78** (1936), pp. 717-722.
59 Nyquist's paper was considered as somewhat of a 'snow job', and his contemporaries would needle him about it at the various AIEE meetings. Finally, over a few drinks in someone's hotel room he is said to have commented that Bell Labs had not been especially anxious to pass on the information to their competitors, so he really had not tried to simplify the paper. Story attributed to R.S. Glasgow, see Thaler, G.J.: *Automatic Control and Classical Linear Theory*, Benchmark Papers in Electrical Engineering and Computer Science, (Dowden, Hutchinson and Ross, Stroudsburg, 1974), p. 104
60 Bedford, L.H., Puckle, O.S.: 'A velocity-modulation television system', *J. Institution of Electrical Engineers*, **75** (1934), pp. 63-92
61 A number of people have commented on the important role played by scientists and engineers who had undertaken graduate research work, most notably Charles Susskind: 'American contributions to electronics: coming of age and some more', *Proceedings IEEE*, **64**(9) (1976), pp. 1300-5; it should be noted that both Nyquist and Bode had a PhD

in physics, at a time when, in engineering in particular, nobody aspired to the PhD degree unless they intended to pursue an academic career

62 For example, the workman who wound the helical tubes that were used as the pressure sensitive elements in a range of instruments of The Foxboro Company was given strict instructions to stop work and walk away from his machine if anyone tried to watch him. Interview with Archie Hanna, The Foxoboro Company, 27th April, 1989. For a discussion of the use of patents and trade secrets as a means of protecting innovations see Eric von Hippel, *The Sources of Innovation*, (Oxford, OUP 1988), Chapter 4

63 Stevenson, A.R., Howard, A.: 'An advanced course in engineering', *Trans. AIEE*, **54** (1934), pp. 265-268

64 Dyches, H.E., Hellmund, R.E.: 'The Pitt-Westinghouse graduate program', *Trans. AIEE,* **53** (1934), pp. 103-4

65 W. Bernard Carlson reports that General Electric's support of the co-operative course in Electrical Engineering was lukewarm and they did little to ensure that students who took the course remained with the company—see Carlson, W. Bernard: 'Academic Entrepreneurship and Engineering Education: Dugald C. Jackson and the MIT-GE Cooperative Engineering Course, 1907-1932,' *Technology & Culture,* **29** (July 1988), pp. 536-567

66 Ott, P.W.: 'The use of mathematics to engineers', *General Electric Review,* **38** (1935, p. 138

67 Arnold Tustin in interview with the author, July 1976

Chapter 4
Theory and design of servomechanisms

In the same year, 1934, that Harold S. Black's paper describing the negative feedback ampliier was published, two papers by Harold Hazen were published in the *Journal of the Franklin Institute*: one on the theory of servomechanisms and the other on the design of a high performance servomechanism.[1] The central point of the papers was Hazen's recognition that negative feedback made the overall behaviour of a servomechanism dependent primarily on the difference between the input and output, and that the effects of non-linearities and variation in parameters of the amplifying devices used in the forward path were significantly reduced. However, unlike Black he did not recognise that feedback also reduced noise and distortion in the system.

These papers were the outcome of some eight years work on the design of following mechanisms for equipment for calculating machines designed by Vannevar Bush and Norbert Wiener,[2] who worked in the Electrical Engineering Department at MIT, which Hazen first joined as a student in 1920. The Department was then under the direction of Dugald C. Jackson who believed in the involvement of staff and students in industrial work.[3] As well as being head of the Electrical Engineering department Jackson was also a senior partner in the consulting firm of Jackson and Moreland and it was through the interests of this firm that members of the Electrical Engineering Department, in particular Bush, became involved in calculating machines.

4.1 Network analyser

Bush's first approach was to investigate the use of analogue simulators for analysing the performance of electrical networks. In the 1920s, the idea of interconnecting power generation sources to form a network was still in its infancy. During the First World War, engineers realised that an increase in the load factor, and hence an increase in available power, could be obtained by connecting together distribution networks which had a diversity of load. After the war demand dropped but interconnection schemes continued to be promulgated, albeit at a slower rate.[4] The firm of Jackson and Moreland were consulting engineers for the project to bring Canadian hydro-electric power into the New England and New York areas and, in 1924, Bush worked with Ralph Booth of Jackson and Moreland on the project.[5] The practical problems involved in network interconnection were manifold, ranging from consideration of overall stability in the event of a sudden loss of a line caused, for example, by a lightning strike, through to how to maintain synchronism and prevent hunting when a number of generators, each with their own frequency control, were connected in parallel. The formulation of the network equations was well understood but obtaining practical solutions was difficult because of the quantity of numerical calculation required, so typically engineers resorted to experimental methods. Several analogue computing devices — in essence miniature power systems — were developed. These included a dc short-circuit calculating table (Lewis, 1920 and Fortescue, 1925) which provided information on current magnitudes but was unable to provide any information on phase angles. The first miniature system to employ alternating current was constructed in 1917 (Gray, 1917) but this was also limited to a static representation of the system. From 1919 onwards, several systems were developed using three-phase generators, motors, static loads, and lumped three-phase artificial lines. Schurig (1923) produced a system which used machines of 3.75 kVA rating with a voltage of 440V, while Evans and Bergvall (1924) developed a system using 200 to 600 kVA machines and voltages of 2300V. The larger system was necessary for stability investigations involving consideration of mechanical momentum and electrical damping since the normal relationship between such factors cannot be maintained as the scale of the model is reduced.

In 1924, Bush got Hazen for his undergraduate project to develop a model system for network analysis (this was for Hazen the beginning of a twenty year period of close involvement with Bush). He used 5 kVA generators to represent 50 000 to 100 000 kVA units. However, when Hazen began his work, scale effect problems were not clearly understood and when General Electric Company engineers tried the system it was found to be unstable: the low mechanical momentum of the small machines in relation to the electrical characteristics resulted in very large transient power flows in the model. Bush suggested replacing the rotating machines by static phase-shifting transformers. Hazen, together with Hugh H. Spencer (another undergraduate), pursued this approach and the scheme was found to work well.[6] On graduation Hazen joined the General Electric Company 'on test', but soon found himself working with Spencer in the office of the Chief Consulting Engineer Robert E. Doherty, on the Canada to New England transmission line. Doherty suggested that Hazen and Spencer should follow up the work they had done as undergraduates and both returned to MIT[7] where they continued to work on the network analyser system and showed how it could be used for the analysis of transient behaviour as well as for steady state studies.

Hazen then decided to do graduate work at MIT under the guidance of Bush. He was appointed as a research assistant in 1925 and an instructor in 1926. For his master's degree he continued to work on computational devices for network analysis and the outcome was the MIT Network Analyser which, in the early 1930s, was the most advanced system for transmission network analysis in the world.[8]

4.2 Product Integraph

As an alternative to using simulations based on miniature power networks, Bush also pursued direct methods for solving the network equations. He based his approach on John Carson's work on transients in electrical networks.[9] The solution of the network equations by Carson's methods involves lengthy calculations to evaluate convolution integrals of the form

$$E\omega \cos(\omega t + \theta) \int_0^t \cos(\omega\lambda) a(\lambda)\, d\lambda \qquad (4.1)$$

The general form is

$$f_3(x) \int_a^x f_1(x) f_2(x)\, dx \qquad (4.2)$$

where f_1, f_2 and f_3 are given functions.

Bush had been interested in calculating machines from an early age and in 1912, while still at college, he invented a surveying machine that incorporated a mechanical integrator and a spring operated servomechanism.[10] Later, in 1920, he developed a harmonic analyser.[11] By 1927, he had become convinced that if engineering was to progress the general need for calculating machines must be satisfied:

> Engineering can proceed no faster than the mathematical analysis on which it is based. Formal mathematics is frequently inadequate for numerous problems pressing for solution, and in the absence of radically new mathematics, a mechanical solution offers the most promising and powerful attack wherever a solution in graphical form is adequate for the purpose. This is usually the case in engineering problems.[12]

Working under bush, Herbert Stewart had developed a machine called the Product Integraph, to evaluate integrals of the above form. Mechanical integrators were not new but, as Bush and Stewart remarked, they 'usually evaluate the definite integral between given fixed limits. There has been a need for a machine which would continuously evaluate and plot the integral as a function of a variable upper limit'.[13] They continued with an explanation of how by means of 'an expedient fairly well described as "back-coupling",' the Product Integraph could be used to solve differential equations. For example, it could be used to solve equations of the form

$$\phi(x) = \int_a^x f(x)\phi(x)\, dx \qquad (4.3)$$

for an unknown ϕ when f is known.[14] The idea of 'back-coupling' came from the work of another of Bush's graduate students, King E. Gould, who used the Product Integraph to investigate the temperature distribution along a thermionically emitting filament.[15]

As part of the graduate course on operational circuit analysis Bush set a problem involving the analysis of a resistance-inductance (RL) coupled amplifier. Using the Product Integraph, Hazen derived the output of the amplifier, including in the solution consideration of the actual nonlinear vacuum tube characteristic. Bush, wanting to try something more complicated, suggested Hazen analyse a simple oscillator circuit. Hazen recalled that he;

> Played with that at home for an evening or two and, by golly, if we had a second integrator we could handle the oscillator circuit, so I made a little sketch of the mechanical wheel and disc integrator as a simpler one than the Watt hour meter and took it to Bush. He came back next morning with a sheaf of 'Bush scrawl' showing how it would not only solve the equation I had now, but a whole family of second order differential equations.[16]

Hazen was assigned as his next task the design of a wheel and disc integrator which was to be added to Product Integraph. The modification led to the two-stage Product Integraph and eventually the Differential Analyser. Neither Bush nor Hazen were aware at that time of William Thomson's proposal for using wheel and disc integrators to solve the general second order differential equation. Neither were they aware of James Thomson's integrator design.[17].

In the original single-stage Product Integraph graphs of the two known functions $f_1(x)$ and $f_2(x)$ were mounted on a table which was driven at constant speed (Figure 4.1). The graphs were arranged so that the x axes of the graphs were parallel to the movement of the table. Fixed across the table at A and B were two centre-tapped potentiometers. Two operators moved pointers attached to the slides of the potentiometers along the plotted curves; the outputs from the potentiometers were fed to the potential and the current coils of a modified Thomson direct-current, integrating, watt-hour meter. The integral of the product of the two functions $f_1(x)$ and $f_2(x)$ is given by the rotation of the watt-hour meter which moved a pen, located at C, thus continuously recording the value of the integral.

Success depended on a number of factors: the accuracy of 'tracking' achieved by the operators, the accuracy of the potentiometers, the dynamic performance of the watt-hour meter, and the performance of the mechanism used to follow the movement of the watt-hour meter disc (it was important that this mechanism did not load the disc). Construction of a mechanism to follow the movement of

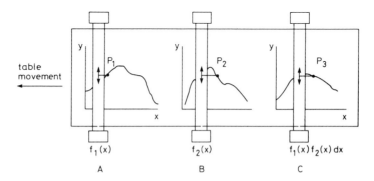

Figure 4.1 *Arrangement of input and output table for the Product Integraph*

Adapted from Figure 1 in V. Bush, and H.L. Hazen: 'Integraph solution of differential equations', *J, Franklin Inst.*, 1927, **211**, pp. 575–615

the watt-hour meter disc presented a fundamental servomechanism problem — automatic follow-up — a problem that was appearing in many different industries and applications. For example, the industrial instrument manufacturers were producing recorders, based on the use of a sensitive galvanometer, that required the pen to follow accurately the movement of the galvanometer without loading it. The problem is to devise an electro-mechanical positioning mechanism which will accurately follow a rapidly changing signal, a mechanism which, as Hanna writing in 1945 put it, 'not only must accelerate at the proper rate and move at the right speed, but also must be in the correct place'.[18] The problem is, of course, directly analogous to the amplifier problem solved by Black. In Black's case the rapidly varying signal to be followed was derived from speech or other sounds, in Hazen's case from the position of mechanical components.

The follow-up mechanism of the Product Integraph (see Figure 4.2) consisted of a Bakelite table on which metal contact strips were fixed, mounted 0.25 inches below the watt-hour meter disc. Three platinum tipped screws were attached to the disc and adjusted to ride just above the contact strips. Contact was established

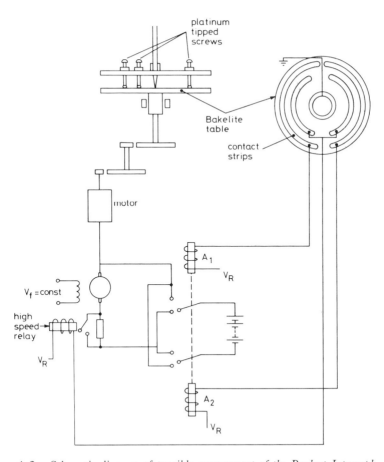

Figure 4.2 *Schematic diagram of possible arrangement of the Product Integraph*

by wetting the tips of the screws with mercury. The centre contact on the table formed the ground return; the outer contact was connected to a two-way relay (A_1, A_2); the remaining contact operated the high-speed relay. The table was driven, through a gearbox, from a dc motor. When the watt-hour meter disc rotated slowly, the outermost contact operated relay (A_1, A_2), energising the motor and thus turning the table. If the speeds of the motor and the watt-hour meter disc differed, then the contact screw moved into the gap between the contact strips and onto the other contact strip. When this occurred the direction of the motor was reversed. The Bakelite table thus followed the movement of the contact mounted on the watt-hour meter disc but with a superimposed oscillatory movement. If the meter speed exceeded the normal motor speed a resistance in the armature circuit was short circuited thus increasing the motor speed. The movement of the Bakelite table was used to drive the pen P_3 (Figure 4.2). The oscillatory motion of the follow-up mechanism reduced the friction in the meter support. The use of an oscillating contact with feedback to obtain proportional action from a relay system was common practice at this time.[19]

This single stage integraph formed the basis for the two-stage unit. The watt-hour meter was retained and a wheel and disc integrator (the second stage) and additional recording shafts added. For successful operation of the device:

> It is essential that these integrator shafts — in the first stage, the watt-hour meter rotor; in the second, the wheel shaft — be free from all friction and load torque, and hence they cannot directly furnish energy to drive the recording shafts. A servo-motor follower mechanism is therefore used to drive each recording shaft, and this not only reduces the necessary energy output of the integrator shafts to a negligible value, but, as mentioned above, practically eliminates bearing friction on these shafts at the same time. This mechanism is really the key to the success of the machine from the practical point of view.[20]

The servo-followers used were simple. An electrical contact detected the position of the shaft to be followed and the signal from this contact turned the follow-up motor on and off. In this simple form the follower exhibited velocity lag. The lag error is irrelevant when the device simply integrates a function. However, if the result is to be fed to a second integrator, the output of which is fed back to input of the first integrator, any error becomes cumulative. Hazen and Bush[21] identified three principle sources of error:

- backlash
- lag in the servomechanism
- lag due to the inertia of the watt-hour meter rotor.

Backlash was eliminated by careful design. The velocity lag in the servo-system was eliminated by adding a second servo-motor to the system (see Figure 4.3). If the speed of the main motor A matched the watt-hour meter disc speed, the auxilliary motor B merely oscillated about a mean position. Meter speed fluctuations caused motor B to make unequal oscillations, causing a change to the voltage supplied to the main motor armature and thus correcting the speed of the main motor. The resulting servo-system was of type 2 and hence had zero velocity error. There was another method of reducing lag: the motion of the auxiliary control motor was fed forward and added to the motion of the main motor. The angle

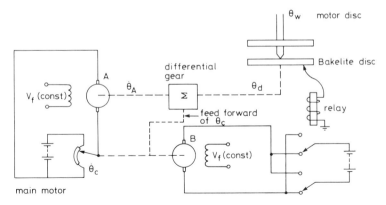

Figure 4.3 *Schematic diagram illustrating the addition of a motor to eliminate velocity error*

added was proportional to the speed of rotation of the watt-hour meter disc and was adjusted to compensate for the error associated with the mechanical time constant of the watt-hour meter, in effect cancelling the inertia of the watt-hour meter disc. This was similar to the method for eliminating velocity lag used by the Admiralty Research Laboratory in the UK for a oil-hydraulic searchlight stabilisation system (see Chapter 1).

4.3 Differential analyser

In 1928, Bush was awarded the Levy Gold Medal by the Franklin Institute in recognition of his work on the Product Integraph (his co-workers Stewart, Gage and Hazen were given honourable mention). He was anxious to build a larger version—a 'real machine'—and with Hazen began work on the design of a six-integrator machine, the Differential Analyser, which, by 1930, was successfully used in the solution of many complicated problems.[22] Hazen's contribution to the Differential Analyser project was mainly design and construction of the wheel and disc integrators with their torque amplifiers and frontlash units.

Again, the crucial problem was to avoid loading the output shafts of the integrators. The solution emerged by chance. In 1928, while on vacation in his home town of Three Rivers, Michigan, Hazen came across copies of two papers by C.W. Nieman.[23] One of these papers described a two-stage torque amplifier which Hazen saw could be used to provide the coupling between the wheel of the integrator and the mechanical shafts of the rest of the analyser. The torque amplifier worked well when carefully adjusted: the nuisance was that without careful adjustment oscillations could occur, which according to Bush, were 'presumably caused by a small part of the output being fed back in one way or another into the input'. His solution was to add a vibration damper to the output shaft. Arthur Porter, who worked on the Differential Analyser built at the University of Manchester, recalls that they faced similar problems caused by transverse vibrations of the metal bands of the torque amplifier which they solved by coating the drums with Vaseline. To compensate for the decreased friction they had to increase the number of turns on the drum.[24] The second of Nieman's papers described a device to eliminate backlash which he called a lashlock. The lashlock formed the basis of the frontlash unit, an adjustable differential gear mechanism used in the Differential Analyser to compensate for backlash in the gear trains and the elasticity of the connecting shafts. The frontlash unit was invented by Hazen and Bush during one weekend, and Hazen was pleased to see that Bush 'struggled just like the rest of us, just like a dog with a bone'.

The six-integrator Differential Analyser was a success, and in the early 1930s machines were built at the Aberdeen Proving Ground, Maryland, at the General Electric Company, at the universities of Pennsylvania, California and Texas, and at the universities of Manchester and Cambridge in the UK.[25] On the basis of the success of the machine Bush planned to build a larger, faster differential analyser, and in 1935 succeeded in persauding the Rockefeller Foundation to finance the design and construction of such a machine. This machine was completed in 1942 and was immediately put to use in support of the war effort.[26]

From the early 1930s Hazen, increasingly involved in teaching, gradually withdrew from participation in the work on the Differential Analyser. He did, however, remain interested in a problem common to the whole range of analogue computing machines: the design of a follow-up servomechanism, activated by an electrical signal, with a fast and accurate response. The impetus came from two problems, one of which arose in connection with yet another computing machine, the Cinema Integraph, and the other from the design of an automatic curve follower to reduce the labour and increase the accuracy of inputting data to the Differential Analyser.

4.4 Cinema integraph

Norbert Wiener suggested the 'Cinema Integraph' as an alternative to the Product Integraph, for evaluating the convolution integral and integrals of the form

$$\int_A^B f(\lambda) \, {\sin \atop \cos} (n\lambda) \, d\lambda \tag{4.4}$$

as a function of n, where A and B are any fixed limits. Evaluation of this integral, providing that f converges with sufficient rapidity, provides the approximate evaluation of

$$\int_0^\infty f(\lambda) \, {\sin \atop \cos} (n\lambda) \, d\lambda \tag{4.5}$$

Wiener's method, based on the passage of radiation through apertures whose shapes represented the functions to be multiplied and integrated, was investigated by King E. Gould.[27] A line source of radiation of uniform brightness (SS' in Figure 4.4) is placed at a distance x from a mask shaped to represent a function $f(\lambda)$, with the distance x made sufficiently large so that the intensity of the radiation falling on the face of the mask from any point on SS' varies by less than 1%. Placed an equal distance behind this mask is a second mask shaped to represent a function $g(\lambda)$ (drawn to twice the scale of $f(\lambda)$). The amount of radiation which passes through a vertical strip of width δ at point a of mask 1 is proportional to the ordinate of $f(\lambda)$ at λa. Similarly, the amount of radiation which passes through

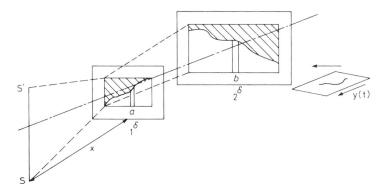

Figure 4.4 *Arrangement of masks in Cinema Integraph*

a vertical strip of mask 2 at point b is proportional to the ordinate of $g(\lambda)$ at λb, but since the radiation incident on mask 2 is dependent on $f(\lambda)$, and if $a = b/2$, where a and b represent the distance from the origin for masks 1 and 2 respectively, then the radiation transmitted through mask 2 is the product $f(\lambda)g(\lambda)$, and hence the total transmitted radiation over the range $\lambda = 0$, $\lambda = \lambda_1$ is

$$I_3 = \int_0^{\lambda_1} f(\lambda)g(\lambda)\, d\lambda \qquad (4.6)$$

The evaluation of the integral involving the variable parameter t can be performed by moving the second mask horizontally so that the radiation passing through a strip at point k_a of mask 1 falls on the strip corresponding to $(t \pm \lambda_a)$ on mask 2. By attaching a table to mask 2 and moving both the table and the mask as t is varied, the value of the integral as a function of t can be recorded, providing some mechanism is available to instantaneously measure the light transmitted.

Though simple in concept, implementation required the solution of many practical problems. Gould omitted any precise figures for the accuracy obtained, but it is known that he experienced great difficulty in excluding extraneous radiation and in accurately measuring the radiation transmitted. Development of the photo-electric cell and the high impedance vacuum tube led Bush to try a different approach. The system was designed by Truman S. Gray (1931) and used a null balance technique for the measurement of the transmitted light. He was able to obtain an accuracy of between 2% and 5% but hand adjustment to obtain the null balance was slow and laborious: an automatic balancing system would have been of great benefit. However, automation was not achieved until the late 1930s when Hazen's high speed servo-system was used.

The significance of Gray's device when it was built in 1930–32 was that it represented a computing mechanism which would benefit from the development of a fast, accurate position control mechanism, and was a device that provided an electrical, not a mechanical, input for the servo-system. As inevitably happens, once a reliable computing device of reasonable accuracy becomes available, the demands on it grow rapidly and it is used to solve problems that could not have been imagined when the machine was designed. This happened with the differential analyser, and studies were made of ways of reducing the set-up time and the solution time of the machine. Little could be done to reduce the setting-up time because setting up involved the physical interconnection of shafts and gears. The solution time was largely dependent on the speed at which the machine could be run — the motor speed controlled the rate of change of the independent variable, and the limiting factor on this speed was the ability of the operator to trace the curve of the input function with the required degree of accuracy — hence a balance had to be achieved between speed and accuracy. It was in an attempt to speed up the machine while still retaining accuracy that Hazen began work on designing and building an automatic curve follower. An essential feature of such a unit was a high-performance servomechanism.

4.5 High speed servomechanism

One of Hazen's 1934 papers ('Design and test of a high perormance servomechanism') gave a detailed account of requirements for, and design of, the servomechanism used in the Cinema Integraph.[28] The input signal to the servomechanism was the difference between the outputs of two photo-electric cells and had a power level of the order of 10^{-10} watts. The servo output had to operate a shutter to adjust the light balance and the recording unit — a total power requirement of the order of 10 to 100 watts giving an overall power amplification requirement of 10^{11} or 10^{12}. The servomechanisms had to have a non-oscillatory response, the ability to follow rapid changes in the input signal, and had to convert electrical energy into mechanical energy.

Hazen conjectured that the essential elements were an electrical power amplifier and an electric motor but he did not rule out the possible need for mechanical power amplification. He considered the 'dynamic' elements introduced by each of the possible units, concluding that a resistance coupled dc amplifier would not introduce friction or inertia effects, an electric motor would introduce inertia and damping, and a mechanical amplifier would introduce only inertia. He noted that this inertia, together with the inertia and friction of the load (suitably reduced by the amplification factor of the torque amplifier) could be added to the inertia of the motor. For the purposes of analysis he reduced the system to that shown in Figure 4.5.

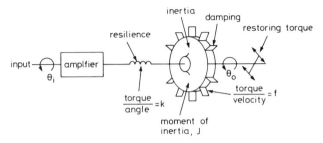

Figure 4.5 *Schematic diagram of Hazen's basic servomechanism*

Adapted from H.L. Hazen.: 'Theory of servomechanisms', *J. Franklin Inst.*, 1934, **21**, pp. 283–331

The subsequent stages of the design were based on the use of a 'figure of merit' expressed as

$$M = \frac{\omega_m^2}{\theta_m} \quad (4.7)$$

where ω_m = maximum speed of the servo, and θ_m = steady state error at speed x_m. In the other 1934 paper ('Theory of servomechanisms') he had already shown that

$$M = \frac{\tau_m}{4\gamma^2 J} \quad (4.8)$$

where τ_m = maximum driving torque, J = total moment of inertia (including load inertia) and γ = 'factor characterising the oscillatory tendency of the output' that is in modern terminology ζ, the damping ratio.

In the early 1930s Hazen could not turn to a manufacturer's catalogue for a suitable servo-motor. He had to start from first principles. He worked out that a high torque to inertia ratio was necessary to obtain the high speed and acceleration required. But should he use an electric motor or a mechanical system for amplification? He calculated the electromagnetic force that could be developed in copper and the force that could be transmitted by a unit mass of steel one centimetre in length, concluding from these calculations that 'a device which develops forces mechanically could be made much lighter i.e. with much less mass, than one which develops forces electromagnetically'.[29] Thus he decided to use a mechanical torque amplifier with a small, low inertia, d.c. motor.

By requiring that the motor provide the damping necessary for an overall aperiodic response for the systemn, he was following the current practice which was to produce components with inherent regulatory characteristics rather than to use feedback to modify the overall *system* behaviour. He designed and built a motor with the following characteristics:

- armature resistance   1590 ohms
- moment of inertia 0.34×10^{-5} kg m^2
- torque constant   1.12 Nm/A
- mechanical time constant   0.0146 s
- damping coefficient 0.465×10^{-3} Nms^{-1}
  (including eddy current damper)

The inclusion of damping as part of the motor design procedure suggests that the work must have been well advanced before Hazen discovered Minorsky's work and became aware that the overall system characteristics could be modified by using the time derivatives of the error, for he simply comments in the design paper that these methods were not used.[30] Whereas in the later theory paper he states that

> the servo operation can be made aperiodic or oscillatory in any desired degree by the damping effect introduced by the component of restoring torque depending on the first derivative of θ....Physically the torque corresponding to the [damping] term...is readily introduced, when a dc vacuum-tube amplifier is used, by an inductance component of inter-stage coupling.[31]

108 *Theory and design of servomechanisms*

The papers as published had, in Hazen's words, 'the appendices at the front'. He had begun the work in 1931 as an attempt to build a practical high performance servomechanism and such a device built with the help of Gordon S. Brown was exhibited at the Chicago World Fair in 1932. One of Hazen's students, Harold A. Traver, described a curve-follower in his S.M. thesis of 1933[32] and during 1932–33 Hazen worked on an analysis of continuous control systems. It was while doing the latter that he found the paper by Minorsky (1922) with its analysis of ship steering control. Hazen wrote up the work on the theory of continuous control and the design of a servomotor as one paper and showed it to Bush early in the summer of 1933. Bush suggested that during the summer Hazen visited him at his summer cottage at South Dennis (Cape Cod) and that they have a look at the paper in detail. They

> went to work, chewed it over — it was about the third or fourth thing that we'd done together... We discussed the fact there would be off/on relay type controls and definite correction controls... This was early summer and he [Bush] said 'I think you want to go back and put some stuff together on the overall treatment of these as well as the continuous control. That might just be something if did that'. So I went back to work on it.[33]

4.6 Theory of servomechanisms

In the theory paper ten years of experience are codified and presented with great clarity, and we see Hazen change from the practical, experimental, engineering research worker and designer to the teacher. In preparing for the paper, Hazen carried out an extensive search of the literature and lists 32 papers in his bibliography. The greatest number of references are to ship and aircraft applications (steering of ships — 6, stabilisation of ships — 5 and aircraft stabilisation — 4); there are 10 references to applications in the process industries, nearly all concerned with applications of the photo-electric cell; and 4 references to speed control governors for electrical supply; the remainder are references to Nieman and to Bush and Hazen's previous work.[34] Although 30 out of the 32 references are to feedback control applications Hazen still felt it necessary to explain in detail in the introduction the difference between open and closed loop control systems, and also to give a definition of a servomechanism which he described as

> a power-amplifying device in which the amplifier element driving the output is actuated by the difference between the input to the servo and its output.

When he wrote the paper Hazen was unaware of Black's work on the feedback amplifier, information about which was made public for the first time in January of 1934. He was vaguely aware of Nyquist's work at Bell Telephone Laboratories but 'did not recognise at the time the intimate and fundamental interconnection between this and the transient analysis approach', associating it only with communication network theory.[35] However, he reached the same conclusions as Black about the effects of feedback, explaining that for a simple vacuum tube amplifier the linear response 'was due to the constancy of the parameters within the amplifier. Any departure from constancy of these parameters affects the relation between input and output directly'.[36] For a servomechanism, however, the 'only function of the servo amplifier element is to apply sufficient force to the servo output to bring it rapidly to correspondence with the servo input. Such an amplifier can be a relatively crude affair'.[37]

The theory paper deals with the whole range of servomechanism types which Hazen classified as relay, definite correction, and continuous control. The order chosen reflects both the order in which they developed and the practical significance of each in 1934.

In analysing relay systems he assumed certain forms for the output signal from the relay and from this calculated the equation of motion and the terminal conditions for specific segments of the output path. He then used classical operator techniques to solve the equations in terms of the known terminal conditions. The solution gives the conditions which have to be satisfied for the original assumption about the solution to be valid. For example for an ideal relay (that is, no dead space and no time delay) the assumption made was that the output would settle to a steady oscillation. On this assumption the solution of the system equation gives

$$\ln \frac{1 - \frac{\omega_0}{\omega_s}}{1 + \frac{\omega_0}{\omega_s}} + 2\frac{\omega_0}{\omega_s} = 0 \qquad (4.9)$$

Eq. 4.9 is satisfied only for vanishingly small values of ω_0/ω_s and since ω_s is finite then ω_0 must be vanishingly small therefore the output oscillates with infinitesimal amplitude and infinite frequency. Hazen drew the implication that any deviation from the assumed conditions, for example by adding a time delay, will result in the output oscillating with a finite amplitude and frequency of oscillation.

The importance of Hazen's work on relay type controllers lay not so much in the results obtained, although they were of use to designers, but in the approach to the solution and in the encouragement it gave to others. A.L. Whiteley is just one of the people who recall eagerly seizing up and ideas contained in the two papers.[38]

Hazen did not attempt any analysis of the definite correction servomechanism, except to comment that if such a device was to follow a high speed input then either the maximum amount of correction that could be applied at any one instance must be high, or the interval between applying successive corrections must be small, or both. At the time this case was largely of theoretical interest since the most common use of definite correction servomechanisms was in the process industries where the rate of change of the input signal was usually very small. He also noted that such devices had a long life, whereas there was a danger with a relay system, because of its tendency to oscillate in search of a balance, of high wear and hence a short life.

As an example of continuous control Hazen chose the system used in the Cinema Integraph. This system reduces to the standard second order system. He used operator techniques to solve the system equations and he considered three cases: critically damped, underdamped and oscillatory. The important ideas which Hazen extracted from the analysis were:

110 *Theory and design of servomechanisms*

(a) the use of relative damping factor, that is damping coefficient, to characterise the form of the response;
(b) the importance of the time constant in determining the speed of response;
(c) the establishment of a 'figure of merit' as a basis for design.

As in the 'Design' paper Hazen discussed the use of derivatives of error instead of damping to obtain the desired form of transient response. In 1934–5 some of Hazen's students attempted to use inter-stage coupling networks to introduce the first three time derivatives of error into the system. They found that it was difficult to get derivatives of the required magnitude and Hazen reported that 'None of the mechanisms having the more complex control signals has evidenced the complete stability and independence of small variations in the parameters observed in the uncorrected mechanism'.[39]

4.7 Summary

The work of Harold Hazen represents a different tradition — academic research — from that described in the previous two chapters. The work on the differential analyser is a paradigm for academic engineering research: difficulties in solving a practical problem (in this case the solution of complex equations), thorough investigation of the difficulties and a recognition of an underlying general problem (the move from a network analyser to a general differential equation solver), exploration of different possible solutions (the various calculating machines including the cinema integraph), development of a working prototype (the differential analyser), and lastly, the extraction of coherent body of information for transmission to others. This latter was Hazen's major achievement.

Hazen's papers of 1934 are so clear, understandable, and presented in a way which is so familiar to present day engineers that it is easy to underestimate the originality of his contribution. They are more than a workmanlike account of current practice, for when Hazen wrote them there was no fundamental understanding of the problems, no corpus of design methods, and a few standard components with the necessary performance for the construction of a servo-system. These papers explained clearly the crucial features of a servomechanism and provided a methodology for the analysis and design of such systems. Hazen's work marks the change in emphasis from relay systems to continuous control and the beginnings of a method for the design of a control system with a specified response. It is this which is Hazen's most significant achievement.

The basic problem solved by Hazen, the accurate following of mechanical movement, is a direct analogue of the telephone repeater amplifier problem of Black. However, it was complicated by the difficulty in providing accurate and reliable amplification of mechanical movement. In the beginning, Hazen attempted to use existing technology and compensate for its defects. Success came when he was able to make use of the development of reliable electronic devices, in particular the photo-electric cell and the vacuum tube. The photo-electric cell enabled mechanical movements to be converted into electrical signals with much less disturbance than any other device then available. The vacuum tube made available an amplifier capable of handling the low level signals from the photo-electric cell. Black's work had not yet been disclosed, but Hazen was able to use a dc amplifier designed by Brown. This amplifier did not incorporate negative feedback but, since Hazen was trying to automate a manual null balancing system, there was an overall closed loop. Hazen did, however, appreciate the need for damping and he determined the amount of damping required to give the desired response.

Hazen's work on the design of an automatic curve follower, done with Brown and Jaeger, was important in several ways. As with the previous designs, some method of correcting the velocity lag was needed. The method they used was to insert a coil in series with the armature of the motor. The current through the coil generated a force which moved the photo-cell shielding slit relative to the centre line of the pick-up head and hence introduced an offset proportional to the velocity. The system was difficult to adjust to obtain stability and they experimented with the use of derivatives (or approximate derivatives) of the error obtained by using various interstage coupling networks — transformer, capacitance and inductance — in the amplifier.[40] Both techniques were to be extensively investigated in subsequent years.

Through personal contacts by Douglas Hartree and Arthur Porter there was a rapid transfer of these ideas to England. Hartree, Professor of Mathematics at the University of Manchester was a frequent visitor to MIT and was one of the first recipients of a set of drawings of the Bush Differential Analyser. Porter spent a year in the Electrical Engineering Department of MIT before working with Hartree on the applications of the Differential Analyser. First Porter and then P.M.S. Blackett and F.C. Williams of the University of Manchester designed automatic curve followers for the differential analyser which had been constructed at the University of Manchester. Porter used a very simple on-off system in which a relay was operated by the photo-cell. The system worked but the 'jitter' about the mean position caused by the relay switching disturbed the operation of the differential analyser. Blackett designed a continuous controller and Williams solved the velocity lag problem by offsetting the tracking photo-cell.[41]

Work on the curve follower also led to the realisation, at some time in the late 1930s, that the mechanism might have applications in the control of machine tools, in particular the replacement of template operated lathes and other machines.[42] The US Navy also realised the relevance of Hazen's work to the problems which were arising in the development of fire-control equipment to cope with the increased speeds of ships and aircraft. And in 1936 they asked Bush, as the then head of the Electrical Engineering Department, to provide a course on servomechanisms for the Navy. The original intention was that the course should be taught by Bush, with the assistance of Hazen and Caldwell. On Bush's departure to be President of the Carnegie Institute in Washington DC in 1938, Hazen took over the planning of the course and was anticipating teaching it. However, administrative changes consequent upon Bush's departure resulted in Hazen becoming the head of the Electrical Engineering Department and having to pass responsibility for the course over to Gordon Brown.[43]

The course commenced in September of 1939 with an enrolment of four naval lieutenants (all eventually became admirals) and under the guidance of Brown they studied Hazen's and Minorsky's papers in depth. Brown was assisted by John W. Anderson and George C. Newton (then a third year student), and later by A.C. Hall. In the second term Brown began to get together a laboratory and this was the beginnings of the Servomechanisms Laboratory. Although Brown seized this opportunity eagerly and made the most of it, the task of teaching a new course to four experienced and clever naval officers cannot have been easy, particularly as he must have been aware of the close interest that Hazen, as Head of Department, was taking in the course.[44]

By the end of the 1930s there was a growing awareness of the importance of dynamic analysis in the design and application of automatic control and several design techniques were emerging but before consolidation could occur the developed world was enveloped in war. The war disrupted the normal patterns of development, and the adoption of stringent secrecy measures restricted publication of new ideas. Under such conditions, with large numbers of people working on common problems, parallel invention and overlap was inevitable and although at the end of the war accounts of the major development were published, much of the detail remains buried in archives.

4.8 Notes and references

1. Hazen, H.L.: 'Theory of servomechanisms', *J. Franklin Institute,* **218** (1934), pp. 279–331 (referred to hereafter as Hazen 1934a); 'Design and test of a high performance servomechanism', *J. Franklin Inst.,* **218** (1934), pp. 543–80 (referred to hereafter as Hazen 1934b)
2. A more detailed account of Hazen's work is given in Bennett, S.: 'Harold Hazen and the theory and design of servomechanisms', *Int. J. of Control,* **42** (1985), pp. 989–1012. See also Owens, Larry: 'Vannevar Bush and the differential analyser: the test and context of an early computer', *Technology and Culture,* **27**(1) (1986), pp. 63–95
3. See Carlson, Bernard W.: 'Academic entrepreneurship and engineering education: Dugald C. Jackson and the MIT-GE cooperative engineering course 1907–1932', *Technology and Culture,* **29**(3) (July 1988), pp. 536–567
4. An excellent detailed account of the development of electricity distribution systems is given in Hughes, T.P.: *Networks of Power: Electrification in Western Society 1880–1939,* (Johns Hopkins Press, Baltimore MD, 1983) see also Cohn, N.: 'Recollections of the evolution of real-time control applications to power systems', *Automatica,* **20**(2) (1984), p. 145.
5. Bush, V., Booth, R.D.: 'Power system transients', *Trans. American Institute of Electrical Engineers,* **44** (1925), pp. 80–97, disc. 97–103
6. See Hazen, H.L., Spender, H.H.: 'Artificial representation of power systems', *Trans. American Institute of Electrical Engineers,* **44** (1925), p. 72
7. The phrase 'on test' was a relic of the early years of the General Electric Company's predecessor, the Thomson-Houston Company who had used graduate engineers to carry out quality control tests on equipment for an extended period before becoming fully qualified engineers. By the 1920s, this had changed into a cooperative programme between MIT and GE, however, the terminology of being on test was retained. See Carlson *op. cit.* (n.3), p. 549. See also Wise, George: 'On Test': Postgraduate Training of Engineers at General Electric 1892–1961', *IEEE Trans. on Education,* **E-22** (November 1979), pp. 171–77
8. The analyser was built with the support of the General Electric Company, see Hazen, H.L., Schurig, O.R., and Gardener, M.F. (1930)
9. J.R. Carson worked for the Bell Telephone Laboratories and his work was done to meet the needs of engineers working on telephone transmission systems. The main results were published in 'General expansion theorem for transient oscillations of a connected system', (Carson, 1917) and 'Theory of transient oscillations of electrical networks and transmission systems', (Carson, 1919)
10. US Patent No. 1,048,649 issued 3rd December 1912; the harmonic analyser is described in Bush, V.: 'A simple harmonic analyser', *J. American Institute of Electrical Engineers,* **39** (1920), p. 903; see also Bush, V.: *Pieces of the Action,* (Morrow, New York, 1970); Owens, *op. cit.* (n.2 above) gives details of the surveying machine
11. See 'A simple harmonic analyser', (Bush, 1922)

12 Bush, V., Hazen, H.L.: 'Integraph solution of differential equations', *J. Franklin Institute*, **211** (1927), p. 615.
13 Bush, V., Gage, F.D., Stewart, H.R.: 'A continuous integraph', *J. Franklin Institute*, **211** (1927), p. 64. Gage's contribution was to suggest that Stewart interpret the integral electrically rather than mechanically which led Stewart to use the watt-hour meter, see Owens *op cit.* (n.2 above), p. 69
14 Bush, V., Gage, F.D., Stewart, H.R.: 'A continuous integraph', *J. Franklin Institute*, **211** (1927), pp. 63-84
15 Bush, V., Gould, K.F.: 'Temperature distribution along a filament', *Physics Review*, **29** (1927), pp. 337-345
16 Hazen in interview with author 1976; Owens *op cit.* (n.2 above) reproduces Hazen's sketch
17 Hazen in interview with author. Owens suggests that Stewart might have been aware of Thomson's work and that Bush knew of the general nature of Thomson's designs. Hazen's recollection was that he came across them while preparing a term paper in 1927
18 Hanna, G.R., Osbon, W.O., Hartley, R.:A.: 'Tracer-controlled position regulator for propeller milling machine', *Trans. American Institute of Electrical Engineers*, **64** (1945), p. 201
19 Bennett, 1985 *op cit.* (n.2 above), p. 996
20 Bush and Hazen 1927 (n.12 above), pp. 585-588
21 Bush and Hazen 1927 (n.12 above), p. 589
22 Bush, V.: 'The Differential Analyser', *J. Franklin Institute*, **212**(4) (October 1931), pp. 447-488; see also Owens *op cit.* (n.2 above)
23 Nieman, C.W.: 'Bethlehem torque amplifier', *American Machinist*, **66** (1927), pp. 895-897. Nieman, C.W.: 'Backlash eliminator', *American Machinist*, **66** (1927), pp. 921-924. Hazen recalled that he was given the papers by his former Sunday School teacher, Adam Armstrong, who had continued to take an interest in his career.
24 Arthur Porter in interview with author, 1975
25 Goldstine, H.H.: *The Computer from Pascal to von Neumann*, (Princeton University Press, Princeton, 1972); Berkeley, E.C.: *Giant Brains, or Machines That Think* (New York, 1949); Differential Analysers were also built in Ireland, Germany, Norway and the Soviet Union; parts of the University of Manchester Different Analyser are preserved in the Science Mustum, London
26 Owens, *op cit.* (n.2 above), p. 65
27 Gould, K.E.: 'A new machine for integrating a functional product', *MIT J. Mathematics and Physics*, **3** (1928), p. 309
28 Although published first the theory paper was written after the work on the second paper had been completed. During 1931 Gordon S. Brown, then a graduate student, worked on the design of an automatic curve follower as did another of Hazen's students Harold A. Traver. Hazen wrote up the work early in 1933, including in it some theory on continuous control. He showed it to Bush who suggested it should be extended to include a study of on-off controllers, a suggestion to which Hazen acceded.
29 Hazen (1934b), p. 549
30 See Minorsky (1922)
31 Hazen, 1934b, pp. 324-5
32 See Wildes, K., Lindgran, N.: *The history of electrical engineering and computer science at MIT 1882-1982*, (MIT Press, Cambridge, MA, 1984)
33 Hazen in interview with author
34 The referencees are: aircraft and ship applications, 'New Compass and Path Indicator', 1929; 'Automatic Yacht Steerer', 1933; The Gyroscope stabilising equipment..., 1932; Chalmers, 1930; Ferry, 1932; Frisch, 1931; Green and Becker, 1933; Haus, 1932; Huggins, 1930; Minorsky, 1922, 1930; Saujvaire-Jourdan, 1931; Schilovsky, 1932; Sperry, 1932; Viterbo, 1931; process applications, 'Kent recording and controlling...', 1931; Alfriend, 1933; Bayle, 1931; Bernarde and Lunas, 1933; Gulliksen, 1933; Hardy, 1929; Harrison, 1931; La Pierre, 1933; Osbourne, 1931; Stein, 1930; and speed control governors 'Hydraulic turbine governors and frequency control', 1931; Jones, 1930; McCrea, 1929; and Sporn and Marquis, 1932
35 Hazen, private communication
36 Hazen 1934a, p. 282
37 Hazen 1934a, pp. 282-3
38 A. L. Whiteley in interview with author

39 Hazen, H.L., Jaeger, J.J., Brown, G.S.: 'An automatic curve follower', *Review of Scientific Instruments*, 7 (1936), pp. 353–57
40 Hazen, H.L., Jaeger, J.J., Brown, G.S., (1936)
41 Porter, A.: Ph.D. Thesis, University of Manchester, 1937; Blackett, Williams (1939); Williams (1939)
42 For a detailed account of MIT's involvement in numerical control of machine tools see Noble, David: *Forces of Production: A Social History of Industrial Automation*, (Alfred A. Knopf, New York: 1984) and Reintjes, J. Francis: *Numerical Control: Making a New Technology*, (Oxford University Press, New York, 1991)
43 Brown, recalling the period in his address at the memorial service for Hazen said, 'Later he passed to me the opportunity to exploit all that he had created in the field of servomechanisms and control...For his expression of confidence in me, for his willingness to help me whenever necessary, and for the sacrifice he made in forgoing the joy of teaching a course he would have loved, I am eternally grateful'. (Brown, G.S., Address at the Harold L. Hazen Memorial Service 25th February 1980)
44 The publication of the 1934 papers marked the end of Hazen's career as an engineer personally involved in technical work. From assuming responsibility for the Graduate Study and Research, in 1937, until his retirement in 1967 Hazen became much more involved in teaching and administration than in active research. He was appointed Head of Department in 1938 and served for 14 years during which time the size of the Department doubled. In 1952 he was appointed Dean of the Graduate school, a post he held until his retirement. He was also active in enhancing engineering education abroad through his work with engineering institutions in Japan and his trusteeship of Robert College, Istanbul and the University of Petroleum and Minerals, Dhahran, Saudia Arabia. And he was, from December 1942 until 1946, head of Division 7 (Fire Control) of the National Defense Reseach Committee (NDRC)

Chapter 5
Wartime: problems and organisations

The exigencies of war raised myriad control problems: some simple and easy to solve, some of great difficulty and complexity. The problems addressed included voltage stabilisation for aircraft generators[1], theoretical investigations of aircraft stability, developments in aircraft instruments[2], governors in power stations[3] and process control for the Los Alamos atomic weapon plant. However, most of the control system advances arose out of work on fire-control applications, that is the detection, tracking, and prediction of the future position of a moving target, and the aiming and firing of guns. In this chapter I first describe the problem of anti-aircraft fire-control and the attempts made at finding solutions, and then I examine the way in which the UK and the USA organised to solve this and other control problems.

5.1 Anti-aircraft fire control

The problem of fire control was not new. As described in Chapter 1, the major navies of the world had introduced fire-control techniques at the beginning of the century. Extensive work on developing such equipment was carried out during the First World War and was continued in subsequent years. However, during the 1930s a new and different fire-control problem was recognised: how to direct gunfire at fast moving aeroplanes. Contrary to the secrecy which surrounded naval fire-control systems, this problem, defence against attack from the air, was much discussed in the inter-war years.[4] There was considerable popular apprehension that bombing would not be limited to military targets and that, as Stanley Baldwin warned on 10th November, 1932, 'the bomber [would] always get through'.[5] An official British history of armament development comments that given the level of public anxiety that was being expressed at the beginning of re-armament in the mid 1930s, it is surprising that the British Army had made no significant improvements to AA armament since the end of the First World War, particularly since the War Office had reviewed the problem from time to time.[6] Given the mood of fatalism at the highest levels, the lack of improvements is perhaps not so surprising.

The two main forms of defence were fighter aeroplanes and anti-aircraft guns. The effectiveness of both was crucially dependent on being able to locate the attacking aircraft. In clear conditions and in daylight attacking aircraft could be seen. During darkness, fighter aircraft still relied on visibility, but anti-aircraft gunnery used sound locators to establish an approximate position and then searchlights to make the enemy aircraft visible. Once the searchlights had located a target an observer tracked the aircraft using a telescope. The movement of the telescope was transmitted by magslips to the searchlight operators who moved the searchlight by hand to keep its pointers aligned with the pointers on the magslip receivers. The telescope data was also transmitted to the predictor (known as the director in the USA), and the output of the predictor was transmitted to the gun operators who manoeuvred the guns to keep the predictor and gun direction pointers aligned.

In poor weather, defence was 'increasingly difficult', H.E. Wimperis, then Director of Scientific Research at the Air Ministry, explained in 1934, 'on account of high speed, higher ceilings, less noisy airscrews and engines, and the ability to fly with an automatic pilot in clouds and fog'.[7] Though the problem was, undoubtedly difficult, there was a growing willingness to tackle the difficulties. Professor F.A. Lindemann, in *The Times* of 8th August 1934, made a strong appeal for investigations into new modes of defence against aircraft to be made, and in September Lindemann and Winston Churchill visited the Prime Minister, Stanley Baldwin, to urge him to support such work. Lindemann had in mind the formation of a small independent committee, financed by the Government, which would investigate possible methods of defence against aircraft. The Committee for the Scientific Survey of Air Defence (the Tizard Committee) was formed and met for the first time on 28th January 1935.[8] However, it was not the independent committee envisaged by Lindemann, but a department committee of the Air Ministry. After making strenuous efforts to get a committee independent of the Air Ministry formed, Lindemann eventually accepted an invitation to join the Tizard Committee. During the the following years there ensued a merciless battle between Henry Tizard and Lindemann,[9] despite which the Tizard Committee was successful in persuading officials and military officers to begin a thorough review of possible methods of defence.

Arising from that review came support for Robert Watson Watt for his experiments on the use of radio waves for the detection of aircraft. The work was referred to as RDF (Radio Direction Finding) in an effort to conceal its real purpose. The reflection of radio waves from object had been observed by many people and the principle of pulsed radio signals for measuring distance was used by Appleton in the UK, and by Breit and Tuve in the USA, for measuring the height of the ionosphere.

Secret investigations into the use of radio waves for military purposes took place in a number of countries: USA, UK, France, Germany, Italy, Japan, Russia, Hungary and Holland.[10] For example, in 1922 in the USA, Hoyt Taylor and Leo Young, working at the Naval Research Laboratory, observed a distortion in received radio signals caused by the reflection from a wooden steamer on the Potomac River. In the summer of 1930, they detected the presence of an aircraft using radio direction finding equipment. As a consequence of this discovery, the Naval Research Laboratory was ordered to carry out further investigations. In 1934, staff at the laboratory demonstrated a continuous wave system and, in 1935, produced a pulsed radar. The development of the latter was given high priority. By 1938, radar systems were being supplied to both the US Navy and US Army for trial.[11]

5.2 Organisation in the UK

In the UK, during an experiment made on 26th February 1935, an echo from an aircraft flying in the 50 metre beam from the BBC's station at Daventry was detected in a receiver eight miles away. H.E. Wimperis at once took steps to provide facilities at Orfordness for an intensive investigation of methods of locating aircraft. Members of the Tizard committee discreetly suggested to a number of young research workers in the universities, both staff and students, that they might like to join the Bawdsey Research Station at Orfordness (and later other similar stations).[12] The aims of the research were initially the detection of aircraft at the maximum possible range. Gradually the programme grew to cover systems that would provide more or less continuous determination of the positions of hostile and defending aircraft, for use for Ground Control of Interception (GCI), the control of AA guns, Airborne Interception (AI), and eventually bombsights and bombing radar.[13]

The urgent need in Britain to establish an operational system led to a decision to construct a chain of stations using continuous wave radar to 'floodlight' the sky rather than waiting for further technical developments in pulsed radar to provide a directional beam system. By March 1938, a chain of five stations—the Home Chain—protecting London and the Home Counties was operational. In parallel, Tizard began the task of persuading senior military personnel to consider how the information obtained from the radar stations might be used to remedy a major weakness of existing air defences—the detection of attacking aircraft. He was successful in persuading the Air Ministry to carry out experiments—the Biggin Hill experiments—on interception of bombers by fighters. The co-operation of scientists and service personnel produced effective methods of interception based on the use of radar to detect aircraft and radio to communicate directional commands to pilots, the so-called Ground Controlled Interception (GCI) system. GCI was so effective that it was no longer considered necessary to have fighter squadrons continually airborne. The experiments demonstrated the value of having scientifically trained personnel observing operations and formed the beginnings of Operational Research, techniques and applications for which were developed extensively during the war.[14]

Information from the Home Chain radars could be passed to the AA gun and searchlight crews to alert them, and provided the general direction of the raiding aircraft, an improvement on the existing system of aligning the searchlights on the basis of telephoned messages from sound locator operators.[15] For land based systems the overall arrangement of the AA system at the beginning of the war is shown in Figure 5.1. The radar operators manually tracked the target in bearing, elevation and range; the bearing and elevation data was transmitted by magslip units to pointers on the searchlights and the searchlight operator matched the searchlight pointers to the magslip pointer. Once the target was illuminated the telescope operator centred his cross-wires on it and switched off the radar operators' signals, thus connecting the pointers on the searchlight to his telescope. The data from the telescope was also transmitted to the predictor and hence to the gun operators. A separate range finder operator, using optical range finding techniques, determined spot measurement of the range and these were also fed into the predictor. The output from the predictor was transmitted using magslips to the gun operators who manipulated the gun controls to match the gun position to

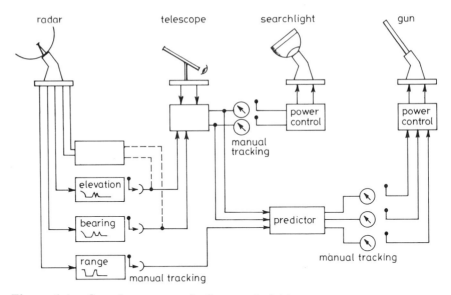

Figure 5.1 *General arrangement for fire control of AA guns*

the receiver magslip pointers. The range information was converted by the predictor into a fuse setting for the shell. The whole system could involve up to 14 operators.

Visual methods of tracking provided an accurate measure of the angles of elevation and azimuth; the major errors arose from use of optical range finders. During 1937 a group attached to Bawdsey Research Station began investigating the use of radar for ranging and gun laying. At first they thought that it would not be possible to achieve the necessary accuracy but during the summer of 1937 they were able to produce an accuracy of 100 yards at ranges between 3,000 and 14,000 yards, and later that year accuracies of 25 yards were claimed. Another advantage was that radar provided a continuous measurement of range not just the spot measurements produced by optical range finders. As a result a decision was made to develop what became known as the GL Mark I radar set. This set provided a continuous measurement of range but only intermittent measures of bearing (at about 30 second intervals) and did not measure elevation. A contract for the production of 500 sets was placed with Metropolitan-Vickers (Metro-Vick) (transmitters) and A.C. Cossor (receivers) early in 1939 and by the end of the year 59 sets were in service. Early in 1940 work stated on the GL II set which was to incorporate elevation (height-finding). The development work was done by the Gramophone Company and the Metro-Vick Company. Based on the method of measuring elevation used in GL II, L.H. Bedford of Cossor devised a means of modifying the GL I sets to measure elevation and the sets were modified before the GL II sets went into service. Further improvements in accuracy (typically about 2 degrees for GL I and GL II sets) could only be achieved by using higher frequencies than the 50 MHz currently used.[16]

Work on experimental systems for what was to become known as GL III began during 1941 and production orders were placed in autumn of 1942. However, the system was overtaken by technical developments as it became clear that future GL equipmnt should be based on automatic following. A political decision was made to use the American SCR 584 system and in the autumn of 1943 the GL III programme was drastically curtailed.[17]

In setting up the Bawdsey Research Station to pursue research on radar, H.E. Wimperis was following a system that had evolved in Britain over a 30 year period. The use of research establishments under the control of a Director of Scientific Research in the sponsoring ministry with staff appointed as civil servants was the normal way of operating. It had grown up in an *ad hoc* manner in response to various perceptions of problems and could be seen as being a practical, adaptable system. Except for a large increase in the number of staff employed, the pattern of organisation was little changed during the war.

This form of organisation provided a close linkage between the service and the research scientists, but could result in lack of co-ordination between the three services. During 1935 representatives of the Navy and the Army were briefed on the developments in radar and both services began work in their research and development establishments — the Admiralty Signals School (ASE) and Air Defence Experimental Establishment (ADEE).[18] As the war progressed, each arm of the services began to develop its own radar and fire-control systems and to build up links with different private contractors. There were attempts at a high level to co-ordinate research and development programmes but the traditional rivalries between the services and compartmentalisation on the grounds of security meant that there was little interchange of ideas between the staff in the various establishments.

The position was further complicated when the arrangements for carrying developments into production are considered. Each service had different ideas as to when research and development work should be passed to design staff and whether or not the service should employ its own design staff or contract all design work to private companies. Even within one service, 'the business of designing a fire control system was', according to John F. Coales of the ASE, 'almost impossible with the Director being designed at Teddington [Admiralty Gunnery Estalishment], the mountings at Vickers or at Bath [Directorate of Naval Ordnance] and the radar at Portsmouth [Admiralty Signals Establishment].'[19] Add to this the involvement of various commercial companies — Metropolitan-Vickers, British Thomson-Houston and Vickers — and the fact that the ARDE operated under the control of the Ministry of Supply rather than one of the service ministries, and the potential for duplication of work and incompatibility of systems is readily apparent. The ARDE, as the only organisation with an overall systems responsibility, should have had an advantage, but their work was hampered by having to satisfy the conflicting requirements of the different armed services.[20]

At various times the issue of co-ordination of research and development activities was raised. For example, in 1942, Oliver Lyttleton, the Minister of Production, expressed to the Prime Minister his dissatisfaction with the present arrangements and recommended a full enquiry with a view to achieving a closer integration and coordination of research and development activities. The Prime Minister passed the matter to Lord Cherwell (Prof. F.A. Lindemann) for comment and received the response that an enquiry was not necessary for while better arrangements could be conceived, changeover would cause loss of time and possibly outweigh any advantages.[21] An enquiry limited to coordination on RDF between the three services was carried out by Lord Justice du Parq who reported on 11th August 1942 that it was desirable for the services to maintain their own research establishments because it was essential that research workers must have close contact with the serving personnel.[22] Proposals made at lower levels to improve coordination and cooperation also met difficulties. For example, a proposal that originated in the Admiralty Research Establishment for an Inter-Services Fire Control Group to develop predictors for use by both the Army and the Navy made little progress, because with the exception of Major Kerrison, Army representation was to be minimal.

A problem common to all the establishments involved in radar work, and in fire-control generally, was the design and development of automatic control devices, and in March 1942 an informal group which came to be called the Servo-Panel was formed. The idea originated with A.K. Solomon, an American scientist who was a section head at ADRDE. It received strong support from F.A. Vick, Assistant Director of Scientific Research, Ministry of Supply, and from J.D. Cockcroft, then Chief Superintendent of ADRDE Malvern. The decision to form the panel was taken at a meeting held on 20th March 1942 attended by representatives of the Services, the research establishments and industry. The industries represented were the British Thomson-Houston Company, A.C. Cossor, Evershed and Vignoles, Ferranti, Elliott Bros., Stothert and Pitt, Sperry Gyroscope and Vickers-Armstrong. Cockcroft was appointed as the first chairman. In October 1942 Douglas Hartree took over as chairman with K.R. Hayes as deputy and A. Porter as secretary. These three largely controlled the operation of the panel for the next four years.

The original intention of the group was that an Interdepartmental Committee on Servomechanisms should be formed but the Admiralty was unwilling to cooperate fully so instead it became an informal panel under the auspices of the Ministry of Supply. It eventually achieved full interdepartmental status in 1944.[23] The Servo-Panel primarily acted as might a committee of a learned society: it organised meetings, provided a library and generally attempted to facilitate the exchange of information. Because it had no executive, or even advisory powers, with respect to the various ministries and the services, it was considered by some industrial members to be simply a 'talking shop',[24] a somewhat harsh, but understandable judgement, because in its early meetings the panel concentrated on nomenclature[25] and linear design techniques at a time when industrial contractors were struggling to get control systems working in the presence of noise and non-linearities. Some indication of its activities can be obtained from the list of meetings organised and the sub-committees formed which are shown in Table 5.1. The Servo-Panel was formed too late in the war to have had much influence on the development of systems that became operational during the war, many of its members did, however, play an important role in disseminating control system design techniques after the war.

5.3 Organisation in the USA

On first examination the arrangement made in the USA for the organisation of research and development immediately preceding and during the war was much simpler, with clear lines of command. Overall responsibility was exercised by the

Office of Scientific Research and Development (OSRD) whose Director, Vannevar Bush, reported directly to the President. The formation of the OSRD was a consequence of an initiative taken by a small group of scientists and engineers led by Bush and including Karl T. Compton, James B. Conant and Frank B. Jewett who, in the spring of 1940, believing that the USA would get drawn into the war, solicited the President's support for an organisation that would harness scientific and technical expertise. They argued that the existing organisation — the National Research Council of the National Academy of Sciences which had been formed in 1918 to provide advice to the government on scientific matters — was a reactive body, and what was needed now was a proactive body that would not only make its own assessment of service rquirements, but also get the weapons made, with or without their cooperation. Service leaders were not, and could not be expected to be, sufficiently acquainted with modern science and technology to know what could be achieved, and so the scientist and technologist must become acquainted with the needs of the services.[26]

The President (Franklin D. Roosevelt) General Marshall and Admiral Stark agreed to the proposal and the National Defence Research Committee (NDRC) was formed on 27th June 1940. The committee was to oversee all areas of warfare, except problems of flight which were to remain within the purview of NACA (in the UK the Aeronautical Research Council and the RAE seem also to have remained somewhat independent from other agencies and the various Ministries). Roosevelt also made it clear that NDRC was not to replace any of the excellent work already being carried out in the research laboratories of the Army and Navy. As might be expected, these caveats led to boundary disputes between NDRC (and its successor, OSRD) and the research organisations operated by the military. These disputes were a constant source of irritation and friction throughout the war.[27]

The NDRC was initially organised into five sections of which section D — Detection, Controls and Instruments — headed by Karl Compton is of the most intersecting for this study. Section D was subdivided into four sub-sections, of which D-1 Detection under Alfred L. Loomis and D-2 Controls under Warren Weaver are the most relevant.

After just one year and a day the NDRC was subsumed within the newly formed OSRD. Bush became the director of the OSRD and Conant became chairman of the NDRC which became a committee of the OSRD. Conant reorganised the NDRC into divisions and Division 7, Fire Control, under Harold Hazen, was the one largely concerned with control systems. Division 7 operated with the following sections:

Section	Title	Section Head
7.1	Surface Systems	Duncan J. Stewart
7.2	Airborne Systems	S.H. Caldwell
7.3	Servomechanisms	E.J. Poitras
7.4	Optical Range Finders	Thornton C. Fry
7.5	Fire Control Analysis	Warren Weaver
7.6	Seaborne Fire Control	I.A. Getting

The Divisions operated in many different ways and were largely autonomous. For example, Division 7 operated mainly through contracts that were awarded

Table 5.1 *Servo-panel meetings*

Meeting No.	Date	Authors	Title
1	13th May 1942	Hartree, D., Porter, A.	The differential analyser and its application to servo-problems
2	12th June 1942	Bedford, L.H. Taylor, L.K.	Servo-motor developed at Cossors' Servomechanisms developed at Ferranti's
3	24th July 1942	Oxford, A.J.G. Jofeh, L.	The automatic ranging unit for S.L.C. An electrical method of analysis of servomechanisms
4	21st August 1942	Tustin, A., Morris, D.G.O.	Talk and demonstration dealing with the stability of cyclic systems and the application and construction of the Metadyne
5	2nd October 1942	Ludbrook, L.C.	Practical problems involved in servo-systems with special reference to the Ward-Leonard system in position control servos
6	30th October 1942	Warren, N., Bell, J.	The A.R.L. hydraulic servo-system with magslip control
7	4th December 1942	Williams, F.C.	Low power split field servos
8	8th January 1943	Whiteley, A.H.	Some servo developments in the U.S.A.
9	29th January 1943	Inglis, C.C. Tustin, A. Somerville, T.B. Barnett, P.S.	Tank gun stabilisation Gyro-electrical control systems Polarised twin vibrator regulator Brief decription of a hydraulic method of stabilisation
10	19th February 1943	Ashdown, G.L.	An electro-hydraulic remote control system
11	19th March 1943	Robinson, B.W.	Some aircraft servomechanisms
12	16th April 1943	Williams, R.R. Tustin, A.	The servo selsyn and D.C. impulse control systems Short talk on the vernier selsyn
13	21st May 1943	Uttley, A.	Servomechanisms used in R.A.F. synthetic trainers and The human link in the cyclic control system
14	25th June 1943	Daniell, P.J.	The interpretation and use of harmonic response diagrams (Nyquist diagrams) with particular reference to servomechanisms
15	23rd July 1943	Bell, J.	Symposium on the technique of testing of servomechanisms Methods of measuring (a) sensitivity, (b) dead sector, (c) running lag, (d) power or torque ratio, (e) stability, (f) the ratio of rate of doing work to stored energy
		Whiteley, A.L. Morris, D.G.O.	Testing techniques used by BTH Methods of harmonic response testing used at Metro-Vick
16	27th August 1943	Jofeh, L.	Electrical models in servomechanism design

No.	Date	Author(s)	Title
17	17th September 1943	Hamon, B.V.	Servomechanism development in Australia
18	15th October 1943		Discussion of interim report of the Nomenclature Committee
19	26th November 1943	Hayes, K.A. Hyde, A.D.	Use of servos in fire control A new method of analysing servomechanisms Meeting held at C of M.S. Bury
20	7th January 1944	Marchant, E.W.	The testing laboratory at Liverpool University for small servomechanisms
21	4th February 1944	Craik, K.J.W.	Some characteristics of the human operator in control systems
22	3rd March 1944	North, J.D.	The electro-hydraulic system used in Boulton-Paul gun turrets
23	24th March 1944	Brown, G.S.	Some activities of the servomechanism laboratory, MIT
24	4th May 1944	van Leeuwen, J.J.S.	Some general aspects of stabilisation
25	9th June 1944	Hartree, D.R.	The differential analyser and its applications to control problems
26	7th July 1944	Porter, A.	A study of the performance of a fire control with special reference to smoothing
27	18th August 1944		Discussion of 'glossary of terms used in control systems'
28	29th September 1944	Spraule, C.O.	The Holmes repeater compass and the four pen recorder
29	10th November 1944	Donald, M.B. Callender, A. Griffiths, J.B. Forster, E.W.	Symposium on 'Some industrial control (servo) problems' Control and measurement in the chemical industry Industrial control problems Circuit interlocking Electronic industrial servos
30	Not known	Caldwell, S.H.	U.S. developments in fire control systems
31	9th February 1945	Hayes, K.A. Holland, E.	The specification of a servomechanism The 1939 G.E. searchlight remote control
32	23rd March 1945	Ludbrook, L.C. Ovens, O.	Searchlight control gear mark VIII
33	20th April 1945	Tustin, A. Brailsford Hayes, K.A. Braddick, H.J. Sutton, O.G.	Symposium on 'Servo testing instruments and testing techniques' Method of deriving harmonic responses from unit function responses and vice-versa Instrument for obtaining the harmonic response of equipments Response characteristics of the Hughes 4 pen recorder Instruments developed by R.A.E. The Cambridge Instrument Co. gyrograph
34	Not known	Not known	
35	29th June 1945	Tustin, A. Belsey, F.H.	Symposium on 'Backlash' The effect of backlash in servo-systems Effects of backlash and resilience between motor and load on the stability of a servo-system
36	24th August 1945	Sudworth, J.	Control systems of the German pilotless missiles V1 and V2

to many different groups within industry and the universities; whereas Division 14, concerned with radar, concentrated early all its resources in the one research laboratory — the Radiation Laboratory based at MIT.[28]

5.4 Relationships between civilian scientists and engineers and military and government personnel

The formal structure and line of command within the OSRD and between the OSRD and other arms of government was direct and clear. This did not, however, avoid the problems of parallel development, service rivalries, inter-agency disputes, and jockeying for commercial advantage. Attempts were made by the services to by-pass the perceived bureaucracy of NDRC, and to deal directly with universities and companies. The role of NDRC was not made easier by application of 'need to know' principles for disclosure of information, and the reluctance to allow civilians to observe active operations. Of continuing concern to the civilian members of the NDRC was how well the equipment performed in the field. The services argued that they needed technical improvements to increase accuracy, whereas civilian members of NDRC saw little need for further improvements until training was improved so that the existing technical potential was achieved. At the heart of many of these disputes was a lack of adequate and systematic performance data.[29]

Civilians working for the NDRC did have the advantage of executive authority whereas their British counterparts lacked such authority. In Britain there was strong resistance at the highest level to creating any organisation that would give civilian scientists and engineers any executive power. The Committee for the Scientific Survey of Air Defence had been formed as a departmental advisory committee of the Air Ministry and it had no direct access to the Committee on Imperial Defence or to any Minister, let alone the Prime Minister. Lacking executive power and having no formal status outside the Ministry it relied on force of argument and informal channels of influence.

When, in July 1939, Sir William Bragg, President of the Royal Society, suggested the formation of a small committee of scientists to advise the Committee of Imperial Defence on scientific matters regarding the conduct of war, the suggestion was ignored. Sir William repeated the suggestion in June 1940 and on this occasion the Prime Minister was advised in a memorandum,

> that a Committee should be formed under Lord Hankey's Chairmanship, mainly as a sort of 'safety valve' for Sir William Bragg and his friends in the Royal Society. The Terms of Reference are I think such as to ensure that they do not interfere in the work of the Scientific Departments generally.
>
> I have shown this [the proposals] to Prof. [Professor F.A. Lindemann] and he thought that the Terms of Reference proposed were reasonably innocuous. As you know he has always been opposed to any Committee of this kind; but if a safety valve for the malcontents is necessary (and I genuinely think that it is) this seems the most harmless form.

Churchill agreed, subject to the proviso that secrets relating to the work being carried out were not imparted to a wider circle, and that scientists already working for the Government did not have to spend much time on the matter.[30] The engineering community fared no better: in response to requests from the engineering institutions an Engineering Advisory Committee was formed in April 1941, but once again it was specified that it was to have 'no teeth'.[31] Some advisory groups were formed; for example, the Ministry of Supply formed an Advisory Council on Scientific Research and Development, but this council largely comprised people who were already actively involved in some capacity in government activity.[32]

Although the organisation in the USA was apparently more open and inclusive there were complaints that NDRC favoured the east coast establishment. NDRC also faced strong resistance to its involvement in designing fire-control systems for the Navy. The main suppliers of fire-control equipment to the US Navy included the Ford Instrument Company, the Arma Corporation, and the General Electric Company all of which built up close ties with the Bureau of Ordnance and naval personnel. The close ties and the 'excessive application of security regulations', Ivan Getting argued, produced a 'closed technical society' which could 'easily lead to over-confidence within the group' and to the discounting opinions, criticism and proposals of people outside the group.[33]

In both the USA and the UK, civilian scientists and engineers drawn from the universities and from industries with no previous military involvement, met resistance to their ideas during the war. A particular issue which was raised repeatedly was criticism of and concern about the methods used by the military to evaluate the performance of equipment and systems. So although, by the beginning of the war, the US Navy had the best fire-control equipment of any of the major sea-powers,[34] how effective was it? Recently, R. Garcia y Robertson has argued that long-range gunnery against moving targets — the problem which first generated work on fire-control mechanisms — has always been and is still inaccurate. (He offers as one example the reply given to a Congressional question by US Navy, in 1985: the US Navy claimed, on the basis of a paper study, that they would have a 30% success rate at hitting the US Capitol Building from a range of 13 to 18 miles.)[35]

5.5 The systems approach

The civilian scientists on the Tizard Committee, and also the civilians who became involved in the fire control problem in the USA, recognised at an early stage the importance of examining the problem as a whole and that the way in which various components — target position finding, prediction, gun positioning and fuse setting — were integrated would be vital in finding an effective solution. They were also realistic enough, however, to realise that an overall system based solution, was a long term solution, and that the exigencies of war demanded some short term, piecemeal, solutions.

The problem was thus broken down into several areas: provision of accurate measurements of position using radar (particularly range), improved methods of manual tracking, remote power control of gun positioning, improved predictor design, automatic fuse setting,[36] and automatic following radar. The eventual aim was to have fully automatic tracking of the target and aiming of the guns.

Proposals for a completely automatic fire control system with an automatic tracking radar feeding elevation, bearing and range to the predictor, which in turn was connected directly to the power controls of the guns, were made early in the war. However, the Scientific Advisory Council of the Ministry of Supply urged caution, recommending that the first step should be the automatic guidance of searchlights.[37] The use of radar to direct searchlights received political support at the highest levels and an experimental searchlight control (SLC – known colloquially as Elsie) was put into production during autumn 1940. Effective use of the system was constrained by the War Office rules on the location of searchlights (they had to be spaced at 10 000 yard intervals and in groups of three) and on how they were to be used. ADRDE scientists argued (and sought to demonstrate) that they were most effective if used with a single light at 3500 yard intervals with one light in three radar controlled. AA Command eventually agreed with this view in 1943. Further problems arose from the unreliability of sets – this was largely caused by the hurried way in which they had been put into production. In 1942 authority was given for a redesign and for the redesigned systems to use automatic following techniques.[38]

Work on fully automatic following radars did not begin in Great Britain until 1941. In the United States, with less urgent needs, it was decided in October 1940 to adopt, 'wholly automatic tracking in both azimuth and elevation, and to incorporate Alfred Loomis's novel suggestion of a conical scan.'[39] This decision eventually resulted in the production of the SCR-584 based AA fire-control system, 'the first Allied ground radar to surpass the German Wurzburg radar system.'[40] In the circumstances, the British decision to proceed by piecemeal modification and improvements of existing equipment was wise, for even with a unified design team based on the Radiation Laboratory at MIT it was not until 1943 that the SCR 584 entered service.

Problems relating to feedback control systems also emerged in many other areas of work during the war; for example, extensive work was carried out on the development of gyroscopically stabilised systems and the use of gyroscopes in the 'lead computing sight'. In both the USA (led by Hanna at Westinghouse) and in the UK (led by Sutton and Tustin), work was carried out on gun control systems – stabilised using gyroscopes – for tanks.[41]

The biggest control system requirement at the beginning of the war was the provision of a fast, accurate, high powered position control system that could be used to move guns, searchlights and radar antennae. To meet this requirement techniques for the analysis and design of control systems had to be refined and developed, and a whole range of improved components – motors, amplifiers, transducers – had to be developed. As the war progressed other problems began to emerge: the design of filters to smooth signals obtained from manual tracking and radars, the design of circuits to give the best estimate of the future position of a target, techniques for analysis and design of systems with non-linearities, analysis of servomechanisms with pulsed data (sampled data systems), and for the design of systems with optimum response with respect to some performance criteria in the presence of noise and disturbance. I deal with details of these developments in the next two chapters.

5.6 Notes and references

1 Holliday, T.B.: 'Application of electric power in aircraft', *Trans. American Institute of Electrical Engineers,* **60** (1941), pp. 218–225
2 Imlay (1941), Kotelnikov (1941), Richards (1941), Sudworth (1941), Dunn (1942), part of the interest in automatic controls was for use in pilotless target aircraft, Osborne (1942)
3 See for example McClure and Caughey (1941), Church (1941), Concordia *et al.* (1941), Higgs-Walker (1941), Osborne (1941)
4 A detailed description of the Sperry Company system appeared in R.H. Ward, 'Anti-aircraft Gun Control', *Army Ordnance,* **XI**, 1931, pp. 452–457 and this article was abstracted in the *Aeronautical J.,* **XXXV**, 1931, p. 1087. A description of this system was given by R. Schmitt in 'A new night fire control apparatus for AA Artillery', *Luftwehr,* **5** (1938), pp. 243–246 (in German), see abstract in *J. Royal Aer. Soc.*, (1938), p. 842. In 1931 Guyomar described the problems of fire control in an article, 'Fire-control from Aeroplanes against targets and from the ground against aeroplanes' *Rev. F. Aer.* 25th August 1931, pp. 935–945 (in French) for abstract see *Aeronautical J.* (1932), p. 78
5 Clark, R.W.: *Tizard*, (Methuen, London, 1965), p. 105
6 Postan, M.M., Hay, D., Scott, J.D.: *Design and Development of Weapons; Studies in Government and Industrial Organization*, (HMSO, London, 1964), pp. 279–81
7 Clark, 1965, *op. cit.*, (n.5), p. 110.
8 Members of the Committee were Mr. Henry Tizard (Chairman), Professor P.M.S. Blackett, Professor A.V. Hill, and Mr. H.E. Wimperis with A.P. Rowe as Secretary. A.P. Rowe, Wimperis's assistant, had at sometime during 1934 examined all the files on air defence that he could find and had communicated his fears to Wimperis who in October met with A.V. Hill to discuss problems of air defence. On 12 November 1934 Wimperis drafted a note proposing the setting up of a committee to carry out a scientific survey of possible methods of air defence. See Swords, S.S.: *Techncal history of the beginnings of RADAR*, (Peter Peregrinus, Stevenage, 1986), p. 83
9 For accounts of the antagonism between Tizard and Lindemann see Lord Birkenhead: *The Prof in two Worlds*, and Clark: *Tizard*, 1965. Both men played a major part in Britain's use of science and technology for military purposes: during the Second World War, Lindemann, as Churchill's personal scientific adviser was in the dominant position but Tizard continued to proffer advice and to carry out important work including heading the British Technical and Scientific Mission to the USA in September 1940. This mission passed onto the Americans Britain's scientific secrets including information about short wave radar. It is generally recognised that the mission played a significant role in the development of cooperation between the two countries and their scientific and technical endcavours in support of military operations.
10 See Swords, *op. cit.*, (n.8) for a brief outline of early developments in these countries.
11 A brief account of radar development in the USA is given in Fagen, M.S., (ed.): *A History of Engineering and Science in the Bell System: National Service in War and Peace (1925–1975)*, (Bell Telephone Laboratories, Murray Hill, NJ, 1979), pp. 5–51. See also Swords *op. cit.,* (n.8), pp. 101–120
12 F.C. Williams a junior lecturer at the University of Manchester was recruited in this way, he had been working with Blackett who suggested that he might find the work at Bawdsey interesting (and more importantly as far as Williams was concerned better paid).
13 Postan, 1964, *op. cit.*, (n.6)
14 During the war Blackett played a leading role in the introduction of Operational Research, Clark, *op. cit.*, (n.5), p. 320; see also Waddington, C.H.: *OR in World War II*, (London, 1973)
15 Clark, *op. cit.*, (n.5), p. 382
16 A group headed by P.M.S. Blackett attempted to collect reliable data on the accuracy and effectiveness of blind firing of AA guns. From the data they collected during September and October 1940 over 250,000 rounds had been fired and 14 aircraft destroyed (nearly 19,000 rounds per aircraft). From data collected during 1941 after the addition of elevation measuring units to the GL radars it was claimed that over 100 aircraft had been destroyed at a cost of 4100 rounds per aircraft. Cockcroft (1985), *op. cit.*, (n.17), p. 330; Swords, p. 86

17 Cockcroft, J.D.: 'Memories of radar research', *IEE Proc.*, **132**, Pt A (1985), pp. 331-2
18 The ADEE was officially under the control of the Royal Engineers and Signals Board and the Director of Scientific Research in the Ministry of Supply had no official say in its activities. During 1941 the ADEE was reorganised as the Air Defence Research and Development Establishment (ADRDE) and eventually became the Royal Radar Research and Development Establishment.
19 Coales, J.F.: private communication. Coales took over responsibility for developing naval fire-control radars in 1937.
20 J.F. Coales considers that even within the Admiralty organisation there were too many design and development groups and that the division seriously interfered with fire control developments; private communication to the author.
21 PREM/4/97/9 (1942): *Arrangements for Scientific Research and Development*. The Minister of Production's memo was dated 5th May 1942 and the file passed to Cherwell on 8th July 1942, who responded that 'It is all very well to demand a scientific general staff. Hill [A.V. Hill] and company should...be made to put forward a detailed plan...giving the names of people they propose for principal office...Some of our distinguished friends are not content with such relatively obscure activities [scientific and administrative work]. Ignorant of all knowledge of military and production difficulties, they have the vague feeling they could win the war if only they were given a free hand.'
22 CAB66/27 W.P.(42) 352 Radio Communication and Equipment.
23 The idea originated with Dr. A.K. Solomon, an American scientist who was a section head at ADRDE, and it received strong support from F.A. Vick, Assistant Director of Scientific Research, Ministry of Supply and J.D. Cockcroft then Chief Superintendent of ADRDE Malvern. The decision to form the panel was taken at a meeting held on 20th March 1942 attended by representatives of the Services, the research establishments and industry; the industries represented were: BTH, A.C. Cossor, Evershed and Vignoles, Ferranti, Elliott Bros., Stohert and Pitt, Sperry Gyroscope and Vickers-Armstrong. Cockcroft was appointed as the first chairman, but in October 1942 Hartree took over as chairman with K.R. Hayes as deputy and A. Porter as secretary and these three largely controlled the operation of the panel for the next four years. See Porter, A.: 'The servo-panel — a unique contribution to control-systems engineering', *Electronics and Power*, October, 1965, pp. 330-333; see also Hayes, K.A.: 'Servo-mechanisms: recent history and basic theory', *Trans. Society of Instrument Technology*, **2** (1950), pp. 2-13
24 This view was expressed by A. Tustin, A.L. Whiteley and F.C. Williams in interviews with the author.
25 A glossary of terms was produced in March 1944 with a second edition in March 1946. The second edition was prepared for use in the Military College of Science: *MCS Notes: Glossary of terms used in control systems with particular reference to servo mechanisms*, MCSB/DP/21/1, and was issued as a restricted document
26 The main sources of information on NDRC and OSRD are Baxter, J.P.: *Scientists against Time*, (1947), this is a popular history of the use of science and engineering during the war; there was also a seven volume history *Science in World War II* of the OSRD. The most relevant volumes are Boyde, J.C.: 1948 *New Weapons for Air Warfare: Fire-Control Equipment, Proximity Fuses and Guided Missiles*, (Little Brown, Boston, 1948) and Stewart, I.: *Organising Scientific Research for War: The Administrative History of the Office of Scientific Research and Development*, (Little Brown, Boston, 1948). The book by Boyce drew heavily on accounts prepared by the staff of various divisions and some of the original accounts can be found in the NA Record Group 227 — some of the accounts were heavily edited by Boyce to remove the more caustic comments regarding organisation.
27 Bush throughout the war worried about the organisation and at times considered arrangements closer to those operating in the UK. See Baxter *op. cit.* (n.26), pp. 28-30
28 A brief account by Karl Compton of the formation of the Radiation Laboratory is given in Volume 28 of the Radiation Laboratory Series McGraw-Hill. The volume also contains an account of the organisation of the Laboratory and the preparation of the series of books reporting its work. Under an agreement made during the visit of the Tizard mission to the USA the laboratory was to concentrate initially on three projects, a microwave (10 cm) radar for Aircraft Interception (AI), a precision gun laying radar, and a long-range aircraft navigation system.
29 Squabbles also occurred between different sections of NDRC. Division 7 did not like

the proposal that the Radiation Lab should build a complete fire-control system for the navy, they thought that fire-control was their speciality and that the Radiation Lab should stick to providing just the radar. Ivan A. Getting, *Draft history of Section 7.6 of NDRC*, NARS, Hazen papers, box 62 history folder

30 The initial suggestion was made by Bragg on 13th July 1939 and re-iterated on 8th September 1939, see CAB 66/2 WP(39) 51 *Scientific Knowledge: Proposals by the President of Royal Society*; Bragg wrote again to the Prime Minister's Office on 10th June 1940 suggesting a Cabinet Sub-Committee on cooperation in scientific work; a memo from EHA [E.H. Appleyard?] to the Prime Minister proposing such a Committee was dated 26th September 1940; the Prime Minister agreed it on 27th September 1940, see PREM 4/97/2 1940: *Coordination and application of scientific research*

31 PREM4/97/14 1941–42 *Engineering and Science Advisory Committee*

32 The Committee first met on the 25th January 1940 with the following terms of reference:
'1. To consider and initiate new proposals for research and development, and to review research and development in progress in the MoS establishments in relation to the most recent advances in scientific knowledge
2. To advise on scientific and technical problems referred to them
3. To make recommendations regarding the most effective use of scientific personnel for research and development
4. To report to the Minister of Supply'
The council was chaired by Sir Edward Appleton and included among its members Tizard, Cockcroft, Watson-Watt, Oliphant and ex-officio the Directors of Scientific Research of the Ministry of Supply, the Admiralty and the Air Ministry. AVIA 22/155 (Internal 201/General/309) Pt. I

33 Ivan A. Getting, *op. cit.*, (n.29). In the published version of this account the editor replaced 'extensive' with 'stringent'

34 *ibid.*

35 Garcia y Robertson, R.: 'Failure of the heavy gun at sea', *Technology & Culture*, **28**(3) (1987), p. 557

36 I have not dealt with the development of the proximity fuse but this was of great importance with regard to the effectiveness, in terms of the ability of damage and destroy aircraft, of AA gun fire.

37 P.R.O., AVIA 22/866, M.O.S. Scientific Advisory Council, Minutes of AC100, 28th March 1940, at this meeting proposals for a guided projectile were also discussed

38 Cockcroft, *op. cit.*, (n.17), pp. 334–335

39 Baxter, J.P.: *Scientists against time*, 1947, p. 147. The conical scan technique which enabled azimuth and elevation to be determined using one aerial was adopted in 1941 by Coales for the Navy's 10 cm high-angle radar, see Coales, J.F., Calpine, H.C., Watson, D.S.: 'Naval Fire-Control Radar', *J. Institution of Electrical Engineers*, **93**, (1946), Pt. IIIA, pp. 349–379

40 Baxter, *op. cit.*, (n.26), p. 145. The Wurzburg dish was driven by a Ward-Leonard system which incorporated a feedback signal from the motor armatures, the field supply for the generator was from a thyratron amplifier and data transmission was by means of selsyns, AVIA 39/22

41 G.R. Hanna of Westinghouse had been investigating the use of a gyroscope for use as a means of detecting speed changes in rolling mills, see Woodbury D.O.: *Battlefronts of Industry: Westinghouse in World War II*, (Wiley, New York, 1948)

Chapter 6
Development of design techniques for servomechanisms, 1939-1945

> At the beginning of the war in Europe basic inadequacies lay, first in feedback-control components, second in systems concepts, but above all inadequacies lay in our understanding of the problems and the methods of attack that would be effective...The role of analytical techniques primarily is to provide this understanding.
>
> A.C. Hall, 1953

The emergence of what are now referred to as the classical frequency response design techniques for automatic control systems came from the work done during the war on three problems: high power servomechanisms for remote control of heavy guns, automatic-tracking radar systems, and power operated gun turrets for aircraft.

6.1 Remote power controls for heavy AA guns

The development of power operated gun controls—remote power control or RPC was the term used—took place primarily in the context of heavy AA guns and naval guns. The anticipated advantages of introducing power control were, as Brigadier Douch recalled after the war, 'greater accuracy and the saving of men continuously in action'.[1] There was also some anxiety, which was clearly expressed in a US Army report of 1940 following the field trials of the Sperry Company gun control system, about reliance on a 'complex combination of electrical, mechanical and hydraulic units, which would be expensive to build, difficult to maintain in the field, and would double the load on the gun's power generator'.[2] However, the increasing speeds of aircraft giving rise to high rates of gun movement made the change inevitable.

The problem that has to be solved in order to provide remote power control is, of course, the classic servomechanism problem of forcing an output (a voltage, a shaft position) to follow accurately a continuously changing input signal. In gun control systems this input signal is obtained from an operator's handwheel or the output from the predictor and the output is the gun or searchlight position. The gun (or searchlight) is required to follow the input signal accurately and smoothly with zero velocity lag; that is, the gun should be correctly aimed and be moving at the correct velocity at all times. As will be discussed later, the problem is most difficult when direct input from the predictor is used since manual tracking provided both smoothing of the signal and a good operator, by anticipating errors, also provides compensation for the lags in the gun control system. Harold S. Black had solved the problem for electronic signals, as had Harold L. Hazen for low power electromechanical systems, but for gun control the problem was totally different in scale. K.A. Hayes explained after the war that designers had to achieve 'an accuracy of a few minutes of arc' for systems with loads of 'hundreds of tons-ft^2 inertia, with velocities of the order of 30°/sec and accelerations of the order of 10°/sec^2'.[3]

The perceived need for power control led to urgent programmes of development in both the USA and the UK in the late 1930s. For example, in the UK in 1938, adaptation of the ARL oil-hydraulic system (see Chapter 1) for use with 40 mm Bofors AA guns was begun. The equipment went into service late in 1939.[4] In the USA in the late 1930s the Sperry Gyroscope Company began developing hydraulic remote power controls for the US Army,[5] and the US Navy placed contracts with the General Electric Company for power controls based on the Amplidyne.[6]

In 1938, the British Thomson Houston Company (BTH) approached the Admiralty with proposals for a system based on the use of the Amplidyne produced by the General Electric Company. The scheme was turned down because the Admiralty had already placed contracts with the Metropolitan-Vickers Company (Metro-Vick), although this was not disclosed at the time.[7] The original contract with Metro-Vick, placed in 1937, was for the control of an 8-barrel Pom-Pom gun using the system developed by J.M. Pestarini for the Italian Navy. In this system, a single Metadyne supplied constant current to the armature of motors used to operate several guns, each motor being individually controlled by manual adjustment of the field current. The design work on the system was undertaken by Arnold Tustin under the direction of C.D. Dannatt.[8] Tustin found the large inductance of the field windings gave a large time constant and he redesigned the system so that the individual motors were supplied with a constant field current and partially compensated Metadynes were used to control the armature current of each motor.

Work on this system led Tustin to compare the characteristics of the Ward-Leonard motor-generator system, the Metadyne, and the Amplidyne. Although he concluded that all three systems had merit and the choice should depend on the application,[9] he favoured the Metadyne and he argued that the 100% compensation of the Amplidyne, although giving great amplification, introduced a large lag. He accepted that the Amplidyne did have advantages if 'a modified feedback response locus [was] desired' for 'then it may in some cases be more convenient to employ a fully compensated machine, and to apply the feed-back uniformly through suitable networks external to the machine'.[10] Tustin, of course, was very familiar with the Metadyne. The Metro-Vick Company held the British Patent rights and had used the system for electric traction applications.

On the other hand, A.L. Whiteley of the BTH Company favoured the Amplidyne (for which BTH held the British rights). He argued that the Amplidyne required much less power in the control field, that time lags were never a problem, and that the partially compensated (or uncompensated) Metadyne had an awkward internal coupling between the control field and the output circuit of the armature

132 *Development of design techniques for servomechanisms*

which could set up an insufficiently damped oscillatory mode. Controversy over the relative merits of the two systems existed between Whiteley and Tustin for many years.[11]

The Metro-Vick Company designed a gun control system that could operate directly from the predictor or by hand control. Figure 6.1 shows the general arrangement. A coarse-fine magslip system with automatic changeover is used for measuring position and the smoothed error signal is fed to a thermionic amplifier supplying the control winding of the Metadyne. The designers were well

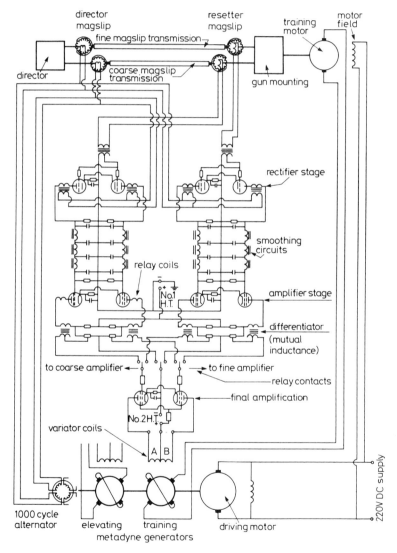

Figure 6.1 *Metro-Vick gun control system*
Adapted from Metropolitan Vickers, C.S. Memo No. 110, Oct. 1940

aware of the problems of time lag. In a report on the system produced in October 1940 they wrote, 'in order to prevent hunting in a closed-cycle control system of this type, time lags of all kinds should be kept to a minimum', continuing 'to keep the time lag in the smoothing circuits to a low value...1000 cycles/sec. AC excitation has been used for the magslips, the time lag of an efficient smoothing circuit being a function of frequency and being shorter the higher the frequency'. The time lags were further reduced by using a 'velocity effect' which was achieved by differentiating the error signal using mutual inductance coupling to the final stage of the amplifier. The report explains that

> The 'velocity' term at all times produces a torque tending to oppose relative motion of director and mounting, the torque being proportional to 'rate of change of displacement' and therefore exactly equivalent to a torque leading the displacement torque by 90 degrees. The resultant torque being the combination of the 'displacement' and 'velocity' terms, therefore leads the misalignment by some phase angle less than 90 degrees, and provided that phase advance is sufficient relative to the lags in the system hunting cannot occur.[12]

There is no reference in the Metro-Vick report to accuracy of following or to the possibility of velocity lag — the emphasis is on attaining stability. However, in an Admiralty Research Laboratory (ARL) report (December 1940),[13] it is claimed that the system provides 'smooth operation, substantially free from hunt, and virtually eliminates lag'. The report criticised the system on the grounds of bulk and because of the need for a special 1000 cycles/sec supply, but still recommended adoption for the 4" Mk. XIX twin mountings.

The Metro-Vick Company was also involved in developing, in cooperation with the Navy, an oil-hydraulic gun control system. The groups involved in the electrical and in the hydraulic system were kept separate, partly on the grounds of secrecy, and partly so that independent approaches to the problems were developed.[14] The company also lobbied hard to get a contract for the Army's 4.5" High Angle (HA) anti-aircraft gun and in September 1941 the Army placed a preliminary contract with Metro-Vick for a pilot study. The Army proposed eliminating the manual tracking operation between the predictor and the power controls by obtaining the input signal for the power controls directly from magslips on the predictor. The scheme which was developed, outlined in a report dated 25th February 1942, is essentially similar to the scheme produced for the Navy described above. The project ran into serious difficulties owing to problems with smoothing and non-linearities.

6.2 British approach to design

Arnold Tustin in a lecture given at a Servo-Panel meeting held on 21st August 1942 outlined the design methods used by engineers at Metro-Vick during the early years of the war. He said that when they started work on the gun control system in 1938 few of them had any experience of considering the dynamics of a system at the design stage: they were used to designing on the basis of steady-state characteristics and they worked largely from design charts. Their first

134 *Development of design techniques for servomechanisms*

approach to the problem of considering the dynamics was the obvious one: they modelled the system using differential equations. They obtained the system transient response either by simulation using the differential analyser (presumably the machine at Manchester University) or they found the roots of the characteristic equation. To reduce the labour involved in this latter method they put the equations in non-dimensional form and plotted transient responses with the dimensionless quantities as parameters. Dissatisfaction with the latter method — it did not allow them to consider elements for which they only had experimental data in the form of the harmonic response — led them to develop an alternative method based on the 'vector respresentation of performance of a cyclic control system'. Tustin explained the method as follows:

> Suppose that in Figure 1 [Figure 6.2a] the mechanical connection between the motor and the pointer M is broken at XX' and the pointer M is given a sinusoidal oscillation of unit amplitude, the unit being supposed small. The sequence of control actions caused by this misalignment results in the motor and the shaft X' oscillating sinusoidally, with an amplitude and phase depending on the frequency. In Figure 4 [Figure 6.2b] the sinusoidal oscillation imposed on pointer M is represented by the vector $\theta = 1$, and the response of the motor in angular movement is represented in magnitude and phase by the vector R. If the frequency of disturbance is successively given values from zero to infinity the end of the vector R follows a curve, referred to as the response locus for the system. Figure 7 [Figure 6.2c] shows the kind of shape that this response locus has for a typical follow-up system using torque control stabilised by phase advance. The response locus for a cyclic system with the scale of frequencies marked on it specifies the performance of the system within the range of linearity.[15] (The figure numbers refer to Tustin's diagrams which have been redrawn in Figure 6.2 and Figure 6.3).

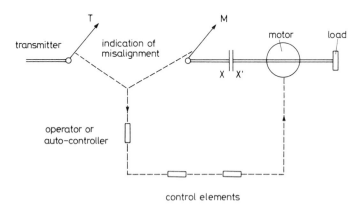

a

Figure 6.2a *Diagrammatic representation of a follow-up control system*
Adapted from A. Tustin Metropolitan Vickers, C.S. Memo No. 48

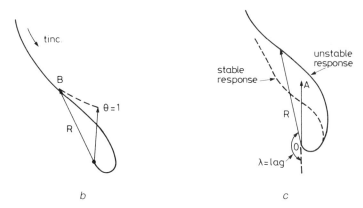

Figure 6.2b,c *Typical response locus of vector 'R' for complete cyclic system*

Adapted from A. Tustin Metropolitan Vickers, C.S. Memo No. 48

Later in the notes Tustin considers the relationship of the response locus to stability: he says.

> A condition $R = 1$ in which the response locus passes through the point $\theta = 1$ thus represents the limiting condition between a stable and an unstable system. In Figure 7 [Figure 6.2c] a system with the response R shown by the full line would be unstable because the locus passes beyond the point A. The curve represented by the dotted line, however,

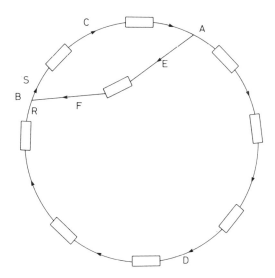

Figure 6.3 *Diagrammatic representation of a cyclical control system with feedback*

Adapted from A. Tustin Metropolitan Vickers, C.S. Memo No. 48

is stable, since the locus cuts the vector $\theta = 1$, and when the response coincides in phase with the disturbance, its magnitude is not sufficient to sustain or build up the motion without the addition of an extraneous driving force'.[16]

The plotting of the response locus is referred to in several earlier reports produced within Metro-Vick and in particular in a number produced during 1941 with the initials DGOM.[17] But in none of the earlier reports is there any general reference to stability. Although it seems hard to believe, the terminology of the reports of the group at Metro-Vick suggests that they were unaware of the developments which had taken place in the USA with respect to the use of frequency response techniques. The first reference to Nyquist in a Metro-Vick report occurs in one dated 12th April 1943 where the phrase 'vector response diagram (Niquist [sic] diagram)' occurs.[18] By this time there were over 20 published papers that referred to the Nyquist diagram but it seems that the engineers at Metro-Vick were unaware of the work until it was brought to their attention by Professor P.J. Daniell of the Department of Mathematics at the University of Sheffield.[19] Tustin claimed that his 'first hunch about what is now the Nyquist criterion was not from Nyquist, or from any advanced mathematics or from complex number analysis or anything like that, but did have a link with having read previously Van der Pol's paper on valve oscillators'.[20] On the basis of Van der Pol's work he realised that for stability, the amplitude ratio of a system had to be less than one when the phase lag was 180°.

The diagram shown in Figure 6.3b appears in the notes of Tustin's lecture to the servo-panel and in two further reports prepared for the servo-panel in 1943. This form of representation of a control system, with the blocks shrunk to dots, is now referred to as a signal flow graph, a term disliked by Tustin who claims 'they are not signals and they are not flows, they are cause and effect diagrams'.[21] (Signal flow graphs are associated with the name S.J. Mason who is credited with developing the techniques. He was apparently unaware of Tustin's work when he wrote his first paper on the subject (1953); however, in a second paper, published in 1956, he extended the method, using, as he acknowledged, a hint given by Tustin regarding the treatment of multiple forward paths.)[22]

Metro-Vick also had a contract to develop the aerial drive system for the GL Mk 3 radar system. This system was to be designed for both manual tracking and for fully automatic tracking. A schematic block diagram of the system is shown in Figure 6.4. The system is interesting in that it proposed using a combination of feedback and feedforward to reduce the velocity lag error seen by the predictor. In order to do this, the rate-of-change of the position error was used to operate a motor which turned a shaft to add or subtract (as appropriate) an angle to the position being transmitted to the predictor. The motor also turned the wiper on a potentiometer which supplied a signal to the input of the amplifier on the main drive, the effect of which was to provide an offset proportional to the required velocity.

6.3 Development of design methods in the USA

In the USA formal techniques for the design of servomechanisms were much

Development of design techniques for servomechanisms 137

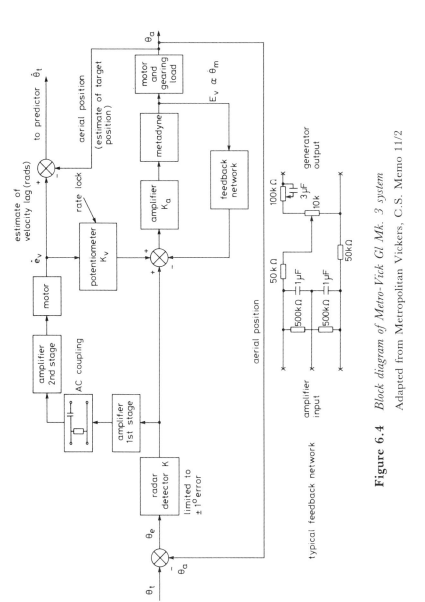

Figure 6.4 *Block diagram of Metro-Vick GI Mk. 3 system*
Adapted from Metropolitan Vickers, C.S. Memo 11/2

138 *Development of design techniques for servomechanisms*

further advanced than in Britain. H.K. Weiss extended Hazen's work in a paper published in 1939. In this paper he dealt with the commonly encountered case in which a torque is applied to a system with inertia and very low friction (typical of many indicating instruments, although Weiss was dealing with the control of pitch of a variable pitch aeroplane propeller). He assumed that the controller was second order, thus obtaining an overall system with a third order characteristic. He expressed the system equation in dimensionless form and provided charts which enabled the values of the real and complex roots and the frequency ratio (damped oscillation frequency/undamped oscillation frequency) to be found from the system parameters (damping coefficient and controller gain).[23] A similar dimensionless approach was used by C.S. Draper and his co-authors in papers published in 1939 and 1940. These papers, ostensibly concerned with instrument dynamics, provided a general summary of the developments in the transient response method of analysis which had been made at MIT since the publication of Hazen's papers in 1934.[24]

During the summer of 1940, G.S. Brown prepared a paper on the behaviour and design of servomechanisms for the sub-committee on machines of the Machine Shop Practice Division of ASME.[25] The paper drew heavily on the work of two students, Lt. Hooper and Lt. Ward, who had attended the Servomechanisms course given during 1939. For their project, under Brown's supervision, they investigated the operation of hydraulic servomechanisms for possible use in the automatic control of machine tools and for remote control of guns.[26]

As originally conceived, the paper covered only the transient analysis and not the frequency response analysis. Brown adopted the use of dimensionless parameters as devices by Draper.[27] For a system with an error equation

$$\epsilon(t) = \frac{1}{J_0 p^2 + f_0 p + K_1}(J_0 p + f_0)\omega_1(t) \tag{6.1}$$

new parameters ζ and ω_{n1} were defined as

$$\zeta = \frac{f_0}{2\sqrt{J_0 K_1}} \quad \text{and} \quad w_{n1} = \sqrt{\frac{K_1}{J_0}} \tag{6.2}$$

The transient response can then be plotted for various values of ζ.

In formulating the error equation as above Brown departed significantly from previous work. The basic idea appears in the thesis written by Hooper and Ward who used it to develop equations of an hydraulic servo-system. They introduced the idea of determining the minimum error of a system. For an arbitrary input signal they expressed the minimum error as a five term McLaurin series, but more importantly, examined the error generated in response to step and ramp inputs and hence derived coefficients that we now refer to as 'error coefficients'.

According to Brown the design of the *system* can begin only when each component part has been analysed such that its performance can be represented in quantitative terms in a suitably simple manner. The designer can then subdivide the problem into sections, representing each section by a block diagram that shows how they are connected.[28] A general servomechanism, Brown observed, can be represented as shown in Figure 6.5 and the error and output equations are

$$\epsilon(t) = \frac{1}{1 + C(p)H_o(p)}[\theta_i(t) - H_o(p)T_o(t)] \tag{6.3}$$

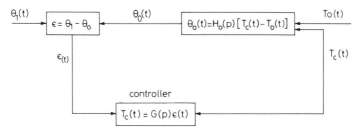

Figure 6.5 *Block diagram showing the elements of a servomechanism*
Redrawn from G.S. Brown, and A.L. Hall: 'Dynamic behaviour and design of servomechanisms', *Trans. Amer. Soc. Mech. Eng.*, 1946, **68**, pp. 503-524

$$\theta_o(t) = \frac{C(p)H_o(p)\theta_i(t) + H_o(p)T_o(t)}{1 + C(p)H_o(p)} \tag{6.4}$$

and the system behaviour is summarised by the *system operator* $[1 + C(p)H_o(p)]^{-1}$. Brown credits John Taplin with the first use of the term system operator (*characteristic polynomial* in modern terminology), in 1937 while a student at MIT. Brown also suggests that Taplin recognised the similarity between the closed cycle servomechanism problem and the amplifier feedback problem dealt with by Black and Nyquist.[29]

The design requirements, Brown noted, are to obtain a stable system with both transient and steady state operating errors within prescribed limits. He illustrated his ideas using a simple servomechanism with three types of controller: proportional, proportional-plus-derivative, and proportional-plus-integral. The choice of proportional-plus-integral gives rise to a third order system operator and hence presents a more difficult design problem. Brown suggests that the difficulties can be overcome by utilising the design charts prepared by Weiss. An alternative and more convenient form of chart for determining the roots of cubic equations was prepared at MIT in the following year.[30] The second part of the paper, contributed largely by Hall, was based on work which was published as a confidential document in 1943.

The NDRC became aware of Brown's draft paper during the autumn of 1940 when they began negotiations with him about carrying out a fundamental study of servomechanisms. Such a study had been proposed at the first meeting of Section D-2 of NDRC at which members recognised that many problems relating to servomechanisms would be common to a wide range of defence problems. They expected that it would take two years and require substantial funds. Nothing came of these initial discussions except that NDRC were made aware of Brown's paper.

As more pressing problems became apparent, the work requested of MIT became more specific and at a meeting on 25th October 1940, the NDRC considered a recommendation from Section D-2 for the award of a contract to MIT for work on 5 projects under the direction of Brown.[31] The contract was to run from 1st November 1940 until 1st September 1941. The projects included a study of relay mechanisms, hydraulic system for follow-up with electrical control, and an Amplidyne generator for the provision of 'automatic guiding of an ultra-short-wave target-locating apparatus to be provided'. For this latter project it was specified that 'the servomechanism shall be capable of keeping the locating apparatus pointed at the target within a precision of 1 mil when the target direction is constant or changing at any rate up to 20°/sec, provided however that the detector signal is not at any time less than 0.1 volt. The complete system shall be substantially aperiodic in operation.'

Two further projects, one concerned with mechanisms for determining first and second derivatives of error for electrical and mechanical quantities, and the other for a general investigation into the relative merits of hydraulic and electric servomotors were considered to be of a broader nature, exploratory in character, with the possibility that they would be expanded considerably should the results justify such an expansion. Contracts were subsequently placed by NDRC with MIT for this work.

Brown did not, however, simply wait around for NDRC to come up with the contracts. He continued working during 1940 on the position control problem using the 8 hp electro-hydraulic servomechanism which he had obtained from the Waterbury Tool Company in 1939. Also, he contacted the Ford Instrument Company and the Sperry Gyroscope Company, both of whom were interested in the use of hydraulic servomechanisms. From the latter he obtained a contract for some fundamental research on servomechanism design with particular attention to hydraulic systems. He began this work in September 1940, although a formal contract was not signed with the Division of Industrial Cooperation, MIT, until 1st December. Brown was granted formal release from NDRC to carry out the work on the Sperry contract.[32] In March 1941 further contracts were placed by the Sperry Company with the Servomechanisms Laboratory for work to be carried out under Brown and in close cooperation with Draper.[33] As a consequence of this second Sperry contract, the NDRC terminated its contract with Brown (for the study of hydraulic follow-up systems) on the grounds that the work was duplicated by that being done under the Sperry contract.[34]

Albert C. Hall, who was then working in the measurements laboratory with T.S. Gray, was assigned half-time and a new graduate student, Duncan P. Campbell, was appointed to work full-time on the Sperry contract. George C. Newton and J.O. Silvey were also in the team, and the Sperry Company seconded Ed Dawson. A little later in the year, Jay W. Forrester joined the group.

Out of this work came radically new designs for hydraulic components and a vast increase in power to weight ratio. 'The hydraulic pump-and-motor assembly was all contained in a seven-inch cubic aluminium box which was relatively light in weight. I remember well my surprise', wrote Brown, 'when I found that, operating at a thousand pounds per square inch, these devices could develop about five horsepower, all in that tiny space, whereas a year and a half earlier I had begun to put together a five-horsepower hydraulic drive system where the power assembly weighed several hundred pounds'.[35] The new form of hydraulic unit was to find its way into the power turrets on the B24 and B17 bombers.

6.4 Albert C. Hall and frequency response design methods

The Servomechanisms Laboratory group, which included Herbert Harris, and

J.A. Hrones, as well as the people mentioned above, quickly saw the difficulties of using the differential equation approach to analyse high order servomechanisms. W.P. Manger, writing in 1946, described the difficulties of the method as:

> The differential equation analysis carried out in detail yields a great deal of information about the system characteristics; the amount of work involved, however, is relatively large, even in the simple example[s]...In...complex case[s] charts displaying the effects of varying the numerous system parameters could not be prepared without an unreasonable amount of effort. If a detailed transient analysis of such a system is required, it is common practice to set up the problem on a differential analyser and determine a large number of solutions by varying the system constants one at a time or (hopefully!) in appropriate combinations. More often than not such a procedure leaves the designer with a large amount of data that are extremely difficult to interpret in terms of optimum performance from the system.[36]

The group began a search for a more powerful and less cumbersome method. They examined the frequency response methods being used by communications engineers; in a report made to NDRC in December 1941, Harris pointed out that improvements in design techniques might result from the adoption of the frequency response approach.[37]

Hall further developed the frequency response technique. Assisted by George C. Newton, he worked with the Sperry Gyroscope Company on the design of a control system for an auto-track radar. The term auto-track, or automatic following was applied to systems in which the radar aerial was made to track the target automatically. They did not imply that the whole system — radar, predictor, gun — were all connected, for even with auto-track radar human operators could be, and were, used to input data to the predictor and even to relay predictor output to the gun input.

Towards the end of 1941, they designed a control system for an experimental auto-track radar. According to Hall, the system 'jittered', because they 'had missed completely the importance of noise', and that 'in attempting to find an answer to the problem we were led to make use of frequency-response techniques. Within three months we had a modified control system that was stable, had satisfactory transient response, and an order of magnitude less jitter. For me this experience was responsible for establishing a high level of confidence in frequency response techniques.'[38]

Hall continued exploring the use of the frequency response approach and described the techniques which he developed in his doctoral thesis presented in 1943.[39] This work was limited to systems with unity negative feedback. He followed Brown in dividing the system into blocks but then, drawing on the recently published work of M.A. Gardner and J.L. Barnes (1942), he used the Laplace transform technique to set up the input-output relationships in terms of 'transfer functions'. He defined a transfer function as the relation between the servo output signal θ_o and the error signal ϵ. In the frequency domain he defined this as being $\theta_o(j\omega)/\epsilon(j\omega) = KG(j\omega)$ where K is invariant with frequency. The frequency response of a servo (with unity feedback) can thus be expressed as

$$\frac{\theta_o}{\theta_i}(j\omega) = \frac{KG(j\omega)}{1 + KG(j\omega)} \tag{6.5}$$

Hall discussed the difficulty of working directly with the frequency response $\theta_o/\theta_i(j\omega)$ because of the complex relationship between it and the system parameters. He suggested using the transfer function $KG(j\omega)$ because of its simpler relationship with the system parameters. He proposed a design method based on making a parametric polar plot of the transfer function — he called the plot the transfer function locus or transfer locus (these terms are now rarely used and the plots are referred to simply as Nyquist diagrams). Using this method, the effect of changing the parameters of the transfer function or of adding compensation components can easily be seen. He also showed how, by means of a simple graphical construction, the frequency response $\theta_o/\theta_i(j\omega)$ could be obtained from the transfer locus plot.

Hall was trying to design servomechanisms which were stable, had a high natural frequency, and high damping. The basic performance criterion he used was that the overshoot ratio — the peak magnitude of the amplitude response $|\theta_o/\theta_i(j\omega)|$ — should be in the range 1.0 to 1.3. Therefore, he needed a method of determining, from the transfer locus, the value of K that would give the desired amplitude ratio. As an aid to finding the value of K he superimposed on the polar plot curves of constant magnitude of the amplitude ratio. These curves turned out to be circles, with the centre at $x = -M^2/(M^2 - 1)$, $y = 0$ with radius $r = |M/(M^2 - 1)|$ and he named them M circles. He also made the suggestion that by plotting $G(j\omega)$ rather than $KG(j\omega)$, the permitted 'system sensitivity' could be determined by changing the scale of the real axis. By plotting the response locus on transparent paper, or by using an overlay of M-circles plotted on transparent paper, the need to draw M-circles was obviated (eventually printed sheets of polar graphs paper with M-circles became available).

The practical requirement behind the development of the design methods was the need to obtain a servomechanism with zero velocity error, and Hall devoted a long section of his thesis to possible methods of achieving this aim. He noted that the use of two motors, each of which introduced an integration term in the loop, would give zero velocity error but that the system would be inherently unstable. The addition of a phase lead network could give conditional stability which would be dependent on the gain but he recommended using one motor and a device that has infinite gain at low frequency with its gain reducing as $\omega \to \infty$. This led him into an extensive coverage of methods of designing compensating filters.

Hall's work exemplifies the change in attitude which was taking place: this was described by Ivan Getting writing after the end of the war:

> The practice before the war in the design of servos was to employ a mechanism *adequate* for the problem. The difficult problems encountered in the war, particularly in the field of fire control, emphasize the necessity of designing the *best possible* servo system consistent with a given kind of mechanisms... If it is desired to design a 'best possible servo', it is necessary to define a criterion of goodness.[40]

The definition of goodness proposed by Hall was the minimisation of the root-mean-square (rms) of error of following and this definition was also adopted, independently, by R.S. Phillips.[41] Clint Lawry of the General Electric Company gave the practical reasons for carefully considering the performance of a system: he explained that a follow-up system must operate smoothly as well as having an error less than a given value, for 'it was of little value... to keep gun sights exactly level in spite of a ship's roll, but to have the gun sights vibrate in a blur because of roughness in the follow-up system. This quality of *smoothness* involves two quantities: first, damping of the system, which must be adequate — not necessarily critical, but with damping ratios of 0.6 or better and second, the filter property of the system, i.e., the follow-up system must not amplify or even pass roughness frequencies even though introduced by the director.'[42]

6.5 The Radiation Laboratory and automatic tracking

Another group, also based at MIT, were working on the design of an auto-track radar system, this was the group in the so-called Radiation Laboratory. The Radiation Laboratory was under the control of the NDRC and its priority task was the design of an auto-track radar system for anti-aircraft gun control. The Laboratory designed the radar system that went into service as the SCR-584. The control aspects of the system are described in great detail in the book *Theory of Servomechanisms* edited by Hubert M. James, Nathaniel B. Nichols and Ralph S. Phillips.[43] In particular the SCR-584 system is used as an example to illustrate the use of the new 'decibel-phase-angle diagrams' — now known as Nichols charts. However, as the authors acknowledge, the system was actually designed using the harmonic locus (Nyquist diagram) and differential equation methods.

The block diagram of the position control system used in the SCR-584 radar (for one axis since similar systems were used for azimuth and elevation) is shown in Figure 6.6 and it is redrawn in a more modern notation in Figure 6.7. The

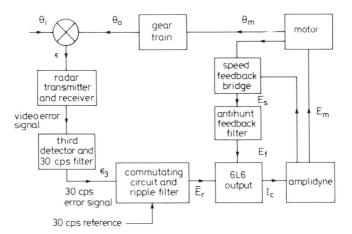

Figure 6.6   *Schematic of SCR 584 radar system*

Reproduced (with partial redrawing) by permission of H.J. James, N.B. Nichols and R.S. Phillips: 'Theory of servomechanisms; Radiation Laboratory Series Vol. 25' (McGraw-Hill, 1947), p. 213

144 Development of design techniques for servomechanisms

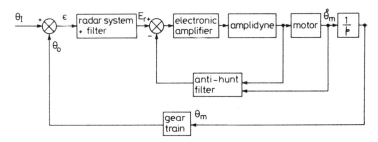

Figure 6.7 *Block diagram of SCR 584 radar system*

radar transmitter and receiver send a modulated video signal to a diode detector. The amplitude of the video modulation envelope is proportional to the angle between the parabolic reflector axis and the target — the angular tracking error. The modulation frequency is 30 Hz and the induction motor which drives the conical scanner also drives a reference generator which provides a 30 Hz signal to the commutating circuit used to separate the elevation and azimuth error signals.[44] The box labelled 6L6 is an electronic amplifier.

A crucial part of the system was the Amplidyne amplifier. The General Electric Company had been using the Amplidyne for several years for both military and civil purposes and had evolved several anti-hunt circuits.[45] Sidney Godet claimed, in a report written in 1941, that the simplest way to stabilise an 'Amplidyne follow-up system' was to feedback to the control amplifier, through an 'anti-hunt' network, a voltage proportional to the Amplidyne output voltage. He proposed the circuit, shown in Figure 6.8, as the 'anti-hunt' network which he claimed provided an improvement on a simple low pass filter, and he gave the following rules for choosing the component values:

(i) R_2 should be made as large as is consistent with other circuit requirements.
(ii) L may be chosen at any convenient value.
(iii) f_b should be determined either by calculation ($f_b = 1/2\pi\sqrt{T_a T_m}$) or experimentally by measuring the natural frequency of the system with no anti-hunt feedback. [T_a = effective time constant of the Amplidyne; T_m = effective time constant of the driving motor].
(iv) Calculate C_1 by $C_1 = 350000/Lf_b^2$.
(v)   Calculate $C_2 = (2000 \sqrt{LC_1})/R_2$.
(vi) Error signal gain should be set so that desired stability is obtained when Amplidyne output is fed back directly to amplifier input.
(vii) With anti-hunt network in place, R_1 should be adjusted until desired stability is again obtained.
(viii) If system is too stable with no external resistance added to choke L, C_1 should be reduced and C_2 recalculated accordingly.

Godet used the harmonic locus (Nyquist) approach for determining system stability and the characteristic of the anti-hunt circuit: he also described how the locus could be obtained by experimental methods.[46]

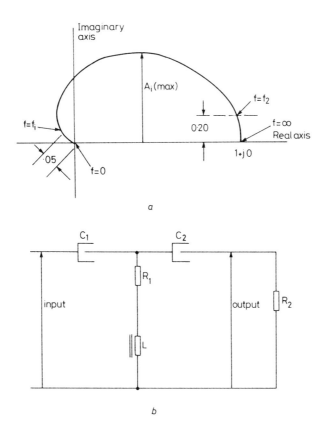

Figure 6.8 *Anti-hunt filter*
Adapted from S. Godet, General Electric Co., Schenectady, circa 1941

In a subsequent report on the Amplidyne system, produced for the Radiation Laboratory, Clint Lawry provided a detailed model of the system using operator notation and referred to use of the differential analyser for obtaining information on the behaviour of the system: he used the Routh technique for analysing the stability of the system.[47] He noted that it was well known that the Amplidyne could not be stabilised by tachometer feedback from the motor alone, and he proposed a combination of tachometer feedback and Amplidyne output voltage feedback, with the combined signal being passed through a suitable anti-hunt network (this was the method adopted for the SCR-584 system).[48] He also argued that if suitable circuits could be found it would be preferable to use a controller in the forward loop that provided derivatives of the error signal, rather than the anti-hunt circuit in the feedback loop.[49] He claimed that in some commercial applications of the Amplidyne, lead circuits were incorporated as part of the electronic amplifier and in these systems no anti-hunt voltage was necessary.[50]

F.C. Williams, who was a strong advocate of the use of velocity feedback, recalled seeing a prototype of the SCR-584 system when he visited MIT in 1942.

'What fascinated me', he recalled, 'was that they made a presentation of it in which they explained how they had exactly calculated using Bode, Nyquist etc., how much phase advance — they were still using phase advance which made me think that they were silly — they wanted to get it stable. I thought . . . very mathematical . . . but I wonder if it works. When we were finally shown the equipment there, right in the middle of the console, was a large knob which said 'anti-hunt' control. So having worked it all out you then went back to the previous method which was to turn a knob until it was working best'.[51] The explanation given in *Theory of Servomechanisms* for retaining the adjustable control was that the Amplidyne and motor time constants varied widely because the allowable manufacturing tolerance on the resistance of the windings was 25 to 42.5 ohms and hence the need to be able to adjust the anti-hunt filter.[52]

The Americans, in return, were critical of the state of knowledge of British workers. Samuel H. Caldwell who with Thornton Fry had attended a meeting of the Servo-Panel sometime in 1942, wrote to Hazen in November 1943 to the effect that, judging by the minutes of a meeting of the Servo-Panel of 23rd July 1943 which he had received, things had not improved, and suggesting that 'someone like Gordon Brown or Al Hall going over there [England] and giving about a week of lectures with everything put on the table and no holds barred' would be a good idea. In March of 1944 Brown did go to England and attended a meeting of the Servo-Panel as well as making a number of other visits.[53]

6.6 Frederick C. Williams

The development of airborne radar sets — Airborne Interception (AI) — had been given high priority even before the war started.[54] For although the Ground Controlled Interception system enabled the fighters to be directed to the vicinity of enemy aircraft, interception was dependent on visual contact which was difficult to attain at night or in dense cloud. A further difficulty was that at night rear gunners in the bombers, having had several hours to accustom themselves to the darkness, were frequently able to spot the fighter before the fighter pilot could spot the bomber.

The development work on the radar sets was carried out by the Telecommunications Research Establishment (TRE) and sets were available late in 1939.[55] These first sets required an operator, mainly in order to determine range, and hence could only be fitted to two-seater fighters, typically Blenheims. The fighters could now locate the bombers but the Blenheim was frequently too slow to catch them. Early in 1940, TRE was asked to investigate making the sets automatic so that they could be fitted to single-seat fighters.

Frederick C. Williams, who was working on the design of the so-called Identification Friend or Foe (IFF) system for installation in aircraft to enable the crew to determine whether the 'blip' on the radar screen was a return from friendly aircraft or an enemy aircraft, was asked to add automatic range finding to the system.[56]

The early radar sets in use operated on 1.5 metres and the signals seen by the operator were an echo pulse (if there was one), a transmitter pulse, and in Williams' words, 'a thundering great ground return'. The automatic system had to ignore the transmitter pulse, search down the trace to find an echo and if one was found, lock onto it. The echo then had to be tested to see if it was the ground return or a friendly aircraft. The ground return was always a wide pulse, so if a wide pulse was detected, the scanner had to return to the beginning of the trace and repeat the scan. The IFF systems fitted to friendly aircraft detected the radar pulses and responded by retransmitting pulses with a definite pattern of wide and narrow. The AI radar scanner had to detect this pattern, release the echo, and then continue to search along the trace until either the echo from an enemy aircraft — permanently narrow — or the ground return was found. Once the automatic scanning system locked onto a pulse the range was displayed to the pilot.

A schematic of the automatic scanning system is shown in Figure 6.9. The strobe forming unit develops a pulse with a trial value of t_2. The discriminator compares this value with the value of t_1 of the echo pulse and the difference, in the form of a current i is passed to the shaping unit which develops a voltage E such that the strobe unit changes t_2 until ultimately $t_2 = t_1$. In the first versions of the equipment the shaping units were constructed from passive networks but later versions incorporated active elements based on the virtual earth amplifier devised by Williams.[57] (It was in connection with this work that Williams met and worked with A.D. Blumlein and E.C.L. White.)[58]

Williams then turned his attention to another problem, that of steering the combined transmitter-receiver aerials of centrimetric radar sets. (The sets had to be made steerable because the centrimetric radar has a much narrower beam width and if it is not steered it cannot lock onto and follow a target. AI radars were different from navigation and bombing airborne radar systems, which had only to be stabilised against the movement of the aircraft.)[59] Steering the aerial was different from the problem of locking onto an electrical signal for, Williams recounted, 'one really had to think about it as a control system because there were mechanical elements involved'. TRE had adopted a split-field, separately-excited, dc motor as the basic power unit for positioning aerials. They used the field control of the motor with the field supplied from a high gain amplifier and the armature winding supplied from a constant current source. After some difficulties, including attempts 'to phase advance what in the first place was a very ragged signal', Williams decided to add an integral tachogenerator to the motor. 'Then one realised that if you just fed the velocity back, the mirror [the radar transmitter-receiver dish] would always be pointing some way behind the aeroplane on account of its velocity so, simple-mindedly, one said...let's put a condenser in the way and call it ac velocity feedback and in the fullness of time, providing the aeroplane is going at a uniform rate, it will latch on to it.'[60]

A schematic of the system is shown in Figure 6.10. Williams (and others) found that the gain of the amplifier could be made very high and yet the system remained

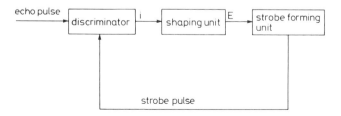

Figure 6.9 *Range-finding strobe circuit from F.C. Williams*

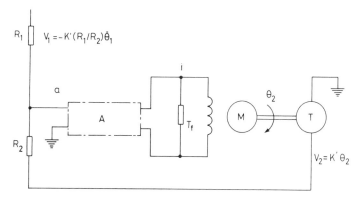

Figure 6.10 *Schematic of the velodyne*
Reproduced (with partial redrawing) by permission of Williams and Uttley, from *J. IEE*, 1946, **93**, (IIIA), p. 1257

stable. The combined motor-tachogenerator was named the Velodyne. It provided an excellent integrator and became widely used within TRE for the development of simulators and computing mechanisms.[61]

The experimental version of auto-tracking radar developed by Williams was demonstrated to service personnel and to staff from other research establishments in April 1942. The work had been carried out without full official sanction (TRE did not seek official permission to work on auto-tracking until March 1942) and demonstrates that even in the crucial stages of the war research workers did have some freedom to develop ideas. The TRE auto-tracking radar was never fully developed although it was installed in a few aircraft and in a motor gunboat.[62] The ASE did begin work early in 1943 on a self-contained, automatic (in the sense that the range, bearing and elevation data from the radar set were to be fed directly to the predictor and the radar was to follow the target) fire-control system for a twin Bofors gun mounting. This system used the automatic ranging circuit devised by Williams and the Velodyne system.[63]

6.7 A.L. Whiteley, the inverse Nyquist technique and 'Standard Forms'

The performance of the TRE system stimulated the interest of other research workers in the application of auto-tracking to AA gun control. The problems of AA control were greater: the airborne radar sets were small and light, requiring drive motors with a power of the order of 100 watts, but the radar aerials and associated equipment for AA control with their much greater range required drives of several horsepower. Arthur Porter, who had joined ADRDE in March 1942 as head of the fire control section, began work on an analysis of automatic following systems for gun-laying.[64] The designs produced at ADRDE were based on the use of the Metadyne and a report on the system (referred to as AF-1) was made in May 1942.[65] This report was brought to the attention of the Scientific

Advisory Council of the Ministry of Supply who, in July 1942 recommended that attention should be given to the development of automatic-tracking radar systems for gun-laying and noted that a lot of work on such systems had been done in the USA.[66] In September 1942 it was suggested that ADRDE should consider carefully the principles of the auto-follow system developed at the TRE.[67] Late in 1942 a contract was placed with Cossor for the development of AF-1 for use in conjunction with a new electric predictor being developed by them and ARL.[68] In July 1943 approval was given for the manufacture of sets for field trials — by this time the design had changed, with the Metadynes being replaced by Amplidynes. BTH, who had cooperated in the design, was asked to manufacture the sets. It initially refused because of the small numbers involved, but eventually agreed when promised that the manufacture of further sets would be accorded high priority.[69] Few sets were made for the delays in ordering equipment and changes in design requirements had already delayed the development to such an extent that Cherwell had urged the use of the new American SCR-584 equipment.[70] This radar set was undoubtedly a major achievement of the Radiation Laboratory at MIT. It provided, together with the M9 director developed at the Bell Laboratories and the proximity fuse, an effective defence against the V1 and V2 rockets.[71] A trial set arrived in England on 7th October 1943 and on 12th November, Cockcroft reported favourably on its performance and recommended its adoption.[72]

Some interesting and valuable control ideas emerged from the design work on the AF-1. BTH had considerable experience in designing electrical position control systems and had two experienced designers in A.L. Whiteley and L.C. Ludbrook. Ludbrook had designed auto-follow for the searchlight radar systems in 1940–41. The aerials of the SLC Mk. 6 radar were mounted on the searchlight and as the radar operators steered them to track the target they also moved the searchlight. In 1941 he produced an experimental system which by-passed the operator so that the radar aerials and searchlight tracked the target automatically. The work provided the basis for the automatic follow-up systems for the AA No. 3 Mk. 4 and Mk. 7 gun laying radars begun in 1942.

Whiteley had worked with ASE on searchlight control systems using magnetic amplifiers and Ward-Leonard drives, and on the contract awarded to BTH in 1942 for the development of an electric remote control aircraft gun turret. It was specified that this system should use the Amplidyne.[73] It was work on these systems that led Whiteley into studying methods of design of servo-systems which incorporated networks in the feedback path. He viewed the use of networks in the feedback path as the most effective and best way of stabilising a servo-system. He stated his views in 1945:

> The aim in Hall's approach to stabilisation is however appreciably different, a local 'regenerative' amplifier being devised, whereas the aim in the writer's approach is a 'degenerative' feedback system, extending (generally) over the whole servo sequence of control media.[74]

In advocating the use of multiple, local feedback, as opposed to networks in the forward path, Whiteley was in agreement with the GE engineers. He may well have been strongly influenced by them, as it is known that he visited them in

150 *Development of design techniques for servomechanisms*

August 1942 in connection with the use of the Amplidyne for the aircraft turret system.[75]

Both methods of design available to Whiteley present difficulties when the system has networks in the feedback loop or has multiple feedback loops. He explained that the

> feedback-back loop can be designed by the use of operational methods, though perhaps with much labour if say the complete servo system yields 6th order differential equations, or more. The Nyquist diagram does not lead to a ready solution because if feedback is used, the feedback operator has a multiplicative relationship with the operator relating to the main servo-system control sequence.[76]

Whiteley therefore began looking for a way of designing servomechanisms incorporating feedback networks, and late in 1942, the idea of using the inverse of the Nyquist diagram occurred to him. He developed the method during 1943[77] and it was formally written up as a BTH research report issued in October 1944.[78]

In this report, Whiteley summarised the advantages of the method:

(1) Stabilising ('anti-hunt') feedback vectors are *additive* to those of the main servo sequence; hence feedback networks can be readily synthesised.
(2) The damping factor $Q = |\theta_o/\theta_i|$ can be immediately read from the diagram.
(3) If the plot relates to an unstable system, the number of unstable modes can be readily reduced.
(4) The method permits a theoretical (and therefore presumably linear) anti-hunt network to be 'married' to a non-linear servo sequence whose frequency characteristics can be determined by test.
(5) The low frequency end of the locus (usually the important part) is at the origin end, and therefore accessible for examination [item (5) is added in Whiteley's hand—undated].

The reciprocal of the Nyquist diagram—the so-called Inverse Nyquist method—was also developed independently by H.T. Marcy and by H. Harris, Jr.[79]

During 1944 Whiteley gave a series of lectures to BTH engineers on the design of servomechanisms and these lectures formed the basis of his doctoral dissertation presented in 1945.[80] And in 1946 he presented an extensive paper on the theory of servomechanisms in which he outlined his design techniques.[81] In this paper he clearly outlined the reasons for classifying servos according the type of error—displacement, zero-displacement, zero velocity and so on. He also introduced the idea of 'Standard Forms' to reduce the labour in choosing parameters for a given type of servo.[82]

In choosing the parameters for his Standard Forms, Whiteley adopted as a criterion an overshoot of about 10%. He was strongly critical of Brown's and Hall's recommendations for damping in servomechanisms (they both favoured underdamping), and he argued that 'a practical form of zero-velocity-error servomechanism even though critically damped, can have an overshoot of more than

50% to a position impulse. The ultimate user is not likely to consider this as excessive damping.' In the discussion of his paper he received strong support from L. Jofeh of the Cossor company who expressed concern about the proposals in some of the contemporary literature.[83]

The standard forms for 3 types of system as given by Whiteley (1946) are shown in Figure 6.11 and Table 6.1. For type A systems (i.e. displacement and zero displacement error systems) the standard forms are based on the criterion of obtaining the maximum range over which the frequency response of the system will be flat. For type B systems (i.e. zero-velocity error) the coefficients are chosen on the basis of obtaining a maximum overshoot of 10% while for zero-acceleration error systems (type C) the coefficients are chosen such that the poles are located on the real axis and spaced in geometric progression. Whiteley notes that in general the coefficients tend to give a well damped response, but he suggests that it is preferable to err on the side of safety at the design stage.

To illustrate the use of the standard forms let us consider a system with an open loop transfer operator $K/[p(1 + p/8)(1 + p/12)]$ for which we wish to choose a value for K and to improve the performance by adding velocity feedback with a gain factor of K_f. The characteristic equation of the combined system is

$$p^3 + 20p^2 + 96(1 + k_f)p + 96K = 0 \qquad (6.6)$$

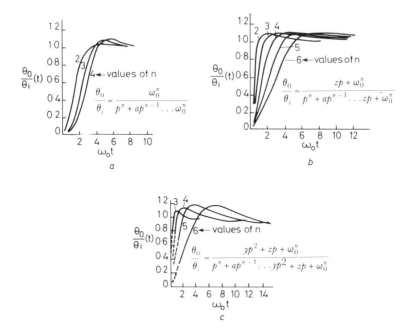

Figure 6.11 Whiteley's standard forms

    a Recovery characteristics, eqns. 54, 55 and 56 in Table 6.1
    b Recovery characteristics, eqns. 57, 58, 59, 60 and 61 in Table 6.1
    c Recovery characteristics, eqns. 62, 63, 64 and 65 in Table 6.1

152 *Development of design techniques for servomechanisms*

STANDARD FORMS

Class	Type and general form of $\dfrac{\theta_o}{\theta_i}$	Basis of characteristic	Maximum overshoot	Standard form; coefficients of p	
A	Displacement and zero-displacement systems: $$\dfrac{\omega_o^n}{p^n + ap^{n-1} \ldots + \omega_o^n}$$	$Q = 1$ for disturbing frequencies up to $f = \dfrac{0.8\omega_o}{2\pi}$ approximately	% 5 8 10	$p^2 + 1.4\omega_o + \omega_o^2$ $p^3 + 2\omega_o + 2\omega_o^2 + \omega_o^3$ $p^4 + 2.6\omega_o + 3.4\omega_o^2 + 2.6\omega_o^3 + \omega_o^4$ Higher orders not checked; use of binomial coefficients suggested.	(54) (55) (56)
B	Zero-velocity-error systems: $$\dfrac{zp + \omega_o^n}{p^n + \ldots + zp + \omega_o^n}$$	Based on maximum overshoot of 10%; no subsequent undershoot	10 10 10 10 10	$p^2 + 2.5\omega_o + \omega_o^2$ $p^3 + 5.1\omega_o + 6.3\omega_o^2 + \omega_o^3$ $p^4 + 7.2\omega_o + 16\omega_o^2 + 12\omega_o^3 + \omega_o^4$ $p^5 + 9\omega_o + 29\omega_o^2 + 38\omega_o^3 + 18\omega_o^4 + \omega_o^5$ $p^6 + 11\omega_o + 43\omega_o^2 + 83\omega_o^3 + 73\omega_o^4 + 25\omega_o^5 + \omega_o^6$	(57) (58) (59) (60) (61)
C	Zero-acceleration-error systems: $$\dfrac{yp^2 + zp + \omega_o^n}{p^n + \ldots yp^2 + zp + \omega_o^n}$$	Based on maximum overshoot; one small subsequent undershoot exists	10 15 20 20	$p^3 + 6.7\omega_o + 6.7\omega_o^2 + \omega_o^3$ $p^4 + 7.9\omega_o + 15\omega_o^2 + 7.9\omega_o^3 + \omega_o^4$ $p^5 + 18\omega_o + 69\omega_o^2 + 69\omega_o^3 + 18\omega_o^4 + \omega_o^5$ $p^6 + 36\omega_o + 251\omega_o^2 + 485\omega_o^3 + 251\omega_o^4 + 36\omega_o^5 + \omega_o^6*$	(62) (63) (64) (65)

*Given as a guide; no actual check available; shows excessive damping.

Table 6.1 *Whiteley's standard forms*

Equating coefficients between this equation and the standard form

$$p^3 + \omega_1 x_o p^2 + \mu_2 \omega_o^2 p + \omega_o^3 \quad (6.7)$$

where μ_1 and μ_2 both equal to 2, we get

$$96K = \omega_o^3$$
$$96(1 + K_f) = 2\omega_o^2 \quad (6.8)$$
$$20 = 2\omega_o$$

And hence we find $\omega_o = 10$, $K = 10$ and $K_f = 1$ (approximately).[84]

For higher orders of type A, which he had not checked, Whiteley suggested the use of the binomial coefficients which leads to a system that has a response composed of equal, critically damped modes. Such a response had been suggested as optimum by Imlay in 1940 in connection with the lateral stability of an aircraft under the control of an automatic pilot and by Oldenburg and Sartorius in connection with regulators and servos.[85]

In 1930, Butterworth suggested standard forms such that the poles of the characteristic function of the system are evenly distributed on the unit circle, in connection with filter design for interstage coupling in pulse amplifiers.[86] The idea of standard forms can thus be extended to cover any type of system and criterion of performance. The crucial issue is deciding what is the appropriate choice for the performance index.

The criterion of performance used in the development of standard forms by Whiteley and Butterworth and in the design charts of Draper *et al.* were based on the use of some parameter of the transient or frequency response of the system, but as we have seen in order to find an analytic method of design we really need some index which will give a measure of overall performance. I shall return to developments in this area in the next chapter.

6.8 The practical problems arising from non-linearities

It became very clear early in the war that production of accurate servomechanisms for power operation required more than the design of a cascade or feedback filter based on linear analysis. In their book *Theory of Servomechanisms*, James, Nichols and Phillips listed the practical problems as '(1) backlash, (2) static friction, and (3) locking mechanisms',[87] and to these they added mechanical resonance and problems of amplifier overload caused by parasitic signals — picked up or generated by non-linearities in the system. Tustin expressed the problems in much the same way in 1944:

> In practice, however, the problem of servo design is much more complicated than that of arranging to produce a desired overall transfer function, and several aspects of design in particular raise acute problems of smoothing and filtering, namely:
> (a) The non-linear and 'saturable' nature of many circuit elements, which results in any supplementary high-frequency signal applied

to those elements causing them to be thereby saturated for more or less of the total time and correspondingly desensitised for response to 'legitimate' signals.

(b) The effect of backlash in gearing, that makes the load mass under some conditions ineffective as part of the smoothing, so that 'jitter' results in hammering of gears, and possible mechanical failure.

(c) The use of A.C. 'carrier' for the signals which take the form of modulation of the carrier, with subsequent rectification, which results in the association with the 'legitimate' D.C. signals of unwanted A.C. of twice the carrier frequent and above.

(d) The frequent presence of parasitic modulation of A.C. signals, due for example to voltage modulation at the source (alternator rotational frequency, throw in bearings and the like).[88]

Other non-linearities were introduced through the use of relay controllers and eventually pulsed (definite correction) controllers. The development of auto-tracking radar systems added other problems relating to noise and signal fading which are considered in the following chapter.

The problems caused by backlash, stiction and locking mechanisms were reduced by improvements to gear production techniques resulting in much higher quality gears, careful design to reduce stiction and by the avoidance of worm gears and other forms of irreversible drive mechanisms. In addition studies of the effects of such non-linearities on closed loop control systems were undertaken. Arnold Tustin, with the assistance of P.J. Daniell, investigated backlash in position control systems.[89] Daniell produced an internal report for Metro-Vick in 1945 outlining an analytical approach for dealing with backlash in gearing.[90] He proposed taking into account only the first harmonic generated from an assumed sinusoidal input to the backlash and he proposed using a Nyquist plot for finding possible limit cycles. Daniell's ideas, which form the basis of the describing function approach to handling non-linearities, were taken up by Tustin and extended further in papers published in 1947.[91] L. Jofeh of the Cossor Company also investigated behaviour of non-linear control systems examining systems involving stiction, saturation and discontinuous control action.[92] Derek Atherton suggests that the describing function approach was used in several countries before the publication of papers on the subject at the end of the 1940s.[93]

Nicholas Minorsky briefly mentioned non-linear problems at the end of a long paper on using operator methods to solve control problems published in 1941. The significance of the paper is that he makes reference to the work of Liapunov on the stability of control systems and seeks to show how his methods can be used to determine the stability of a simple non-linear system.[94]

Obtaining rapid response from the servomechanism meant that the 'stiffness' of the control had to be increased and this led to problems of mechanical resonance. Coales and Whiteley have both commented on the difficulties involved in trying to ensure that the resonant frequencies of all mechanical elements were outside the bandwidth of the system — this task was not made easier by the need to construct systems from existing components. Tustin analysed the auto-follow aerial system for the GL III radar system, considering the effects of resilience in shafting and gearing between motor and load, resilience of the gear case, resilience of the base or mounting and the effects of lateral freedom of movement of the load on its centre of rotation (caused by 'play' in ball races). The approach he used was to model the resilience and to examine the effect it had on the transfer locus of the control system.[95] The problem of dealing with the effects of mechanical resonance and backlash were reduced by adopting 'divided reset' whereby part of the feedback was taken from the motor and part from the load. Divided reset was devised by F.H. Belsey.[96]

6.9 Relay and pulsed servomechanisms

A large number of control systems in use and designed during the war used some form of relay operated control. Relay based controllers had been in use for many years but the only widely available English language theoretical treatment of such systems was Hazen's published in 1934. There were, as Tsypkin and Fuller have pointed out, several German and French papers on the subject; for example, early attempts to deal with relay systems were made by Leaute, Proell, and Houkowsky.[97]

A major study of relay based control systems was carried out by MacColl. The use of relays deliberately introduces non-linearity into the control system but, MacColl argued, if the non-linearity thus introduced was a parasitic, that is a small, incidental defect in a linear system, then in the first instance it can be ignored and the system treated as linear. Recourse to *ad hoc* methods for dealing with the non-linearity can be made as necessary. However, for 'essential', deliberately introduced non-linearities, attempts at analysis have to be made and the only available technique is to form the fundamental differential equations and solve them by whatever means are available. MacColl suggested two approaches: firstly a piecewise solution, that is forming the linear equations describing the motion for the different relay positions and solving these equations; and secondly, the use of phase plane analysis. H.K. Weiss also studied techniques for analysing relay servomechanisms and made use of phase-plane techniques.

The phase plane method arises from the work of Poincaré, who showed that the solution to two first order nonlinear differential equations

$$dx/dt = P(x, y)$$
and
$$dy/dt = Q(x, y)$$
(6.9)

could be sketched as a plot of y against x since the slope of the solution curve (trajectory) at a point in the phase plane is

$$dy/dx = Q(x, y)/P(x, y) \qquad (6.10)$$

Poincaré in his original work described features of the topology of the phase plane such singular points and limit cycles.

Phase plane techniques were extended by Bendixson (1901), Dulac (1923), Lienard (1928), Andronov (1929, 1930), Khaikin (1930), Volterra (1931), Le Corbeiller (1931), Papaleksi (1947) and Nemitzky (1949) and more recently by Minorsky (1962), Blaquiere (1966), Andronov (1966) and Tsypkin (1974). The non-linear functions considered by the mathematicians were usually simple polynomials whereas in control systems the non-linearities typically encountered —

relays, with or without dead zone and hysteresis; dead zone in valves; saturation in amplifiers; backlash in gears; and non-linear friction — are best represented by linear segments.

Closely related to relay control systems are those systems that receive data at specified time intervals. Hazen referred to this type of control as 'definite correction' controllers and they were commonly found in process control applications during the 1930s. There are two types: one in which the amplitude of the signal remains constant and the time for which it is present is varied, and the other in which a pulse of varying amplitude is applied to the control system at specified intervals of time. Two systems in which the latter pertained were of crucial importance during the war.

All radar systems in use, other than the Home Chain system, used pulse techniques such that at fixed intervals of time a pulse was transmitted and at some time interval later an echo from the object detected was received. Thus in the auto-follow radar the input to the control system was actually a pulse. However, the pulse repetition frequency was such that the input could be smoothed and the system designed on the basis of continuous theory.

One of the problems that the NDRC investigated during the war was predictor design. It soon became apparent that the Services had no means of dynamically testing predictors. Contracts were placed for the design and development of dynamic testers, that is devices that could supply as an input to the predictor data representing a course of an aircraft and which could also record the difference between the predictor output and the aircraft course. The basis of the mechanism (see Figure 6.12) was that the data representing the aircraft course was coded

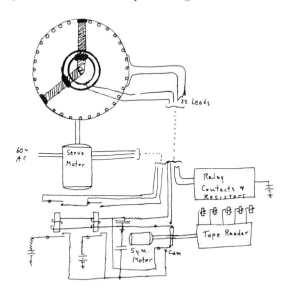

Figure 6.12 *Sketch by George Stibitz for his tape controlled servo*

Reproduced from NARS, Record Group 227, Records of OSRD, Records of Division 7, General Project Files, Project No. 60, Contract No. OEMsr-904 Bell Telephone Lab Misc. Memos, Rpts., Drawings, Photos etc. 1/30/43

as holes in paper tape. A relay system was used to read the tape at fixed intervals (0.1 seconds) and compare the code with the code on a disc on the input drive shaft to the predictor.[98] The error was applied as an electrical voltage to the control circuit for the Thyratron drive for the dc motor as shown in Figure 6.13. George R. Stibitz, who was the NDRC Technical Aide responsible for the project, wrote a whole series of reports over the period 1942 to 1944 on techniques for analysing and stabilising servomechanisms of this type.[99] At some time during the war, possibly late 1942, Witold Hurewicz was asked to undertake a detailed study of the problem.[100] The outcome was Chapter 5, 'Filters and Servo Systems with Pulsed Data' of the book *Theory of Servomechanisms*, by James, Nichols and Phillips. Hurewicz showed how the Nyquist stability criteria could be extended to handle the sampled-data system and developed an approach that was eventually to lead to the z-transform methods for analysing sampled data systems.[101]

6.10 Notes and references

1 Douch, E.J.H., 1947, p. 177
2 Green, C.M., Thomson, H.C., Roots, P.C.: *The Ordnance Department: planning munitions for war, US Army in World War II* (GPO, Washington, 1955), p. 417. This was the opinion of U.S. Army officers following trials in 1940 of an hydraulic system of gun control designed by the Sperry Company; no doubt many British Officers would have agreed. At a later period of the war when civilian technologists became concerned about the ability of the military to maintain in the field and to utilise to the full extent the equipment being demanded, the military response was that the apparatus was not too complicated and that there was a need to improve on the precision of the instrumentation being supplied. NARS, Office papers of H.L. Hazen, SHC to HLH 14th Dec. 1943, HLH to KTC 16th March 1944, KLW 21st April 1944
3 K.A. Hayes' discussion of paper by A.L. Whiteley, 22nd March 1946, *J. IEE*, p. 368
4 PRO WO 185/92
5 Wildes, K.L.: extract from a history of the Electrical Engineering Department at M.I.T., communicated to the author privately.
6 See Chapter 7
7 A.L. Whiteley interview with author 21st and 22nd June 1976
8 The control of the work was in the hands of the research department at Trafford Park, Manchester and C. Dannatt was on loan to them from Birmingham University.
9 CS Memo 20 1941; CS Memo 54A 1943. Metro-Vick held the British rights to Pestarini's patents for the Metadyne and Tustin had worked on the designs for the application to electric traction see Fletcher, G.H., Tustin, A. (1938); Dummelow (1949); the system was also described by Butler, O.I., (1938a, 1938b, 1940). In none of these papers is there any attempt at an analysis of the transient behaviour.
10 CS Memo 48 1942, p. 9
11 Letter from A.L. Whiteley to the author, 14th October 1983
12 CS Memo 108, 8the April 1940 and CS Memo No. 110 14 10th October 1940 the quotations are from the latter.
13 PRO WO/185/92
14 Tustin: the hydraulic unit used was the Vickers VSG unit and a report was produced in August 1943 Metro-Vick, Research Department Report C.405 'The harmonic response characteristics of V.S.G. hydraulic engines'.
15 CS Memo 48 1943, p. 2
16 CS Memo 48 1943, p. 3
17 CS Memo 3 undated and anonymous; CS Memo 17 22nd August 1941; CS Memo 18 23rd August 1941; CS Memo 22 4th September 1941 all with initials DGOM: DGOM would appear to be D.G.O. Morris.
18 CS Memo 80 1943
19 Daniell prepared a report on the interpretation of Nyquist diagrams for the Servo-Panel, Report S2 1945. Percy John Daniell (1889–1946) was Professor of Mathematics,

158 *Development of design techniques for servomechanisms*

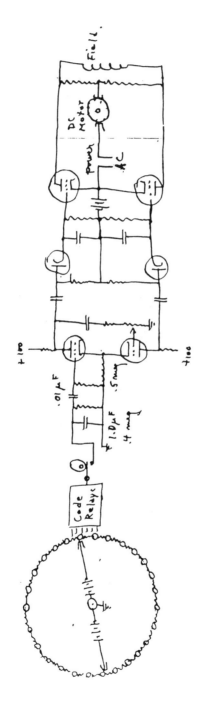

Figure 6.13 *Sketch by George Stibitz for the thyratron control circuit for the tape controlled servo*

Reproduced from NARS, Record Group 227, Records of OSRD, Records of Division 7, General Project Files, Project No. 60, Contract No. OEMsr-904 Bell Gelephone Lab Misc. Memos, Rpts., Drawings, Photos etc. 1/30/43

University of Sheffield from 1923 to 1946, see Stewart, C.A.: 'P.J. Daniell', *J. London Mathematical Society,* **22** (1947), pp. 75-80 for biographical information.
20 Tustin in interview with author 2nd July 1976. Van der Pol wrote several papers in the late 1920s and early 1930s, the paper to which Tustin is probably referring is Van der Pol, B.: 'Nonlinear theory of electric oscillations', *Proc. IRE,* **22** (1934), pp. 151-86
21 The diagrams appeared in CS Memo 80 1943, and CS Memo 95 1943. Tustin's comment was made to the author in an interview with Tustin. In a paper published in 1950 (Tustin, A.: 'Problems to be solved in the development of control systems', *Trans. Society of Instrument Technology,* **2** (1950), pp. 19-27) he discusses the importance of energy concepts and storage of energy in control systems and it is possible that he had in mind the idea that the diagram should be a more general representation of the relationship between components, perhaps something akin to the Bond Graph introduced by Paytner.
22 Mason, S.J., 1953, 1956; and Percival, W.S., 1953, 1955 developed the formal rules for manipulating signal flow graphs.
23 Weiss, H.K., 1939
24 Draper, C.S., Schliestett, G.V., 1939; Draper, C.S., Bentley, G.P., 1940
25 The paper was brought to the attention of the newly formed NDRC who requested that it be withdrawn from open publication because of its relevance to gun control problems and it was issued on restricted circulation by the NDRC in November 1940. The paper was published after the war (with added material) as Brown, G.S., Hall, A.L.: 'Dynamic behaviour and design of servomechanisms', *Trans. American Society of Mechanical Engineers,* **68** (1946), pp. 53-524
26 For an account of MIT's involvement in automatic control of machine tools see Noble, D.: *Forces of Production: A Social History of Industrial Automation*, (Alfred A. Knopf, New York, 1984) and also Reintjes, J.F.: *Numerical Control: Making a New Technology*, (Oxford University Press, New York, 1991). Brown also seems to have drawn heavily on the project report of E.B. Hooper and Ward, 'Control of an Electro-hydraulic Servo-unit' submitted 16th May 1940; see Naval Operational Archives, NRS 627. Hooper and Ward, at the request of Brown re-worded their acknowledgement to him so that he could use some of the results: this re-wording was important in view of possible patents, Hooper believes that one feature of the system which they developed appeared in the controls for the 40 mm mount developed by Sperry with the assistance of Brown. Two other naval Lts Rivero and Mustin were also at MIT as students—they had been enrolled on the Servomechanisms course in September 1939—and they worked with Draper on the design of a lead computing sight. Reminiscences of Vice Admiral Edwin B. Hooper USN (Retired) USN Operational Archives No. 1, pp. 45-7
27 Two other naval Lts Rivero and Mustin were also at MIT as students—they had been enrolled on the Servomechanisms course in September 1939—and they worked with Draper on the design of a lead computing sight. Reminiscences of Vice Admiral Edwin B. Hooper USN (Retired) USN Operational Archives Nos. 1-45 to Nos. 1-47
28 This seems to be the first use of 'block diagrams' as a formal method of representing a control system. Formal rules for the manipulation of block diagrams were not fully described until the 1950s, see Gosling, W.: *Design of Engineering Systems,* 1962, London, Heywood, p. 28. The description and elaboration of the methods were given by Graybeal, T.D., 1951; by Stout, T.M., 1952; Stout, T.M., 1953
29 James, H.J., Nichols, N.B., Phillips, R.S.: *Theory of Servomechanisms,* Radiation Laboratory Series, **25** (McGraw-Hill, New York,1947), p. 16
30 These were produced by Y.J. Liu in 1941, updated by L.W. Evans in 1943 and were privately printed by the Department of Electrical Engineering at MIT
31 NARS Minutes of NDRC, Meeting of 25th October 1940
32 This would appear to have been from 1st December 1940 until the end of March 1941, see MIT Office of President 1930-59 AC4 Box 3
33 Wildes, K., Lindrau, N.: 'The history of electrical engineering and computer science at MIT 1882-1982' (MIT Press, Cambridge, MA, 1984) gives the details of many of the contracts.
34 Letter from Harold Hazen to Karl T. Compton, 19th March 1941 quoted in Wildes, in this Hazen was accepting that Brown was more interested in working with Sperry than under an NDRC contract. The relationship between Brown and the Sperry Company was the cause of considerable friction and ill-feeling with staff working for the NDRC. The complaints were that Brown would not keep NDRC informed of the work done for Sperry and that he also sought contracts from the military directly and

not through NDRC. In view of the complex patent agreements and understandings which MIT had with the Sperry Company (MIT was carrying out other work under contract for Sperry Gyroscope Company—for a radar drive and for lead gunsights—for the radar drive) MIT reserved the right to be the sole judge as to whether Sperry Gyroscope Company should have any equity in any invention; and in the other case Sperry Gyroscope Company were to be granted a non-exclusive licence but on the understanding that no other licences would be granted for three years. MIT Archives Office of the President ACS Box 2/35 letter K.T. Compton to V. Bush 9th January 1940 and box 2/53 memo (undated) N.McL Sage to K.T. Compton.

Considering Brown's status as a consultant to the company, his reluctance to release information is understandable, but so is the stance taken by NDRC staff who had a responsibility to ensure that information which could be of value in the war effort was disseminated to the services. The view of NDRC was expressed by Warren Weaver in a letter to Karl Compton (3rd March 1941): NARS OSRD, Div. 7 Office files of Warren Weaver, 'Compton, Karl T.' Weaver seems to have been particularly irritated by Brown and at a meeting held on 20th August 1942 lost his temper completely with him. The meeting was concerned with the relationship between section D-2 of the NDRC and the Army Ordnance and the outcome was that the Army Ordnance got what they wanted, complete control over the servo-design and the freedom to deal directly with whosoever they wanted. NARS General Project Files, Project No. 35 OEMsr-522 MIT Corres. 1941–45, Diary of Chairman [Warren Weaver, Thursday, 20th August 1942]. They wished the Sperry Company to authorise Brown to discuss matters freely with NDRC and to permit NDRC to release information to the Servies as they felt appropriate. Weaver gave as an example of the importance of NDRC knowing what work was being carried out the information that the Ford Instrument Company already had in operation a hydraulic follow-up system which incorporated a scheme for obtaining a zero steady state error which was better than that which had recently been designed by Brown for the Sperry Gyroscope Company. There was a further difficulty: the Ford Instrument Company was a Navy contractor whereas the Sperry Gyroscope Company was an Army contractor and they had always been instructed that information was not to be shared between the two companies. NDRC Div. 2 were well aware of (and possibly favourably inclined towards?) the Ford Instrument Company as Edward J. Poitras who had worked for the company was attached to section D-2. Brown's views were made plain in a letter to Sage in which he indicated that he considered that theoretical matters—academic work— should be freely reported to NDRC and considered as being financed by the NDRC contract but that the reduction to engineering practice should not be so reported and should be treated as the property of MIT (or the Sperry Gyroscope Company).

35 Brown, G.S., quoted from Wildes, *op. cit.*, (n.33) pp. 15–16; a note by Weaver suggests that although Sperry was satisfied with Brown's designs, for internal reasons it did not put them into production—see NARS General Project Files Project No. 35 OEMsr-522 MIT Corres. 1941–45 Diary of Chairman, Thursday 30th October 1941
36 James *et al.*, *op. cit.*, (n.29), p. 158
37 Harris, H.: 'The analysis and design of servomechanisms', *OSRD Report No. 454*, 1941; the report was restricted. A shortened version of it was published in 1946: (Harris, H. 1946)
38 Hall, A.C.: 'Early history of the frequency-response field', *Trans. Amer. Soc. Mech. Eng.*, **76**, (8) (1954), pp. 1153–4
39 Hall, 1943 it was originally a restricted publication issued by NDRC which was derestricted in 1946 and extracts appeared in Hall 1946 and Brown and Hall 1946
40 Getting, I.A., in James, *op. cit.*, (n.29), p. 21
41 Phillips 1943
42 Clint Lawry, 'An analysis of an amplidyne servomechanism,' *Group B Radiation Laboratory*, 1942, pp. 5–7
43 James, *op. cit.*, (n.29). Staff were retained by the OSRD for a period of six months after the war to write up reports of the wartime work. The work of the Radiation Laboratory was described in a 28-volume series published by McGraw-Hill
44 The radar used the conical scan technique invented by Alfred Loomis
45 Several papers by Alexanderson and others appeared in the GE Review and elsewhere, see Alexanderson *et al.* 1940; Fisher, 1940; Shoults *et al.* 1940
46 The report by Godet is undated but the copy which I have seen has on it a note 'about

41 R.W. Mayer 6/9/81' — Mayer was a five year co-operative student during the war period. Godet circuit is referred to by Lawry in his report of February 1942 so there seems little doubt that Godet's report was written sometime during 1941; see Godet 1941
47 Lawry, 1942. Clint Lawry was a GE employee who worked closely with Sid Godet. Godet, in his earlier paper on the Amplidyne, had used the Nyquist approach for determining stability and for design of the anti-hunt circuit. GE built a differential analyser based on the Moore School of Engineering model but they replaced the mechanical torque amplifiers with a polarised light follow-up system. See Varney, R.N.: 'An all-electric integrator for solving differential equations', *Rev. Scientific Instruments*, **13** (1942), p. 10
48 James, *op. cit.*, (n.29), p. 213
49 Lawry, *op. cit.*, (n.42), p. 14.
50 *Ibid.*, p. 16. He states that 'pure derivatives of error would be extremely helpful in producing the *optimum* system. The error filter has the purpose of producing the proper function of *error* and *derivatives* of error to operate the optimum system.' He refers to a paper by Coates, G.T.: 'Differentiating Networks and Lead Networks', MIT Publication 6.80-2A
51 Interview with Williams, 1975
52 James, *op. cit.*, (n.29), p. 224
53 NARS, Office Files of Harold L. Hazen, Box 58, Dr. Gordon S. Brown. Memo from SHC to HLH dated 29th Nov., 1943. Brown arrived in London on 7th March 1944
54 PRO PREM3/22/5
55 For a brief account of the development of airborne radar, see Swords, S.S.: Technical history of the beginnings of RADAR (Peter Peregrinus, Stevenage, 1986). pp. 243-253
56 As mentioned in Chapter 4, Williams had some experience of designing control systems prior to the war when with Douglas Hartree and P.M.S. Blackett he designed a curve follower for the Differential Analyser.
57 Details of the work and the techniques for circuit analysis were published in Williams (1946a, 1946b). For a fuller account of the work of Williams, see Bennett, S.: 'F.S. Williams, his contribution to the development of automatic control', *Electronics and Power*, Nov.-Dec., **25**, (11) (1979), pp. 800-803
58 A.D. Blumlein was a gifted electronic circuited designer who worked for EMI. He was killed when an aircraft carrying an experimental H_2S radar set crashed on 7th June, 1942.
59 For a detailed description of the problems of stabilising airborne and ship radars see Cady, W.M., Karelitz, M.B., Turner, L.A., (eds), *Radar Scanners and Radomes*, Radiation Laboratory Series, **26** (McGraw-Hill, New York, 1948)
60 Interview with Williams.
61 Williams did not have a major involvement with the development of simulators or computers at TRE, although he did contribute to a number of papers (Williams 1946c, 1947, 1951). He did, however, make a major contribution to the development of the digital computer. In connection with his work on computers Williams made one further excursion into automatic control when with J.C. West he designed a speed control system for a drum store; he recalls that the major difficulty was to persuade the Ferranti engineers that what was needed was not '...the lightest possible structure on the lightest possible bearing...but...the heaviest possible structure on the very best bearings', interview with Williams, 1975
62 Williams thought that he was too early and that the service personnel had not yet fully realised both the need for and the potential of auto-tracking radar systems.
63 Pout, H.W.: 'Precision ranging systems for close-range weapons', Radiolocation Convention 3rd April, 1946, *J. IEE*, **93**, Pt. IIIA(1) (1947), pp. 380-394
64 AVIA 22/1383; see also Porter 1942
65 AVIA 22/1383
66 AVIA 22/155 Meeting 30th July, 1942
67 AVIA 22/1383 21st Stepember, 1942
68 AVIA 22/864
69 AVIA 22/1383
70 AVIA 22/865
71 In August 1944 General Sir Frederick A. Pyle wrote to Vannevar Bush 'I am writing to thank you for the great assistance you have given us in meeting the flying bomb. The equipment you have sent us is absolutely first class and every day we are getting

better results with it. Our percentage of "kills" is not high enough, but the curve is going up at a nice pace, already we are far away ahead of the fighters...The SCR 584 is a wonderful job. We are deploying the SCR 584 with the BTL10 [M9] predictor this predictor is also an outstanding job', Quoted in letter Vannevar Bush sent to Harold Hazen, 31st August 1944, NARS Office Files of H.L. Hazen. Establishing the actual performance of the system is not easy although it seems to be agreed that it was effective against the V1 and V2 rockets. This is to be expected as the rocket flight path was much smoother, and hence easy to predict, than that of a piloted aircraft.

72 AVIA 22/1388
73 Postan, M.M., Hay, D., Scott, J.D.: *Design and development of weapon studies in government and industrial organisation* (HMSO, London, 1964), p. 117. In the USA the General Electric Company also developed an electrical gun turret, see Thompson, J.D.: 'Electric Gun Turrets for Aircraft', *AIEE*, **63**, (1944), pp. 799–802, who stated that the turret typically weighed 700–1000 lb, had a wind loading of 300–400 ft lb of torque and had to have a slewing speed of 10 rpm. During the 1930s there was among military personnel a growing realisation that power control of both searchlights and guns for anti-aircraft defence was necessary. Also, in air-to-air gunnery, with closing speeds approaching 600 mph, it was realised that line-of-sight firing was no longer possible and that gunners would not have sufficient strength to manipulate gun turrets at the necessary speeds. Thus there was a need both for predictor sights and power operated turrets. In both the UK and the USA work began in the late 1930s on various forms of automatic gunsights resulting in the production of gyroscopically based sights which were fitted to aircraft from about 1942 onwards and were used for light anti-aircraft guns, particularly on board ships. There was little official support in the UK prior to the war for remote power control of turrets. Some development of power operated turrets was carried out by private companies; for example, both the Boulton-Paul and Frazer-Nash companies developed turrets based on French designs. The Air Ministry made plans at the beginning of the war for the development of power turrets but these plans were halted in 1940 when all forward development was frozen. It was not until early 1942, when the BTH company approached the Air Ministry with a proposal for an electric turret based on the use of the Amplidyne, that official support for the development of power operated gun turrets was given. Work on stabilised sights, including gyroscopically stabilised sights continued throughout the war, see Baxter, J.P. III, 1947, *Scientists against Time*, (Washington: G.P.O.), 1947 for American work. Postan, M.M., Hay, D., Scott, J.D.: *Design and Development of Weapons; Studies in Government and Industrial Organisation*, (HMSO, London; 1964), pp. 117–120 gives a brief account of the British work. In 1937 the British Navy was carrying out trials on various forms of stabilised gunsights, PRO ADM 212/174 'Priority of researches at ARL 1923–37'
74 Whiteley, A.L.: Theory of servo systems with particular reference to stabilisation, D.Sc. dissertation, University of Leeds, 1945, p. 2
75 Postan, *op. cit.*, (n.73), p. 117
76 Whiteley, A.L.: 'Servo systems—new form of phase-amplitude diagram', BTH Research Laboratory Report 1288-5, 9th October 1944, p. 2
77 Private communication from A.L. Whiteley.
78 Whiteley, *op. cit.*, (n.76) 1944
79 Marcy 1946, Harris 1946
80 Whiteley, *op. cit.*, (n.74) 1945
81 Whiteley, A.L.: 'Theory of servo systems with particular reference to stabilization', *JIEE*, **93**II, (1946), pp. 353–67
82 The idea of standard forms is similar to an idea put forward by Clint Lawry in February, 1942. He suggested as a possible way of improving system performance the 'extension of the preselected root method of stabilising which would indicate what roots to choose for the characteristic differential equation. If the optimum characteristic differential equation was known, then all the anti-hunt quantities could be readily adjusted and the system would be operating at its best.' Clint Lawry, *An Analysis of an Amplidyne Servomechanism*, Group B Radiaton Laboratory, 10th February, 1942
83 Whiteley, *op. cit.*, (n.74) 1945, p. 7. L. Jofeh, discussion of Whiteley 1946, p. 370. Jofeh suggested that the standard forms might not be suitable for systems in which noise was present but that they would provide a starting point for the design.

84 This example is taken from Macmillan, R.H.: *An Introduction to the Theory of Control in Mechanical Engineering*, (Cambridge University Press, Cambridge, 1955), pp. 98–100
85 Imlay 1940, Oldenburg and Sartorius 1944
86 Butterworth 1930
87 *op. cit.*, (n.29), pp. 19–20
88 CS Memo 185, 26th October, 1944
89 Tustin, A.: 'The effect of backlash in positional control systems with phase advance', *Metro-Vick CS Memo 168*, 22nd June, 1944. Tustin, A.: 'The effects of mechanical elasticity in a servo system, especially when resetting from the load', *Metro-Vick CS Memo 193*, 19th December 1944
90 Daniell, P.J.: 'Backlash in reset mechanisms', *Metro-Vick CS Memo 199*, 16th March, 1945
91 Tustin, A.: 'The effects of backlash and speed-dependent friction on the stability of closed-cycle control systems', *J. Institution of Electrical Engineers,* **94,** Pt. IIA (1947), pp. 143–151; Tustin, A.: 'A method of analysing the effects of certain kinds of non-linearity in closed-cycle control systems', *J. Institution of Electrical Engineers,* **94,** Pt. IIA (1947), pp. 152–160
92 Jofeh, L.: 'The effect of stiction on cyclic control systems', *A.C. Cossor Report MR110,* 11th October, 1943; Jofeh, L.: 'Study of the effects of saturation on the behaviour of a type of servomechanism', *A.C. Cossor Report MR165*, April 1945. I have not been able to obtain a copy of these reports.
93 Atherton, D.P.: 'Non-Linear systems — Historical Development', in *Systems and control encyclopedia*, Singh, M.G. (Ed.) (Pergamon Press, Oxford, 1987). He suggests the techniques were developed by L.C. Goldfarb in Russia, W. Oppelt in Germany, J. Dutilh in France and R.J. Kochenburger in the USA. The underlying theory of the technique, but not its application to control systems, was developed through studies in non-linear mechanics, in particular by N. Krylov and N. Bogoliubov (1943).
94 Minorsky, N.: 'Control Problems', *J. Franklin Institute,* **232** (November and December 1941), pp. 451–487, 519–551. In this paper Minorsky also investigated methods for determining the stability of systems with 'retarded control' that is systems with pure time delay in the operation of the control action.
95 Tustin, A.: 'The effects of mechanical elasticity in a servo system, especially when resetting from the load', *Metro-Vick C.S. Memo 1945*, 20th February 1945
96 See F.H. Belsey, contribution to the discussion of the paper by Whiteley (1946), *op. cit.*, (n.81),p. 370. Belsey patented the technique, British Patents 610029 (April 1945) and 611046 (June 1945). Apparently the method was also described in *Metro-Vick Research Report C.565*, 1945 but I have been unable to locate a copy of this report.
97 H. Leaute published two papers, 'Memoire sur les oscillations a longues periodes dans les machines actionnées par des moteurs hydrauliques et sur moyens de prevenir ces oscillations', *J. de l'Ecole Politechnique,* **55** (1885), pp. 1–126; and 'Du mouvement trouble des moteurs consecutif à une perturbation brusques. Nouvelle methode graphique, pour l'étude complète de ce mouvement', *J. de l'Ecole Politechnique,* **61** (1891), pp. 1–33; Proell, R.R.: 'Uber den indirektwirkenenden Regulierapparat Patent Proell', *Zeitschrift des VDI,* **28** (1884), No. 24,25, pp. 457–460, 474–477; and Houkowsky, A.: 'Die Regulierung der Turbinen', *Zeitschrift des VDI,* **40,** No. 30,31, (1896), pp. 839–846, 871–877
98 Claude E. Shannon had worked on a similar tape input system for setting up the differential analyser for his Master's dissertation at MIT. He had been supervised by Bush, S.H. Caldwell and Hitchcok, see Goldstine, H.H. *The Computer from Pascal to von Neumann*, (Princeton, Princeton University Press) 1970, pp. 119–120. Stibitz worked with Shannon during the latter part of the war on the design of a relay operated (digital) differential analyser.
99 See NARS Project No. 60 OEMsr 904 Bell Laboratories, GPF
100 This could have been in response to Wiener sending copies of Hurewicz curriculum vitae to Weaver with the comment that 'Hurewicz, a very distinguished mathematician, who wishes to get into war work'. NARS Norbert Wiener to Warren Weaver, 19th July 1942, Project Report No. 6 Folder 2, GPF.
101 Hurewicz showed that for stability the roots of the characteristic equation had to lie within the unit circle on the z-plane. In a memorandum dated 9th November 1942, George Stibitz concluded that the absolute value of the roots of the difference equation representing a pulsed servomechanism must be less than 1. NARS 'Stability and Errors in Step-Wise Servos', Project No. 60, Contract No. OEMsr 904 Bell Laboratories, GPF

Chapter 7
Smoothing and prediction: 1939-1945

> The sporadic use of machinery in the seventeenth century was of greatest importance, because it supplied the great mathematicians of that time with a practical basis and stimulant to the creation of the science of mechanics.
>
> Karl Marx

> Science is not a matter of thought alone, but of thought continually carried into practice. That is why science cannot be studied separately from technique.
>
> J.D. Bernal

In 1975, Sir Robert Cockburn argued that the wartime achievements were not 'scientific breakthroughs...they were engineering developments dependent on many years of patient research...The contribution of the scientist was...in applied research, development and production and in management and operational analysis'.[1] For much of the work in the control field this is true: engineers quickly developed design techniques based on the fundamental work of Black, Nyquist, Bode and Hazen, they developed an extensive array of electronic circuits for processing signals, and they learned how to manufacture precision mechanical components in large quantities. Even for engineering and applied research these did not represent major breakthroughs. However, the wartime period did produce some changes of fundamental significance for control systems. The primary one was the recognition of the need for, and adoption of, a systems approach to problems. This was partly through a recognition of the underlying commonality between apparently diverse pieces of equipment, and partly through a recognition that in order to achieve an overall goal it was necessary to consider not just the performance of individual units of a piece of equipment designed to achieve the goal but also how the units interact with each other (including interaction with the human operator). What was recognised and understood was that all devices within the system were, whatever other function they had (power amplification, energy conversion), manipulators of signals, and that the overall behaviour of the system depended on both modifying the signal in an appropriate way and on distinguishing between wanted and unwanted signals (noise). Black, in his work on the negative feedback amplifier, clearly thought in terms of manipulating signals — these were signals that could be described in deterministic terms. Norbert Wiener's contribution was to show how we could handle signals that were non-deterministic and in particular how, for certain classes of system, we can determine by gathering statistical data the dynamics of the system so that we can design appropriate control devices. The particular activity that raised problems which stimulated much of the work described in this Chapter was manual tracking of moving objects.

7.1 Manual tracking

Throughout the war manual tracking (referred to as 'laying' when applied to the guns themselves) was of great importance. The tracking task took a variety of forms: the following of radar signals on a cathode ray tube, the aligning of pointers, or the direct tracking of a target (an aircraft, ship or tank) with an optical telescope; good eye and hand coordination was needed in all cases. The accuracy and effectiveness of gunfire was critically dependent on the accuracy and smoothness of the tracking and, as a consequence, considerable effort was devoted to research on tracking techniques.

The standard technique of 'direct' laying in which the gun, radar aerial, or other object directly follows the movement of the layer's hand wheel was satisfactory when the position of the target was changing slowly. However, operators using 'direct tracking' were unable to track accurately targets such as low flying aircraft which have high apparent rates of movement. Experiments were made using 'velocity' laying in which the operator controlled the velocity of the gun not its position. Velocity laying was not successful since operators were unable to maintain accurate position control and there was a tendency for the gun to oscillate about the target position.

A technique referred to as 'rate-aided laying' gave improved performance and Figure 7.1 shows the principles of the method. In turning the hand wheel H the operator changes both the position of the output and also, by altering the distance d, the *rate* at which the output R_p changes. Therefore, in rate-aided laying the operator adjusts the velocity of tracking and at the same time adds a transient correction to the position. In practice the operator learned to move the hand wheel by an amount proportional to the deviation between the target position and the follow-up mechanism; the rate-aiding device thus provided P + I action. Mechanical integrators were used in the early types of rate-aided laying apparatus, but gradually these were replaced by equivalent electrical devices. Figure 7.2 shows units proposed by Arthur Porter in 1943 as an electrical replacement for the mechanical aided laying system. In these units phase advance is provided by means of an RC network and feedback of the generator voltage. Arnold Tustin had previously suggested that a similar result could be achieved by using 'choked' feedback, that is by adding an inductance in the velocity feedback circuit (voltage obtained from the generator).[2]

The 'aided laying' system did just what its name implies — it aided the operator. The operator remained as part of the control loop (in some of the early fire control systems there were 14 operators) and there was considerable interest in expressing the operator's behaviour in some formal, and preferably quantitative, manner. Arnold Tustin carried out an extensive series of tests 'to investigate the nature

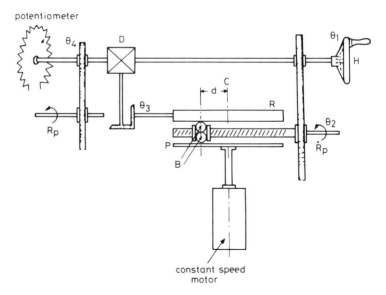

Figure 7.1 *Diagram of a mechanical rate-aided laying system*

of the layer's response' in a number of particular cases, and attempted to find laws of relationship of movement to error. In particular, it was hoped that this relationship might be found 'to be approximately linear, and so permit the well developed theory of 'linear servo-mechanisms' to be applied to manual control'.[3] He tested different operators using a range of tracking apparatus — displacement-speed, rate-aided laying with three different time constants and 'second differential' control (this latter included an acceleration term). The operators tracked a test signal composed of three superimposed sine waves generated by a cam mechanism. Both the test signal and the operators' movement were recorded. A harmonic analysis of both records revealed in the operators' response 'a marked predominance of the frequencies that occur in the target movement, and that all other harmonics

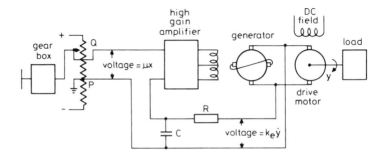

Figure 7.2 *Schematic diagrams of simple aided laying system*
Adapted from Porter ADRDE Minor Report No. 8, 1943 PRO, AVIA 26/1235, Crown Copyright, figures 16a, 16b and 17.

present in this range of frequencies are quite small'. This was a result which Tustin concluded provided 'strong evidence of the approximate linearity of the operator's response'.[4]

Tustin, not content to leave the analysis at this point, attempted to examine the difference between the linear part of the response, that is a signal constructed from the harmonic components in the operator's movements corresponding to the harmonic components in the target signal, and the total operator's movement. The difference, which he called the 'remnant', had its greatest amplitudes between the 18th and 30th harmonics. This result, he argued, was consistent with a linear system close to its stability limit and which was being subjected to small disturbances. These disturbances, he suggested, could arise from irregularities in target motion, variation in friction in laying mechanism, and arbitrary movements by the operator.

Following considerable detailed analysis, including a comparison of his results with those obtained by K.T.W. Craik of the Cambridge Physiological Laboratory, Tustin concluded that the human operator could be represented by a transfer operator of the form $(a + bp)e^{-t\omega'}$ where $t = 0.3$ seconds, this being the delay in response of an average person to a stimulus. He also concluded that aided laying with a time constant of 0.36 seconds gave the best performance — adding cautiously, 'under these conditions'. The problem continued to interest Tustin and he wrote two further reports on the subject as well as returning to a consideration of physiological control systems in his retirement.[5]

Other investigations were carried out on the human operator. J.A. Uttley tried to analyse the behaviour in terms of an intermittent servo. A group at the Royal Aircraft Establishment, according to E.B. Pearson, obtained results somewhat different to Tustin's, which suggested that 'the operator's output may depend on the control system he is called upon to manipulate'.[6]

In the USA there were similar investigations into manual tracking and the response of human operators. Following a visit to Fort Monroe in May 1941, during which he saw a demonstration of existing techniques, Harold Hazen, in a memo to Warren Weaver, explained the crucial part played by the human operator.[7] Men and women, acting as links in the transmission of data between these various stages, were, he reported, major contributors to the 'irregularity and inaccuracy of the final result'. Hazen became convinced that more must be learned about 'the dynamic characteristics of the human being as a servo and therefore his effect upon the dynamic performance of the entire fire control system'. While praising the quality of work currently being done on investigating the performance of human beings he thought that what was needed was not a comparison of the performance of different human operators, but an assessment of how well the dynamics of existing equipment was matched to the dynamics of the person operating it.

He argued for undertaking a fundamental investigation which would be of benefit in the design of future equipment. If it was found that the parameters of the performance of a human being cannot be established with any degree of certainty, 'a definite boundary on the performance of any given automatic control system in which a human link is used' would be set, 'beyond which it is futile to attempt to go.' On the other hand, if the human operator could be modelled with some reasonable degree of accuracy then 'it becomes possible to design an entire automatic control system involving human links rationally', for 'as we have found we must include all of the dynamic characteristics of component parts in the successful design of mechanical automatic control systems'.

Drawing on the techniques used for testing mechanical systems, Hazen laid out a plan for finding a model for the human operator. Such a scheme, he commented, implied that a human being was 'nothing more nor less than a robot' even when applying intelligence to take into account higher derivatives of the path being tracked. The added quality of the human operator, however, is that he 'is endowed with a memory that should permit him to extrapolate from the past history into the future, a feature that is not possessed by any of the simpler mechanisms we have at present'; but to make use of this ability the tracking signal would need to be smoothed. The problem of the variability of operators was also picked up by the Applied Psychology Panel of the Division 7 of NDRC who carried out studies of the tracking performance of trained and untrained operators.[8]

7.2 Smoothing circuits for predictors

In April 1944, H.K. Weiss, an engineer working for the Anti-Aircraft Artillery Board, requested permission to use the records obtained by the Applied Psychology Panel for a study of the power spectrum of the noise. Later that year, in a letter to Weaver he pointed out that what mattered was not the tracking error but the prediction which could be achieved with the data supplied by the tracking instrument: 'What we do care about is how much better (if at all) the prediction which can be made on the first man's tracking is than that corresponding to the second man's tracking'. He noted that only a meagre amount of analysis of predictor behaviour had been done—by Weaver and Stibitz—but that more must be done since tracking studies must take into account the predictability of the data. He wanted analyses to determine the stability of the 'auto-correlation among courses and operators' and to give an 'accurate idea of the manner in which the power spectrum varies with tracking instruments'. Weiss was based at the Camp Davis, North Carolina, testing ground of the Artillery Board supervising the testing of various predictors. He found that the prediction error of the M7 predictor increased with the randomness of the input but that a 'poor' operator produced a less random signal than a 'good' operator and hence the two effects tended to cancel out.[9]

Weiss's work illustrates the growing importance of dynamics and of a systems approach. Attempts to emphasise dynamic testing and the testing of the overall system behaviour led to conflict with service personnel. Weiss, and George Stibitz who, with Duncan J. Stewart of the Barber-Coleman Company, were responsible for the design of a dynamic tester for predictors,[10] all bore the brunt of this antagonism which at times bordered on open hostility. The main complaint of the civilian testers was that the Army did not understand the need for careful, controlled testing and kept making frequent changes to the personnel allocated to testing duties. The Army considered the pace of testing to be too slow.[11]

The importance of a system approach to design was further emphasised when attempts were made to by-pass the human operator by connecting the output from the radar set directly to a predictor. According to Porter, when the output from a radar set was connected to a mechanical predictor the 'howling and screaming ...could be heard a mile away', as the predictor tried to respond to the high frequency components of the radar output signal.[12] Work on investigating the type of data that predictors might be fed with had begun much earlier.[13] Bayliss, a member of Blackett's operational research group began examining the differences between the visual data (which the Sperry predictor had been designed to receive) and the new data obtained directly from the radar system. He found that predictions of future position were much improved if the internal rate calculation mechanisms of the Sperry predictor were disconnected and a rate signal was fed in from the radar set instead.[14] It is not entirely clear from Cockcroft's report what was disconnected and how the rate information was fed to the predictor. The problem was not simply one of adding a smoothing circuit to the predictor: both the Sperry and Vickers (Kerrison) predictors contained smoothing circuits and these could be modified to accommodate different inputs. The smoothing circuit, however, introduces a lag which when using manual input had been compensated for by operator adjustments based on tracer observations. The question raised by the Sperry Company was how to provide an anti-lag unit when using radar input and blind firing.[15]

The Sperry M7 predictor had a mechanism of the form shown in Figure 7.3, which was used to compute the predicted position $p(t) = \tau \, dx/dt$, where τ is the time of flight of the shell. The present position (of one coordinate) $x(t)$ is represented by the rotation of the input shaft and the predicted position $p(t)$ by the rotation of the output shaft. The time of flight of the shell is entered as the speed of the variable speed drive such that $df/dt = K/\tau$. The circuit actually calculates

$$p(t) = \tau \frac{dx}{dt} + \left(p_0 - \tau \frac{dx}{dt}\right) e^{-(t/cM\tau)} \tag{7.1}$$

The circuit acts as a first order lag with a time constant proportional to the time of flight: thus, as George Stibitz concluded, 'any further smoothing obtained from the multiplier, *as it now stands*, will necessarily result in longer lags'.[16] Both the Sperry Company and the United Shoe Machine Company (USMC) were working during 1942 on NDRC contracts to design smoothing circuits for use on the input to the multiplier. The USMC were attempting the design of an electronic smoothing circuit while there were proposals from the Sperry Company and from Edward J. Poitras, the NDRC Technical Aide, for modifications to the electro-mechanical system. These proposals were discussed at a meeting with US Army representatives at Frankford Arsenal in July 1942, at which meeting it was

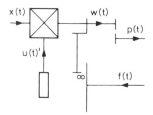

Figure 7.3 *Lag in the Sperry multiplier*

From NARS, Record Group 227, Records of OSRD, Office Files of Harold L. Hazen, 'Sperry Gyroscope Inc.'

suggested that USMC might be helped by the Bell Telephone Laboratories.[17] In October of that year Claude Shannon produced a detailed analysis of all the proposed circuits. He showed that four of the proposals introduced a second order filter and he recommended use of the USMC electrical unit; the fifth circuit— proposed by Edward Poitras—involved an operator entering the rate of change by adjusting the speed of a motor to bring an indicating meter to a zero position.[18] In parallel with these investigations H.K. Weiss was carrying out frequency response tests on the M7. He showed that in azimuth there was a pronounced resonance (with gains of up to 20 times the dc gain) at frequencies in the range 0.10 to 0.5 cycles per second.[19]

Sometime during 1941, P.J. Daniell, Professor of Applied Mathematics, at the University of Sheffield, was awarded a contract by the Ministry of Supply to investigate smoothing networks. Daniell worked closely with Tustin and other engineers at the Metro-Vick works in Sheffield and several reports on smoothing circuits were produced.[20] In one such report dated 26th October, 1944, Tustin, drawing attention to Bode's work on the gain-phase relationship, neatly summed up the position: 'phase advance is anti-smoothing, and smoothing is anti-phase-advance'. He called for work on producing a catalogue of networks which designers could use.[21] Other investigations into smoothing networks were carried out by Porter during 1943-44.[22]

Problems of smoothing were not confined to the inputs to the predictors; similar problems occurred when predictor outputs were coupled directly to the gun positioning control system and in November 1942 Tustin was called to the test range at Monorbier, North Wales, where the gun positioning system designed by Metro-Vick was being tested. The Army had reported inaccuracies in following in certain circumstances. After considerable investigation Tustin found that the gun servo-system was attempting to follow the noise on the predictor output. Tustin explained that if Metro-Vick was given details of the predictor output signals it could redesign the amplifier to 'minimise the response to a particular range of frequencies... and to discriminate between proper target signals and 'pseudo signal'.[23] When Metro-Vick was eventually supplied with data relating to the predictor output, it found that it contained low frequency noise—1/3 Hz in elevation and even lower in bearing, and Tustin reported to Ministry of Supply on 23rd December, 1943 that these frequencies could not be removed by smoothing circuits (in removing the noise the smoothing circuits would also have removed the desired gun aiming information).

When attempts were made to use the Metro-Vick gun control system with the BTL 10 predictor (M9 director), which was just becoming available, further problems occurred. In June and again in July of 1944 Tustin complained to the Ministry of Supply about the 'jitteriness' of the output from the BTL 10 predictor. By October 1944 he succeeded in designing a filter such that the 'jitter' resulting from noise on the input signal was reduced to an amplitude which was less than the backlash of the gears in the gun drives.[24]

7.3 Gun predictor developments

At the first meeting of section D-2 of the NDRC Warren Weaver identified fundamental studies of the essential functions of gun computers and the measurement and analysis of errors in gun computers as two main areas for improving fire control systems (the other areas he mentioned were factors affecting the probability of successful AA fire and servomechanisms for gun aiming).[25] Existing gun computers, for example the Kerrison (Vickers) and Sperry predictors were complex, high precision, electro-mechanical devices which were physically large — the Kerrison weighed 500 lbs. 'Angular and linear variables were characteristically represented by positions of rotating shafts or by lengths of metal bars; resolution and synthesis of vectors and certain multiplications were handled by slide mechanisms; multiplication and integration by ball cage integrators; addition by differential gears; ballistics and other functions were incorporated in two and three dimensional cams.'

Early in 1940 BTL began to design an electrical predictor for the US Army. The BTL team was led by Hendrik Bode and included Blackman and C.A. Lovell (later also Claude E. Shannon). In November 1940 the work came under the direction of the NDRC who hoped that the use of electrical networks instead of mechanical would lead to a simpler, cheaper, and more easily constructed units.[26] The experimental version was known as the T10 gun director and the version that went into service was given the designation M-9. There was some suspicion of the Bell proposal, for what did a research laboratory for a telephone company know about fire control? As Warren Weaver explained in his final report, Bell had developed non-linear potentiometers which could be applied to predictor calculations and also that the fire control problem is analogous to certain problems in communication engineering, namely:

> if one applies the term 'signal' to the variables which describe the actual true motion of the target; and the term 'noise' to the inevitable tracking errors, then the purpose of a smoothing circuit is to minimise the noise and at the same time to distort the signal as little as possible.[27]

At about the time they were considering supporting the BTL predictor project, the NDRC received another proposal for a scheme for predicting the future position of an aircraft. The proposal was submitted by S.H. Caldwell and contained the following statement prepared by Norbert Wiener:

> The proposed project is the design of a lead or prediction apparatus in which, when one member follows the actual track of an airplane, another member anticipates where the airplane is to be after a fixed lapse of time. This is done by a linear network into which information is put by the entire past motion of the airplane and which generates a correction term indicating the amount that the airplane is going to be away from its present position when a shell arrives in its neighbourhood. The principles of design are those of electric networks in general, although the realisation may be by mechanical equivalents to electrical networks. The proposal is, first, to explore the purely mathematical possibilities of prediction by any apparatus whatever; second, to obtain particular characteristics which are suitable for apparatus of this type and are physically realisable; third, to approximate to these characteristics by rational impedance functions of the frequency; then to develop

172 *Smoothing and prediction: 1939–1945*

> a physical structure whose impedance characteristic is that of this rational function, and finally, to construct the apparatus.
>
> Very considerable headway has already been made in the mathematical discussion of possibilities and in the obtaining of rational functions to fit these possibilities.[28]

Wiener had worked on this problem with the help of Richard Taylor of the Department of Electrical Engineering of MIT for about three weeks and the results they had obtained were sent to Warren Weaver. This would appear to be an undated report entitled 'Principles Governing the Construction of Prediction and Compensating Apparatus'. With this report in the files is another undated report by Wiener entitled 'Memorandum on Circuit Theoretical Lead Mechanisms'.[29] In the latter, which appears to be the earlier of the two, there is a discussion of methods for the design of networks to produce phase leads with an example of the requirements for a network to approximate the function $e^{\pi i \omega/2}$ over the range $-1 \leq \omega \leq 1$. For sinusoidal input with $\omega = 1$ this network produces a lead of 90°. It also produces 'large, parasitic amplification' outside the working range of frequencies. It is interesting to note that in this paper Wiener very clearly states the interdependence of amplitude and phase in minimum phase networks, and stresses the need to ensure that in seeking to obtain a particular phase characteristic, serious distortion of the amplitude characteristic is not also introduced. In the memorandum 'Principles...' there is a brief review of network design and a circuit diagram for a differential analyser simulation of a specific network followed by a discussion of the practical problems:

> A difficulty which is to be anticipated is that the manual operation of the telescope or other devices actuating the lead mechanisms may be jerky. It is suggested to avoid this that this manual operation be not direct but through the intermediacy of a filter apparatus designed to take out the irregularities of cranking...Another difficulty in the design of such apparatus is that the object observed may itself have frequencies which will be excessively amplified by the lead mechanism.

Wiener comments that if the path of the plane always contains such frequencies then prediction is not possible, but if they only occur infrequently, such as when the pilot takes avoiding action, it is possible to conceive of an apparatus in which the prediction apparatus is disconnected when the frequency component of the flight path exceeds a preset amount. He also notes that the usual harmonic components of the path of the aircraft and the time of flight of the shell will have to be known in order to compute the required lead network. In these reports there is no mention of treating the motion of the airplane on a statistical basis. Caldwell noted that the main difference between this work and that which he was doing with Wiener and Brown on the control of servo-mechanisms was in the amount of anticipation required, large for fire-control and small for servomechanisms. Hence, Caldwell suggested that a patent clause similar to that in Brown's contract should be inserted in that of Wiener.[30]

Gordon Brown apparently became concerned about patent rights. His research student, A.C. Hall, was working on the design of lead networks and was considering capacitor-resistor networks which could be cascaded and which were similar to the circuits being tested by Wiener on the differential analyser. Hall prepared a report, dated 4th December, 1940, in which he claimed that investigations into the performance of derivative circuits had been underway since 1st July, 1940 and that he had produced both design criteria and 'quantitative, non-dimensionalised performance curves'.[31] Hall's report acknowledges that the scheme was proposed by Wiener, Caldwell and Taylor but in a covering letter to the report, written by Brown and dated 9th December 1940, their participation is minimised.[32] In the letter Brown states: 'The servomechanisms group has known for a long while that one of its most important projects was the analysis and investigation of circuits and mechanisms that would provide a 'lead', or establish a rate or a derivative of a function, or in other words make available means to anticipate the future value of some quantity or function'. He went on to stress that Wiener was a mathematician who 'is but meagrely informed on the techniques necessary to reduce to practice the [mathematics]' and that 'since mathematical analyses are not viewed as patentable material...If any of Dr. Wiener's efforts are reduced to practice in the field of servomechanisms while Wiener is collaborating with me...all patents should become the property of the body supporting the particular servomechanisms project.' (Brown's work was being supported by the Sperry Company). Brown was, however, anxious not to be excluded from access to Wiener's work: 'I believe that the servomechanisms group has real justification for working with Dr. Wiener and for being allowed to use freely the results of the joint work'. Brown was concerned that if NDRC operated its normal policy of allowing access to work only on a strict 'need to know' basis through the technical aides he might lose access to Wiener's insights and network designs.[33]

Early in 1941 Wiener and John Bigelow realised that they could not achieve perfect prediction and that the best they could achieve was a statistical prediction of the future position of an aircraft. In an undated handwritten note Wiener explained,

> To anticipate the future of a function is impossible without restrictions. The restrictions which make it approximately possible are that motions of high frequency or short period are neglected.
>
> It is easy to consider the anticipation operator as a function of the frequency ω. It is $e^{ia\omega}$, a being the time of anticipation. It cannot be approximated over $(-\infty, \infty)$ by a polynomial in operators of the form $([1-\alpha\omega]/[1+\alpha\omega])^n$, but it can be approximated in the least square sense over the range $(-\Omega, \Omega)$. The approximating polynomial may be realised by a Lee-Wiener network, or for large time lags, by an equivalent set-up of integrators. Other network syntheses may be used.
>
> This method is preferable to a synthesis in terms of successive time derivatives, which on the frequency scale is equivalent to a power series development in ω. Such a development grossly exaggerates high frequencies, and is relatively inoperative for low frequencies. The synthesis we suggest is far closer to an extrapolation from values taken at time 0, -1, -2, etc., and automatically suppresses high frequencies. The higher derivatives are as hard to obtain and even more inaccurately given than the desired extrapolation itself.[34]

The ideas are elaborated further in an undated nine page memorandum entitled 'On Linear Prediction'.[35] Both documents probably date from January 1941, for Wiener and Bigelow produced a report entitled 'Analysis of the Flight Path Prediction Problem, Including a Fundamental Design Formulation and Theory of the Linear Instrument' in late February 1941.[36]

By June 1941 Wiener and Bigelow had designed an electrical filter that would produce a good prediction. They described their approach to the Bell group (which included Bode, Blackman, Lovell, and Wente) at a meeting organised by NDRC and held on 4th June 1941 at the Bell Laboratories.[37] Wiener explained that the method he proposed

> [was] to predict non-uniform curvilinear performance of the target ...[This] involved a knowledge of the probable performance of the target during the time of shell flight and that his method proposed to evaluate this probable performance from a statistical correlation of the past performance of the plane. This involves a statistical analysis of the correlation between the past performance of a function of time, and its present and future performance. Such a statistical analysis is important in the study of time series, and also in applications to the design of impulse filter networks intended to produce the maximum distinction between a given signal and an accompanying random disturbing noise.

And Bigelow explained the basis of carrying out this process as:

> (1) The assumption of an available statistical knowledge or history of the functions of time.
> (2) The establishment of a formal integral equation [of the Wiener-Hopf type] expressing the difference between the function as correlated by this knowledge, and the value of the function after a variable time interval.
> (3) The minimisation of the square of this difference equation by factoring into expressions having an appropriate separation of singularities on the upper and lower half planes, and
> (4) The summation of the principal singularities in the lower half plane, thus giving the frequency characteristic of the linear predictor with the least squared error.[38]

In formal terms following, Norman Levinson,[39] the problem can be stated as:

> Consider a function of time $f(t)$ which is the sum of a function $g(t)$ [which could be the co-ordinates of a moving airplane] and a noise $f(t) - g(t)$. How best to determine $g(t+h)$ for some $h>0$ from a knowledge of $f(t-\tau)$ for $\tau<0$? Method of solution: choose $k(\tau)$, $\tau>0$, so that

$$\lim_{T\to\infty} \frac{1}{2T} \int_{-T}^{T} \left[g(t+h) - \int_{0}^{\infty} K(t-\tau)f(\tau)d\tau \right]^2 dt \qquad (7.2)$$

> is minimised.

This leads to an integral equation of the Wiener-Hopf type

$$\int_0^\infty \phi(t-\tau)K(\tau)d\tau = \Delta(t+h) \qquad t \geq 0 \qquad (7.3)$$

where ϕ is Wiener's auto-correlation function for f,

$$\phi(\tau) = \lim_{T \to \infty} \frac{1}{2T} \int_{-T}^{T} f(\tau+t)f(t)dt \qquad (7.4)$$

and χ is the cross correlation function for g and f

$$\chi(t) = \lim_{T \to \infty} \frac{1}{2T} \int_{-T}^{T} g(t+\tau)f(\tau)d\tau \qquad (7.5)$$

The solution of equation 7.3 to find $K(\tau)$—the impulse response of the circuit required for prediction—is no easy task.[40]

7.4 Background to Wiener's work

Wiener had first become interested in probability in 1919 shortly after he got a job at MIT. During the summer of 1919 a visitor, I.A. Barnet, suggested to Wiener that he might consider working on a 'generalisation of the concept of probability to cover probabilities where the various occurrences being studied were not represented by points or dots in a plane or in space but by something of the nature of path curves in space'.[41] Wiener worked on this problem for two years studying the work of P.J. Daniell and G.I. Taylor in particular.[42] Taylor's work was particularly significant in that it used ideas concerning correlation between a function and its derivative. However Taylor had studied the complicated phenomenon of turbulence in fluid flow and Wiener, seeking some simpler physical behaviour to study, chose Brownian motion.

> [Wiener] conceived the idea of basing the theory of Brownian motion on a theory of measure in a set of all continuous paths...this idea proved enormously fruitful for probability theory. It breathed new life into old problems...More than that, it opened up entire new areas of research and led to fascinating connections between probability and other branches of mathematics.[43]

During the 1920s electrical engineers, in their attempts to deal with dynamic behaviour, had increasingly been using Heaviside's operational calculus. There was some concern at the lack of mathematical rigour in Heaviside's techniques and between 1915 and 1930 many people worked to place the operational techniques on a firmer mathematical basis. Around 1924 Vannevar Bush started to teach a course on operational techniques and in preparing a book for the course he sought Wiener's aid. The book *Operational Circuit Analysis* published in 1929 contains an appendix written by Wiener which provides a rigorous justification of the Heaviside operational calculus.[44]

Work with Bush on this book led Wiener to study classical harmonic analysis on a very general basis and he realised that Heaviside's work could be translated word for word into the language of this generalised harmonic analysis. The result was the publication in 1931 of a major paper entitled 'Generalised Harmonic Analysis' (referred to subsequently as GHA).[45] The importance to engineers of GHA was that it extended the range input signals (functions) which could be used in harmonic analysis: Fourier analysis is restricted to a special class of periodic functions of finite amplitude whereas in GHA Wiener extended the techniques to deal with functions which are periodic, almost periodic, and also functions with continuous or mixed 'power spectrum'. In formal terms Wiener's generalised harmonic analysis applies to those measurable functions $f(t)$ for which eqn. 7.4 exists for every τ. This is the Wiener auto-correlation function and the restriction implies that for $\phi(0)$ to exist the indefinitely continuing function $f(t)$ must have associated with it an average power, and for $\phi(\tau)$ to exist in general $f(t)$ must have some form of regularity. Physical phenomena which exhibit random behaviour but have some gross, unchanging attributes can be modelled by a function with the above characteristics. For example noise in an electrical resistor can be modelled by such functions.[46]

In GHA Wiener showed that the spectral density of a signal was related to its auto-correlation function by the Fourier transform, that is the spectral density $s(\omega)$ of a function $f(t)$ is given by:

$$s(\omega) = \int_{-\infty}^{\infty} \phi(\tau)e^{-j\omega\tau}d\tau \qquad (7.6)$$

A. Kintchine independently derived the above relationship and as a consequence eqn. 7.6 is frequently known as the Wiener-Kintchine equation.[47]

7.5 The Wiener predictor

Wiener was well aware of the generality of the work he was doing, pointing out to George Stibitz (who was the Technical Aide handling the contract) in October 1941 that the theory could be applied to the design of networks for television and facsimile transmission systems — in this case the time shift is a lag whereas in the predictor it is a lead. In the report of this meeting, Stibitz — as he was to in several subsequent reports — expresses concern about using minimisation of the square of error as a criterion of performance in the case of anti-aircraft predictors.[48] He also expressed doubts about the assumption that the statistics of the noise and the message would be independent. Stibitz explains what Wiener was trying to do in the following way:

> ...in the predictor problem, a coordinate system is chosen and the target path is given as three voltage functions representing the three coordinates. [Wiener chose distance along target path, curvature in the horizontal plane and height, and assumed the coordinates are independent]...suppose that one of the target coordinates is represented by a voltage function of time, say $f(t)$. It is required to obtain a voltage function $e(t)$ out of an electrical network which is as

'near' as possible to $f(t+\alpha)$, the value of $f(t)$ at a time α units in the future.

Wiener has selected the mean square error, that is, the average of $[e(t) - f(t+\alpha)]^2$, as a criterion of the goodness of prediction. This particular criterion is chosen because it is plausible and simple to handle mathematically. It cannot be the 'best' fit in all coordinate systems, but may be satisfactory with the present choice... it is required to find the network which will minimise the mean square error

$$E = \lim_{T \to \infty} \frac{1}{2T} \int_{-T}^{T} [f(t+\alpha) - e(t)]^2 dt$$

In this expression $\lim_{T \to \infty}$ means 'in the long run'.

It may help to fix in mind the operation of Wiener's predictor if we imagine that we have followed a very large number of flights of the targets we wish to predict. Now we shall sort out of this group all those flights which are alike up to time t, and we notice where the target was a time $t + \alpha$ in each of them.

If now, we look at any particular spot in space, we can find the distance between that spot and each of the target positions for time $t + \alpha$. There will be a certain spot in space for which the mean square of all these distances is least. It is that spot which Wiener's predictor is designed to find.

Stibitz goes on to explain that the proper network is related to the auto-correlation of the message, to explain what the auto-correlation function is, and to explain the response of networks to a suddenly applied unit voltage: '$K(t)$ is called the indicial voltage transfer function of the network'. He then states that in the ideal case, that is no noise the indicial function of the network must satisfy the equation

$$\varphi(\tau + \alpha) = \int_0^\infty \varphi(\tau - \sigma) dK(\sigma)$$

and in the presence of noise the equation

$$\varphi_1(\tau + \alpha) = \int_0^\infty [\varphi_1(\tau - \sigma) + \varphi_2(\tau - \sigma) dK(\sigma)$$

for all $\tau > 0$.

Physically, this means that if a voltage equal to the combined auto-correlations for the message and noise for the τ is applied to the input of the network, then the output voltage must equal the auto-correlation of the message alone at time $\tau + \alpha$. [p. 5]

Stibitz then notes that the problem no longer requires the prediction of an unknown function 'since the auto-correlation functions $\varphi_1(\tau)$ and $\varphi_2(\tau)$ are statistical functions which are presumably known for all values of τ. In this step, therefore, we have reduced the problem of predicting an unknown function to that of designing a network to respond in a definite manner to a definite input'. [p. 6]

Stibitz describes the model predictor that Bigelow was constructing at MIT as containing:

> a sequence of 5 amplifiers and 5 networks in tandem. Each network consists of a series resistance and a shunt capacitance.
>
> Five linear combinations of the resulting voltages at successive steps are obtained by potentiometers; these combinations correspond to networks characterized by polynomials in $(1/1 + Da)$ up to the 5th degree.
>
> The proportions of these separate polynomials which are combined in the final output are determined by the statistical characteristics of the message to be predicted and of the noise present in the signal. [Appendix III]

At the meeting held at the Bell Laboratories on 4th June (referred to above) following Wiener and Bigelow's presentation, Blackman asked about the 'possibility of spurious information in the form of cranking errors entering the apparatus and affecting the correlation and hence the prediction'. Wiener replied that this had been considered and could be dealt with. Unfortunately this was not the case. The transients involved in starting and stopping the tracking completely masked the behaviour of the system at the higher frequencies and made Wiener's stochastic approach much less effective than the BTL approach. This was a major factor in abandoning the Wiener-Bigelow project. The assumptions involved in using the class of functions of GHA was that the apparatus would be insensitive to the fact that the input began at $t = -10$ seconds rather than $t = -\infty$ and yet would still accurately predict the position at $t = +20$ seconds.

Bigelow continued work on building a model predictor system throughout the rest of 1941 and gathered information on airplane tracks. Wiener wrote the manual *Extrapolation, Interpolation, and Smoothing of Stationary Times Series with Engineering Applications*, which became known to engineers during and immediately after the war as the Yellow Peril on account of its yellow cover and difficult mathematics. The manual was issued on a very restricted circulation at the beginning of February 1942 and was eventually published in 1949.[49]

The difficulties of translating the ideas contained in the manual into a practical AA predictor began to grow and in an undated letter (23rd or 24th February 1941) to Warren Weaver, Bigelow began to express doubts as to whether the system could be developed in sufficient time to be used in the present war. A particular concern was the effort involved in calculating, from the statistical data, the coefficients for the various networks. Bigelow proposed going ahead with setting up a demonstration of the system but also expressed the view that the Radiation Laboratory group at MIT were 'in *very* bad need of some good servo-under-noise-input theory'. He had discussed the problem unofficially with A.C. Hall who wanted some help to try to apply Wiener's theory.[50] Wiener had written to Weaver on 22nd February 1941 expressing delight at the reception the manual was receiving and making the request that Norman Levinson be allowed to give a restricted course on the basis of the manual.[51]

In February 1942 Stibitz, concerned about the problem of transient tracking errors, proposed that the statistics used to represent the noise should be based solely on the time during which the trackers were on target and data obtained while they were settling onto the target should be ignored. He realised that this was in fact overly restrictive and proposed that a function be developed that weighted the data according to the time from the beginning of the run. He also called for urgent steps to be taken to obtain auto-correlation data for tracker's errors.[52] He expressed these views to Wiener and Bigelow at a meeting on 21st May 1942, but although Wiener recognised the importance of transient errors and statistical variability he was still optimistic that the predictor would be adequate.[53]

A demonstration of the Wiener-Bigelow predictor was given to Section D-2 (represented by Warren Weaver, E.J. Poitras, T.C. Fry, and G.R. Stibitz) on 1st July 1942. Stibitz was impressed by the accuracy of the prediction, suggesting that for a 2-second period the errors were less than one quarter of those of the tangential predictor and one tenth of those of the memory point mechanism. He was convinced that if prediction for curved flight was to be introduced it must be based on Wiener's method.[54] Weaver thought it was a 'miracle' but 'was it a useful miracle'? He thought that 'for a one-second lead the behaviour...is positively uncanny'.[55] The anonymous official report of the demonstration was more cautious, stating that the circuit was able to predict over a period of 2 seconds and that over 1 second the predicted value was 'astonishingly close'. However, it raised doubts about the ability of the apparatus to separate the signal and the noise: 'are these various errors sufficiently separated in frequency so that Professor Wiener's network can predict the flight errors and at the same time filter out (or at least not predict) the long period and short period tracking errors?' and also whether the prediction time could be extended to a practical length. (The 2-second length of the model predictor was the equivalent of 8 to 10 seconds real time, however, the prediction time required for an AA director was in the order of 20 to 30 seconds). As a result of the demonstration, Section D-2 began to collect realistic data on tracking and other errors which was supplied to Bigelow.[56]

Following the demonstration Weaver wrote to Wiener to say that he thought the theoretical work had been successfully carried out and that the theory would find wide application. He then went on to say that the pertinent question was whether the theory could be applied to the AA problem. Put simply, the input to the predictor comprised four components: (1) signal, (2) short period tracking errors (1 to 3 seconds), (3) long period tracking errors (10 to 20 seconds), and (4) general noise. The signal (1) can be subdivided into (1a) a smooth curve being the normal flight path of the aircraft and (1b) statistical flight error made up of bumpiness and voluntary or involuntary deviations caused by pilot action. The fundamental question was whether the behaviour of (1b) was such that the frequencies were sufficiently different from (2) and (3) to allow the signal to be separated and predicted. If they were not, then the only solution would be to seek to redesign the tracking system to change the period of the tracking errors.[57]

The doubts that had been expressed by Stibitz and Weaver proved correct and by November 1942, Bigelow admitted defeat. In discussion with Warren Weaver he agreed that Wiener's prediction method was not suitable for practical application at this time. He also expressed the view that a rigorously linear system would not be successful and thought that work on non-linear prediction should be pursued.[58] Wiener continued work through November analysing the performance of his system using real data and comparing the results with other methods. He wrote to Weaver on 23rd November, 1942 to say that the best his system could produce over the memory point method was a 10% improvement in prediction and that he thought this was not significant. This was based on using 10 seconds of past data and predicting 20 seconds ahead.[59] The contract was ended and Wiener and Bigelow submitted their final report in December 1942, in which they said that 'an optimum mean square prediction method based on a 10-second past and with a lead of 20 seconds does not give substantial improvement over a memory-point method, nor over existing practice'.[60]

The work was not wasted. Its importance was quickly and widely recognised: John V. Atanasoff of Iowa State College who was carrying out a fundamental study of predictor techniques was sent early reports of Wiener's work and wrote to Weaver saying that it is a 'brilliant and important piece of work'.[61] P.J. Daniell in his 'digest' of Wiener's manual concluded, 'any future theory of statistical fluctuations and of prediction problems will certainly be built on the fundamental ideas expressed in the manual. The technique is a logical extension of Heaviside calculus necessary for such studies.' He did have some reservations. 'There is a difficult problem to be thought out which is to find the proper connections between the methods of the manual and the practical criteria used at present in the design of predictors and filters.'[62]

The reduction of Wiener's results to practical usable form was considered vital to the war effort and several mathematicians, and mathematically adept engineers were involved in interpreting the work. Among these were P.G. Bergman of Princeton University and Norman Levinson of MIT both of whom produced glosses on Wiener's manual.[63] Stibitz continued to work on the problem and proposed an extension to the theory to cover non-linear predictions.[64] He also tried to use the method to determine the optimum response for a sampled data system with input contaminated by noise.[65] Wiener's work was also picked up by R.S. Phillips who was working on the auto-follow radar systems at the Radiation Laboratory and was seeking criteria on the basis of which servomechanisms could be designed that would in some sense minimise the errors caused by noise on the input.[66] It was also an important influence on Claude E. Shannon's seminal work, 'Mathematical Theory of Communication', first published in 1948.[67]

There were also some practical gains from Bigelow's work: in particular he designed an isolation amplifier containing some novel ideas which were picked up by George Philbrick and incorporated in his operational amplifier designs.[68]

Interest in applying Wiener's techniques to fire-control problems continued. In 1944 the USA Naval Research Laboratory expressed an interest in undertaking a broad study of fire-control problems associated with the use of automatic tracking radar. They consulted Wiener and Brown at MIT and discussed the problems with John Bigelow, who spelled out the practical problems and limitations of Wiener's method, and in particular the need for extensive, reliable data, frequency separation of noise and data, linearity, and the limitation of defining optimum performance by the least squares criterion. Bigelow went on to point out that implicit in all the other methods of prediction are the limitations and assumptions that he spelled out for the Wiener method.[69]

During 1941 and 1942 work had continued on developing the Bell Telephone Laboratories' predictor. At an early stage of the development the decision was made to construct a predictor that would be the electrical counterpoint to the existing mechanical predictors. Thus the standard mechanical inputs were converted to electrical voltages, the prediction calculations were performed by, in effect, an electrical analogue computer and the outputs converted back to mechanical movements. Bell originally considered two methods of prediction: the standard existing method which was to predict the future position by simply taking the derivatives of the coordinates and multiplying them by the time-of-flight of the projectile, and by the so-called memory point system of prediction. The problem with the former is that temporary rate variations can lead to significant errors but attempts to smooth the signal used to determine the derivatives automatically leads to the introduction of a lag in the predicted position. The memory point method uses a set of releasable clutches by means of which a reference point of position and rate can be fixed and held, at the point of firing the prediction of the future target position is made on the basis of the average present rate and the stored rate and this is used to extend the average vector between that reference point and subsequent position of the target in space.[70] The Bell predictor went into service in the autumn of 1943 and working in conjunction with the SCR 584 radar was considered to be very successful. Because of the urgency of getting a predictor into service the Bell group did not take into account changes in course, curvature of flight path, performance characteristics of the aircraft or statistical properties. Later in the war, they began to consider predictors for curved flight.[71]

Wiener was greatly influenced and motivated by physical phenomena — be it the eddies in the Charles River or the behaviour of an electronic network. Like Oliver Heaviside with whom he was fascinated,[72] his physical intuition led him to take bold steps, as P. Masani has explained:

> The fact that ergodicity has to be postulated in order to go from time-averages to the expectations and other well known averages of probability theory never bothered Wiener, for his scientific philosophy permitted the free creation of bold and ideal hypotheses.[73]

Masani has remarked on the coincidence of the equivalence of the mathematical theory that emerged from the very practical problem tackled by Wiener and that developed by A.N. Kolmogorov and reported in his paper 'On stationary sequences in Hilbert space' which appeared in 1941 (the results had been reported in 1939).

Well before the final tests on the model predictor Wiener had begun thinking about other applications of the theory: in July 1942 he asked Weaver if he could have permission to talk to Arturo Rosenblueth of the Harvard Medical School about feedback mechanisms in human beings. During the rest of the war he worked on ideas about computing machines and communication. This work was described in his book 'Cybernetics', published in 1948.[74]

7.6 Notes and references

1 Sir Robert Cockburn, contribution to discussion on Research Establishments, p. 489, *Proc. Royal Society London*, A342, 1975
2 CS Memo 48, p. 13
3 CS Memo 169, p. 1
4 CS Memo 169, p. 4
5 CS Memo 184; CS Memo 203; a summary of the work was reported in Tustin, A.:

'The nature of the operator's response in manual control and its implications for controller design', *JIEE*, **94**(IIA), (1947), pp. 190–207 — it was in the discussion of this paper that Porter commented that the difficulties of analysing a system containing fourteen human operators were such that it was easier to comprehend a fully automatic system (p. 203).

6 Uttley, J.A.: 'The human operator as an intermittent servo', *Report of the 5th meeting of the Manual Tracking Panel of the Servo-Panel*, 17/8/1944; a survey of the work carried out during the war is given in Bates 1947; the Royal Aircraft Establishment work is mentioned by Pearson, E.B., in the discussion on 'Applications in the Military Sphere', servomechanisms conference published in *J. Institution of Electrical Engineers, 1947*, **94**, Pt. IIA, p. 205

7 Harold L. Hazen to Warren Weaver, Chairman Division 2, NDRC, 13th May, 1941, NARS RG 227, Div. 7, Offices Files of Warren Weaver, 'MIT, General'

8 These studies were directed by Samuel W. Fernberger and Thornton Fry. See B.O. Williams, p. 214

9 NARS Office Files Harold L. Hazen box 57, Folder Applied Psychology Panel Letter 15th April, 1944, Warren Weaver to George R. Stibitz; Applied Mathematics Panel letter 3rd June, 1944, HKW to Warren Weaver

10 The Barber-Coleman Company developed the so-called Dynamic Tester which provided an imaginary target position, fed the position to the gun director, and compared the correct gun settings with those output by the predictor. Stibitz designed two new relay computers to do this job and these were built at the Bell Laboratories. See B.O. Williams, *op. cit.*, (n.8), pp. 214–228

11 NARS Office Files Harold L. Hazen box 57, A.A. Board, contains several items of correspondence expressing the difficulties faced, see for example letter DJS to Harold L. Hazen 1st July, 1944; difficulties were not confined to relations with Camp Davis, similar problems were reported with regard to the Wright Field, see for example Folder Army Air Force, letter SHC to Lt.Col. W.G. Brown 12th October, 1943

12 Porter, private communication

13 When J.D. Cockcroft, E.G. Bowen and R.H. Fowler met the Fire Control Committee of NDRC on 4th November, 1940, Warren Weaver informed them of the studies he had been carrying out on the need for smoothed data input for predictors. As a result of this meeting instruction relating to the output from the GL III radar set were given. AVIA 10/2 (AR/200/313/2)

14 Cockcroft, J.D.: 'Memories of radar research,' *Proc. IEE*, **132**, (Pt.A), (1985), p. 330 p. 330

15 Report of visit to Sperry Company, 19th June, 1941 by Col. Wallace, Prof. Shenstone and Prof. Fowler. AVIA 10/3 (AR/200/313/3)

16 George R. Stibitz, letter to Joseph Harrington, Jr., of the United Shoe Machine Company, 8th May, 1942. Attached to the letter is an analysis of the M7 multiplier mechanism from which the sketch shown in figure 7.3 is taken. NARS OSRD Div. 7, Office Files of Harold L. Hazen, 'Sperry Gyroscope Inc.'

17 NARS Diary of EJP, Section 2, Division D, 20th July, 1942. OSRD, Division 7, General Project Files, Project No. 51, Correspondence

18 NARS Claude E. Shannon, 'Smoothing Circuits for the M7 Director', OSRD, Div. 7, General Project Files, Project No. 51, folder no. 2

19 NARS H.K. Weiss, memo to the President, Anti-Aircraft Artillery Board, 16th October, 1942, OSRD, Div. 7, General Project Files, Project No. 51, folder no. 2

20 Percy John Daniell (1889–1946) Professor of Mathematics, University of Sheffield from 1923 to 1946, see Stewart, C.A.: 'P.H. Daniell', *J. London Mathematical Society*, **22** (1947), pp. 75–80; CS Memo 25, 97, 185

21 CS Memo 185

22 Porter gave a lecture, 'A study of the performance of a fire control system with special reference to smoothing', at a meeting of the Servo-Panel on 7th July, 1944. A brief summary of this talk is to be found in PRO AVIA 22/2427 Servo-Panel

23 Memo Tustin, 9th April, 1943, PRO WO/185/92

24 PRO WO/185/92

25 NARS Warren Weaver, 'Foreword to report on A.A. Director T-10', undated (about 1944) 'Division & (3)', Office Files of Warren Weaver, Records of Division 7

26 The original idea occurred to David B. Parkinson of the Bell Telephone Laboratories in a dream (10th May, 1940). He talked the idea over with his supervisor Clarence

A. Lovell and early in June 1940 they prepared a report which was sent to Dr. Kelly and Dr. F.B. Jewett. NARS Warren Weaver, 15th March, 1944, Division 7 (3), Office Files of Warren Weaver. The report, rewritten and condensed by Kelly was sent by Jewett to General J.O. Mauborgne, Chief Signal Officer, United States Signal Corps on 12th June, 1940. BTL received positive encouragement from the Signal Corps. A formal letter of support from the Signal Corps was sent to Bush on 30th November, 1940. See letter from Major General R.B. Colton to Dr. M.J. Kelly, 6th October, 1944, Project 2 Correspondence, Project Files, Division 7. The letter contains transcripts of internal Signal Corps memos relating to the BTL proposal.

27 *op. cit.,* (n.25)
28 NARS S.H. Caldwell, General Project Files 6 Box 79, 22nd November, 1940. Wiener, like many of his colleagues at MIT actively sought involvement in the war effort '[I] hope you can find some corner of activity in which I may be of use during the emergency' wrote Wiener at the end of a letter to Vanevar Bush dated 21st September, 1940, General Project Files, Pro. 6, Folder 1, Letter to Dr. V. Bush, 21st September, 1940. Wiener continued throughout the war to seek involvement in the war effort: he wrote to Warren Weaver in July 1942 that he was 'vegetating here, chopping wood, walking n miles a day, and haunting the RFD box in the hope of further orders from D2' (Box 79, Project Rep. 6, Folder 2, Div 7, Norbert Wiener to Warren Weaver, 19th July, 1942); in April 1943 he wrote to G.R. Harrison with a proposal for an instrument for determining the efficiency of bombing patterns (NW to Dean G.R. Harrison, 15th April, 1943 Box 57 Correspondence of H.L. Hazen, AMP Panel). It was not until after the use of the atomic bomb that his attitude to the application of science to war began to change, culminating in his public announcement in 1947 of his refusal to cooperate with governments (see Heims, S.J.: *John von Neumann and Norbert Wiener* (MIT Press, Cambridge, MA, 1980), pp. 187–9 and 332–334)
29 NARS Project No. 6 MIT Reports, General Project Files, Records of Div 7
30 NARS S.H. Caldwell, Proposal to section D-2 National Defense Research Committee, 22nd November, 1940. Division 7 Projects, Project 6(1), Box 79, p. 2. The question of patent rights was of major concern to MIT and it is presumably the implied suggestion that Wiener's work might become subject to patent restrictions that prompted Brown to try to establish some priority. The position was complicated as Brown was also under contract to the Sperry Company and MIT's relationship with that company regarding patent rights was complex and involved.
31 NARS A.C. Hall, Memorandum 4th December, 1940, General Project Files, Project No. 6 MIT Reports, p. 3
32 Gordon S. Brown to N.M. Sage, 9th December, 1940. Sage was Director of Industrial Cooperation at MIT, copies were sent to Hazen, still then head of the Electrical Engineering Department, and to Caldwell.
33 NARS General Project Files 6, Letter Brown to Sage 9th December, 1940
34 *ibid.*
35 *ibid.*
36 The report was dated 24th February and is referred to Bigelow's note of a meeting held on 4th June, 1941, NARS General Project Files 6. I have not been able to trace a copy of the report.
37 Caldwell was the official investigator administratively responsible for the work, Wiener was responsible for the theoretical work and Bigelow, a young electrical engineer had been recruited to aid Wiener in reducing his ideas to practice.
38 The report entitled *Analysis of the Flight Path Prediction Problem, Including a Fundamental Design Formulation and Theory of the Linear Instrument*, DIC Project 5980, was referrerd to by Bigelow in his report of the meeting at Bell Laboratories (see note 17). I have not been able to locate a copy.
39 Norman Levinson, 'Wiener's life', *Bulletin of the American Mathematical Society,* **72** (1966), p. 27
40 The Wiener-Kolmogorov theory of filtering and prediction remained little used until Kalman and Bucy showed how to solve the problem recursively. See K. Astrom, 1970
41 Wiener, N.: *I am a Mathematician: the later life of a prodigy* (Doubleday, Garden City, NY, 1956), p. 35
42 Daniell, P.J.: 'A general form of integral', *Annals of Mathematics*, Series 2, **19** (1918), pp. 279-294; 'Integrals in an infinite number of dimensions', *Annals of Mathematics,* **20**

(1919), pp. 281–8, and 'Further properties of the general integral', *Annals of Mathematics*, **21** (1920), pp. 203–220; Taylor, G.I.: 'Diffusion by continuous movements', *Proc. London Mathematical Society*, **20** (1920), pp. 279–289

43 Kac, M.: *Scientific American*, **211** (1964), p. 105
44 Bush, Vanevar: *Operational Circuit Analysis*, (Wiley, New York, 1929)
45 Wiener, N., 'Generalized Harmonic Analysis', *Acta Mathematica*, **55** (1931), pp. 117–258
46 See, for example, Rice, S.O.: 'Mathematical analysis of random noise', *Bell System Technical J.*, **23** (1944), pp. 282–332 and **24** (1945), pp. 46–156
47 Kintchine, A.: 'Korrelationstheorie der stationaren stochastischen Prozesse', *Mathematischen Annalen*, **109** (1934), pp. 604–615
48 Stibitz, G.R.: NARS Report on visit to Prof. Norbert Wiener, 28th October, 1941. General Project Files, Project No. 6 MIT Reports. Stibitz was a Technical Aide with division 7. Another group with D.J. Stewart as the aide was investigating the basic statistics of gun control.
49 Wiener, N.: *The Extrapolation, Interpolation, and Smoothing of Stationary Time Series*, NDRC Report 370, 1st February, 1942. Copies were sent to England for R.H. Fowler, G.H. Hardy, Littlewood, A. Milne Thompson, R.V. Southwell, G.I. Taylor and J. McNaughton Whittaker. The report was eventually published as, *Extrapolation, Interpolation, and Smoothing of Stationary Times Series with Engineering Applications*, (MIT Press, Cambridge, Mass, 1949)
50 Bigelow had sent Hall his copy of Wiener's report and asked if Hall could receive an official copy.
51 In the letter Wiener asks that Levinson be cleared for security purposes. Unknown to Wiener, Levinson may have already been cleared for Bigelow, in a 1944 report, refers to a report, 'Prediction of Stationary Time Series by a Least Squares Procedure', produced by Levinson about March 1942 under a U.S. Army Air Corps Meteorological contract. NARS Diary of Conference on Methods of N. Wiener, 3rd October, 1944, Project No. 6, Folder 3, 1942–46, General Project Files
52 NARS Stibitz, G.R.: 'Note on Predicting Networks', February 1942, Project No. 6 MIT Reports, General Project Files. Accompanying this note is a longer paper, dated 8th February, 1942, 'Prediction Circuits a la Wiener'
53 NARS Diary of George R. Stibitz, 21st May, 1942, Project No. 6, General Project Files
54 NARS Diary of George R. Stibitz, 1st July, 1942, Project No. 6
55 NARS Diary of Warren Weaver, 1st July, 1942
56 NARS Section Report D-2 No. 6, 31st July, 1942, Proposal Correspondence folder, Office Files of Harold Hazen
57 Letter Warren Weaver to Norbert Wiener, 22nd July 1942, Project No. 6. It is interesting to note that Tustin concluded that the 'wander' that is the tracking errors generated by manual tracking could not be separated from the true signal and as a consequence he recommended the use of auto-follow. CS Memo No. 97, 27th September, 1943
58 NARS Diary of Warren Weaver, 10th November, 1942. General Correspondance, Office Files of Warren Weaver
59 NARS Letter Norbert Wiener to Warren Weaver, 23rd November, 1942, Project No. 6, General Project Files
60 To be strictly accurate Wiener was not given a further contract as the existing contract (NDCrc-83, Suppl. No. 3) was to end on 30th November, 1942. The extract is quoted from Masani, P.R.: *Norbert Wiener 1894–1964* (Birkhauser Verlag, Basel, 1990), p. 190 [page 8 in report]
61 NARS John V. Atanasoff to Warren Weaver, 1st November, 1941, NDCrc-143 and 165 Iowa State College, Corress. 1941–1946, Project No. 12. Atanasoff was attempting to develop a predictor that worked on using weighted averages of past rates of change, the weighting factor decreasing exponentially into the past. His group built model but the method proved to be too inaccurate. NARS 'Electrical Directors and Predictors', October, 1942. Directors (General) folder, Office Files of Harold L. Hazen
62 Daniell, P.J.: 'Digest of Manual on the extrapolation...'. SRD1 PR 328 about March 1943, the report was produced for the Ministry of Supply, Servo-panel: I am indebted to Professor A.M. Walker of Sheffield University for the loan of a copy.
63 Bergman, P.G.: 'Notes on the Extrapolation...', issued by NDRC, 14th December, 1942; Levinson, N.: 'Prediction of stationary time series by a least squares procedure', produced under US Army Air Force Corps Meteorological contract about March 1942,

see NARS, Div. 7, General Project Files 6, Folder 3, 1942-46 report of conference on the methods of N. Wiener 3rd October, 1944
64 Stibitz, G.R.: 'Statistical method for certain nonlinear systems', 9th December, 1944, and 'Noise spectrum for certain nonlinear mechanisms', 11th December, 1944 both in NARS Div. 7 Office Files of H.L. Hazen, Folder: Stibitz
65 Stibitz, G.R.: 'Optimum response for tape servo', 1943 NARS Div. 7, Office Files of H.L. Hazen, Folder: Stibitz
66 See Chapters 6, 7 and 8 in James, H.J., Nichols, N.B., Phillips, R.S.: *Theory of servomechanisms*, Radiation Laboratory Series Vol. 25 (McGraw-Hill, New York, 1947). See Phillips, R.S.: 'Servomechanisms', RL Report No. 372, 11th May, 1943
67 Shannon, C.E.: 'The Mathematical Theory of Communication', *BSTJ*, 1948, (July, October), reprinted in Shannon, C.E., Weaver, W.: *The Mathematical Theory of Communication* (University of Illinois Press, Urbana, 1949). Bode and Shannon also produced a simplified derivation of the theory, Bode, H.W., Shannon, C.E.: 'A simplified derivation of linear least square smoothing and prediction theory', *Proc. IRE*, **33** (1950), pp. 417-426
68 NARS Diary of GAP, 20th July, 1942, visit to J.H. Bigelow, Project No. 6
69 NARS 'Diary of Conference on Methods of N. Wiener', 3rd October, 1944 (JHB), Project No. 6, Folder 3, 1942-46, GPF
70 NARS Bigelow, J.H.: Report of conference at Bell Laboratories, 4th June, 1941, NARS General Project Files, Project 6, Folder 1, 1940-1941
71 See, for example, NARS Diary of Warren Weaver, Saturday, 18th December, 1943, Project No. 30 Western Electric Company, General Project Files. Weaver describes the results of work carried out by Bode, Blackman and Shannon on the use of first, second and third order differences for predicting curved flight.
72 He became fascinated by Heaviside—his novel *The Tempter*, has a character based on Heaviside—and he considered writing a biography of Heaviside. N. Levinson, 'Wiener's Life', *op. cit.*, (n.39), p. 30
73 Masani, P.: 'Wiener's contributions to generalized harmonic analysis, prediction theory and filter theory', *Bulletin of the American Mathematical Society, op. cit.*
74 Whether or not Wiener received official clearance to talk to Rosenblueth, he did so (secrecy irked him and he could not understand the need for it) and as a result of their discussions produced a paper (Rosenblueth, A., Wiener, N., Bigelow, J.H.: 'Behaviour, Purpose and Teleology', *Philosophy of Science,* **10**, 1943, pp. 18-24) relating to feedback in human beings. For Wiener's own account of his wartime work see the Introduction to *Cybernetics* (Wiley, New York, 1948)

Chapter 8
The classical years: 1945-1955

The feasible operation and the effective control of large complex systems constitute two of the central themes of our tumultuous times.
 Simplicity is essential for scientific progress but systems are basically not simple, and the control of systems is equally not simple. Consequently constant examination of our concepts and methods is essential to make sure that we avoid both the self-fulfilling prophecy and the self-defeating simplifying assumption...it should be constantly kept in mind that the mathematical system is never more than a projection of the real system on a conceptual axis.

Richard Bellman

As administrators realised that the war was drawing to a close they began to consider how to disseminate the information obtained during it and how such information might be best used to aid post-war reconstruction. In the USA Vannevar Bush prepared a report, 'Science, the Endless Frontier', in which he called for the creation of a National Research Foundation to be controlled by scientists, not politicians. A bill to establish a National Science Foundation, a modified version of Bush's proposal with more political control, was introduced to Congress in autumn 1945 and eventually passed in 1947. It was promptly vetoed by President Truman on the grounds that it would 'vest the determination of vital national policies, the expenditure of large public funds, and the administration of important governmental functions in a group of individuals who would be essentially private citizens'.[1]

In the UK, much more modest proposals were put forward by Arthur Porter in a report prepared, in 1945, for the Interdepartmental Committee on Servomechanisms and Related Devices (ICSR) in which he noted that the 'future well-being of industry (and of National Defence) will become increasingly dependent upon the wise applications of automatic control systems',[2] continuing, 'the design techniques so successfully applied to military problems, are applied with equal success to the problems of peace-time industry'. He noted that the 'American industrialist appears to be more "control minded" ' than his British counterpart and he drew attention to the lead which America had in this field, noting the greater amount of research being carried out in America and the teaching of the subject at post-graduate level in a number of American universities.

the ICSR were informed at a meeting held on 19th October 1945 that several colleges had expressed an interest in running courses on servomechanisms (they included Imperial College, University of London, Birmingham University, Manchester College of Technology, Armstrong's College (Newcastle), and Northampton Polytechnic (London)). The members agreed to help them run courses. Figure 8.1 shows the advertisement for one such course.

The government, Porter recommended, should form an Industrial Control Systems Panel for the purpose of organising meetings, encouraging personal contacts, liaising between government and industry, and building up a library. He envisaged that one or more of the professional engineering institutions would eventually take over the role of the panel. He also recommended the formation of a research group to handle industrial problems and to collaborate with the existing research organisations, as well as the formation of an Advisory Service on the design and application of industrial control systems. The Advisory Service was to assist industry, universities, and technical colleges. Coordination between the military and civil activities was to be handled by K.A. Hayes who, as a professor at the Royal Military College of Science, was to be a member of both groups.

The British government not only ignored the recommendation but converted the ICSR into a purely military organisation when, in 1946, it decided that because of a reduction in staff it would no longer support the informal panel. The response from the Ministry to a suggestion by Arnold Tustin that either the Institution of Electrical Engineers or the Institution of Mechanical Engineers be invited to take over the work was that 'one could not be asked without the other and the work could not be divided'.[3] The first public presentation of British wartime work in the field had been made at a meeting of the Society of Instrument Technologists held on 28th May 1945 when Arthur Porter read a paper, 'The design of automatic and manually operated control systems'.[4] This paper was not published until 1947. In 1950, belatedly recognising the need for a civilian specialist group, the government asked the Society to form a Control Section. The Society did so and thus became the first British professional body to offer support to the new discipline of automatic control. The inaugural meeting of the section was held on 28th March 1950, and K.A. Hayes, A.M. Uttley and Arnold Tustin read papers at the meeting.[5]

Meanwhile, the Institution of Electrical Engineers organised a Radiolocation Convention, in March 1946, at which several papers containing information relevant to control systems were presented, and in the same month A.L. Whiteley presented a paper to a crowded meeting of the Institution. The large attendance at this meeting, and the interest shown in the paper, prompted the IEE to organise a conference on automatic regulators and servo mechanisms which was held in 1947.[6] Opening the conference, John Wilmot, the Minister of Supply, stressed the importance of automatic control for economic development and emphasised its widespread potential. In his closing remarks V.Z. de Ferranti, the IEE President, observed that the trial and error period in the design of servomechanisms had now been succeeded by a scientific approach based on mathematical concepts.

However, it was not until 1951 that the IEE formally recognised the importance of automatic control developments, when the Council reported that 'in view of developments in automatic control and servo systems and their importance to the Armed Services and in industry, the Committee (Measurements Section) arranged

NORTHAMPTON POLYTECHNIC
St. John Street, London, E.C.1
(nearest Station: Angel, Underground Northern Line)

A Course of Lectures on

ELEMENTS of the DESIGN of
AUTOMATIC CONTROL (Servo) SYSTEMS

will be conducted at the Polytechnic
on TUESDAY EVENINGS at 6 p.m.
commencing 5th March, 1946.

Lectures to be given by

Dr. A. PORTER, M.Sc., A.M.I.E.E.

(Ministry of Supply)

5th March, 1946	"General Introduction."
12th ,, ,,	"Important Classes of Control Systems and Introduction to Theory of Control."
19th ,, ,,	"Further Theoretical Considerations."
26th ,, ,,	"The Stabilization of Control Systems."
	"The Study of a Control System with Finite Time Lag."
2nd April, ,,	"The Design of Manually Operated Control Systems."

Lectures to be given by

Mr. M. L. JOFEH, A.M.I.E.E.

(Messrs. A. C. Cossor Ltd.)

9th April, 1946	"Position Control Systems."
16th ,, ,,	"The Study of Methods of Stabilizing Position Control Systems."
30th ,, ,,	"Application of Electrical Models for Studying Complex Control Systems."
7th May ,,	"Non-Linear and Discontinuous Phenomena in the Design of Control Systems."
14th ,, ,,	"The Design of an Electronic Voltage Regulator."

Lecture to be given by

Prof. K. A. HAYES, O.B.E., B.Sc., M.I.E.E.

(Military College of Science)

21st May, 1946	"The Testing of Control Systems and Demonstration of Apparatus."

Fee for the Course (or portion of the Course), 21/-
Admission by ticket, obtainable from the Secretary,
Northampton Polytechnic, St. John Street, London, E.C.1

The Polytechnic Refectory will be available for light refreshment until 6 p.m.

Figure 8.1 *Northampton Polytechnic course*

that papers of a high academic level concerning the latest investigations into control techniques should be included in the sessional programme, and this policy will be continued'.[7] In 1955, the IEE changed the name of the Measurement section to Measurement and Control. The Institution of Mechanical Engineers did not form a specialist section for control (the Automatic Control Group) until 1961.

In the USA (as previously noted in Chapter 2), the American Society of Mechanical Engineers had formed a committee called the Industrial Instruments and Regulators Committee in 1936, with responsibility for automatic control. In 1943 this committee became the Industrial Instruments and Regulators Division. The division's scope was extended in 1952 to cover dynamic systems, as well as instruments and regulators. Before 1945, the Automatic Stations Committee of the American Institute of Electrical Engineers was responsible for automatic control systems and its main concern was automatic control of steam boilers. The review of the AIEE technical committee structure which began in 1944 led to the formation of an Industrial Control Devices Committee with sub-committees for industrial electronic control and on servomechanisms. In 1946 the servomechanisms sub-committee became a joint sub-committee of the Industrial Control Devices Committee and the Instruments and Measurements Committee, and in 1947 the Committee for Communications and Basic Science became one of the sponsors of the servomechanisms sub-committee. By 1949, recognition of the growing importance of servomechanisms led to the joint sub-committee being upgraded to full committee status and in 1950 the name was changed to the Feedback Control Systems Committee. The Institute of Radio Engineers also had an interest in automatic control systems and it formed a servo-systems committee in 1951. In 1952 it was renamed the Feedback Control Systems Committee and in 1955 the IRE Professional Group on Automatic Control was formed.

8.1 Technical publications on developments during the Second World War

Before the war, publication of information on process control had been growing, particularly in the USA, and during the war several books on process control appeared.[8] Thomas J. Rhodes of the US Rubber Company in his book *Industrial Instruments for Measurement and Control* provided extensive coverage of process instruments.[9] He was influenced by the work of Ivanoff, Mason, Fairchild and Smith and his approach to the theoretical analysis of control systems was of limited use for design purposes. A much more comprehensive book was that of Ed. S. Smith, *Automatic Control Engineering*, published in 1944.[10] The book was a revised version of the notes, previously circulated privately, for a course of lectures given in 1942. In it Smith covers both Fourier and Laplace techniques and he includes appendices dealing with mathematical techniques for solving differential equations, and a comprehensive bibliography. At the end of the war, Donald P. Eckmann of the Brown Instrument Company produced an excellent introductory text book on process control.[11] It was also during the war that Ziegler and Nichols explained, in a series of papers published in several journals, their techniques for finding the optimum settings for two and three term controllers.[12]

Information about the new techniques emerged in several papers published during the war. In 1944, S.W. Herwald of the Westinghouse Company described the use of block diagrams and operational calculus for the study of transient behaviour of feedback systems with compensating networks. He drew attention to the importance of stability and steady-state accuracy in servo-mechanisms. He also discussed the use of normalised transient response curves and compensating feedback networks.[13]

With the end of the war came a greater freedom to publish, and there was a rush of papers explaining the new analytical techniques that had been developed. The ICSR (the old Servo-Panel) planned a series of monographs to be edited by C.C. Inglis, and P.J. Daniell was asked to write a book on control theory. He died before completing the manuscript and A.M. Walker agreed to complete the book, but before he could do so there was a reorganisation in the Ministry of Supply and the project was abandoned. Eventually a monograph on automatic control was published by the Ministry of Supply (in 1951) and included in it were two papers by Daniell.[14]

The OSRD made the decision towards the end of the war to support the writing of reports of the wartime work and the writing of a history of the organisation. Key workers were offered continued employment by OSRD for a period up to six months after the end of the war. NDRC originally intended to report only on radar developments but quickly realised that many of the auxiliary developments were also of great interest. The main form of publication of the technical work was the 27-volume Radiation Laboratory Series, edited by Louis N. Ridenour. The most important volume for control engineers is the book *Theory of Servomechanisms*, written by H.M. James, N.B. Nichols and R.S. Phillips with one chapter contributed by Wittold Hurewicz.[15] This book had a major influence on a generation of engineers and set the style and content of many succeeding text books. It was not the first book on the subject to appear but was the most comprehensive. However, before it appeared many of the important ideas had been published in other papers.

In 1945, E.B. Ferrell of the Bell Laboratories clearly stated the parallels between electro-mechanical control systems and electric network design. He recommended designing in the frequency domain. He also recommended using the asymptotic approximations on the Bode diagram.[16] Robert E. Graham described this technique in detail as part of a comprehensive survey of the techniques used by the Bell Laboratories. The notation used — illustrated by Figure 8.2 — shows the connection with the work of Black and Bode. In discussing design criteria Graham commented that the 'optimum design of a servo system, for a specified input signal and noise...is a compromise between dynamic error and output noise fluctuations, with stability considerations and parasitic circuit elements restricting the possible choice of loop transmission characteristics'. [p. 627]. He noted in particular the conflicting aims of requiring a large value for the loop gain ($\mu\beta$) and reducing the gain to less than unity at a sufficiently low frequency to avoid problems with parasitic elements. He also drew attention to the problems of input noise and the need to narrow the frequency band to avoid problems associated with such noise, observing, however, that reducing the bandwidth can lead to large dynamic errors.

Obtaining the transient response of a system from knowledge of its differential equation is, according to Graham, a straightforward but tedious business, and the response once obtained 'often is of little help either in guiding the initial design or in predicting the necessary changes, should the trial design be found unstable' [p. 628]. He recommends using frequency response techniques which allow 'the

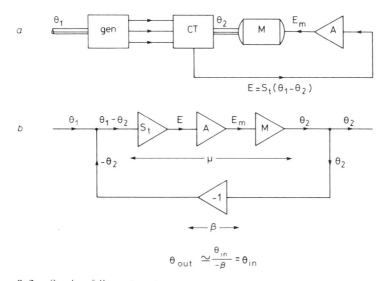

Figure 8.2 *Synchro follow-up system*

Reproduced (with partial redrawing) by permission of Robert E. Graham, from *Bell System Technical Journal*, 1946, **25**, p. 624

additive effects of minor circuit modifications to be evaluated.' Local feedback loops provide an effective means of modifying the system response and he suggests that 'perhaps the simplest and most useful kind of local feedback is negative tachometer (velocity) feedback around motor-driven systems' which increases the bandwidth by reducing the effective time constant of the drive system. [p. 645]. Graham also describes the use of feedforward techniques to reduce the dynamic error of a tracking servo-system.

The use of diagrams of logarithmic gain and linear phase plotted against logarithmic frequency for the design of feedback systems was introduced by Bode in a paper published in 1940.[17] Such diagrams are now known by his name. In the paper he also described the concepts of gain and phase margin as well as elucidating the minimum phase frequency function. The work appeared in extended form in his book *Network Analysis and Feedback Amplifier Design*, published in 1945.[18]

During the war both Hall and MacColl extended the Nyquist criterion to deal with systems with poles at the origin of the complex plane. Servomechanisms containing electric or hydraulic motors often have open-loop transfer functions containing a pure integration term and thus have one or more poles at the origin. Nyquist specifically excluded systems with poles in the right hand half plane and Hall and MacColl developed techniques for dealing with such systems. Nyquist in 1932 did not present a rigorous proof for the class of systems that he considered and MacColl developed, for a restricted class (systems which can be specified as rational functions of a complex frequency variable), a rigorous proof based on the Principle of the Argument. Details of MacColl's proof appeared in his book *Fundamental Theory of Servomechanisms*, which appeared in 1945.[19] His approach was used in Bode's book and in James, *et al.*, and it has subsequently become the standard method of presenting the Nyquist method. W. Frey, in 1946 extended the method to cover open-loop unstable systems.[20]

Several papers published in 1946 by people who worked at MIT drew together and elucidated the techniques for the design of servomechanisms developed during the war. Gordon Brown and Albert C. Hall explained clearly their techniques in a paper based on Brown's wartime report for NDRC and on Hall's dissertation (issued as a restricted report in 1943). In the first part of the paper Brown gave a detailed exposition of the transient response approach to design. In the second part, Hall introduced the idea of a transfer function and explained, using the now familiar notation of $KG(j\omega)$ for the open loop transfer function, that 'the transfer function of a servo system is always the product of two parts, one that is invariant with frequency and a second that is frequency-dependent', where K represents the part of the transfer function which is invariant with frequency.[21] They summarised the design procedure as follows: first determine the transfer function $KG(j\omega)$ approximately which involves finding the amplitude and phase characteristics for the complete servomechanism. Then plot the transfer function as a transfer locus. Decide on the degree of stability required and express this in the form of a value for the M ratio, hence find the gain factor which will make the transfer locus tangent on the desired M circle. Now determine the frequency at which the amplitude ratio (output/input) is a maximum. Using the knowledge thus gained from the gain factor, frequency of maximum amplitude ratio and the zero-frequency behaviour of the transfer locus, consider the following questions:

- Is the servomechanism of the type (zero-displacement-error servo, zero-velocity-error-servo) suitable for the particular application?
- Can the gain factor be adjusted to secure adequate damping?
- Is the sensitivity permitted by adequate damping sufficiently high to minimise the effect of nonlinear factors such as sticky valves, dry friction on the output member, and linear factors such as velocity and acceleration error?
- Is the natural or resonant frequency sufficiently high to provide the speed of response required by the application?

If the answer to all these questions is yes, then the system is satisfactory; if not then some form of corrective action must be taken, either through modification of the system parameters or through the addition of corrective networks. The process of finding and drawing the transfer locus must then be repeated.

Hall presented a greatly extended account of his techniques in a paper published in the *Journal of the Franklin Institute* in 1946[22] in which he explained how the transfer locus related to Nyquist's stability criterion and to the closed loop frequency response. He showed how to construct the M circles and gave a brief mention of N circles (contours of constant phase). For the first time in a paper relating to servomechanism design the Laplace Transform was used and Hall gave a brief discussion about using the final value theorem to find steady state errors. In the latter part of the paper he describes methods of designing compensating filters (both feedback and cascade).

Hall's method of plotting the transfer locus on the polar plane is simple and straightforward but, as Brown and Campbell observed, '[it] becomes tedious when complicated functions are encountered' and it is difficult to identify specific effects of lead and lag circuits.[23] The logarithmic modulus diagrams (Bode diagrams) show such effects more clearly. However, a difficulty with the Bode diagram is the recomputation necessary to obtain the closed loop frequency response when a change is made to the compensating circuit. N.B. Nichols, tired of repeated computation and re-drawing, devised the eponymous chart (see Figure 8.3). The basic Nichols chart is a gain-phase plot with coordinates of $10 \log_{10}|G|$ (where G is the open loop transfer function) and phase of G. Also shown on the chart are contours of constant $|W|$ ($\equiv M$) and constant angle of W ($\equiv N$) where $W(s) = G(s)/(1 + G(s))$. The open loop function $G(s)$ is plotted on the gain-phase plane and the magnitude and phase-angle of the closed loop response can be read off from the intersection with the contours of constant gain and phase. The effects on the closed-loop response of re-shaping the open-loop response can be seen easily. Details of the method were published in the book *Theory of Servomechanisms* (James, Nichols and Phillips) and a tracing containing a plot of the lines of constant M and N in the W plane was given away with the book.

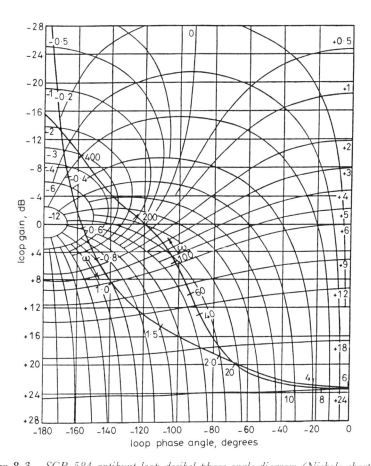

Figure 8.3 *SCR-584 antihunt-loop decibel-phase-angle diagram (Nichols chart)*

Reproduced (with partial redrawing) from H.J. James, N.B. Nichols and R.S. Phillips: 'Theory of servomechanisms; Radiation Laboratory Series Vol. 25' (McGraw-Hill, 1947), p. 219

Further support for frequency response methods came from Herbert Harris who, after describing the steps necessary to use the transient analysis methods — set up the differential equations for the system, obtain the characteristic equation, find the roots and hence the equation of motion — strongly recommends the frequency response method.[24] He says that if the characteristic equation is of degree 5 or higher then time consuming numerical methods have to be used to find the roots. He also observes that the parameters of the system are hidden in the coefficients of the characteristic equation which makes it impossible to see how to change the parameters to improve the performance. The frequency response method, he argues, eliminates the need to find the roots of the characteristic equation and it shows clearly the effect of parameter changes on the overall performance of the system.

He draws attention to the convenience of using the error equation θ_0/e which he refers to as the 'loop transfer function' and points out the similarity to the $\mu\beta$ used by Bode and Nyquist. He notes that the frequency response techniques allow the mixing of experimentally obtained gain-phase values with values obtained by calculation from substituting $j\omega$ for p or s in the block operator for a component.

Harris carried out a detailed study in which he compared series or cascade compensating networks with feedback compensating techniques. He concluded that if correctly applied both techniques give systems with comparable performance. In the discussion of the paper Harold Chestnut of General Electric, after congratulating Harris on an excellent account, argued that series compensation is seldom used alone in production servos. He reasoned that close tolerances on components can seldom be obtained except at considerable cost and hence there will be variations in component parameters for which series networks cannot compensate. He admitted that feedback systems requiring tachometer and other signal sources can add weight and expense. Chestnut's contribution is typical of the combination of deep and sound practical and theoretical knowledge that he and his colleague Robert Mayer expounded in their valuable and informative book, *Servomechanisms and Regulating System Design*, published in 1951.[25]

Harris also makes brief mention of the inverse Nyquist plot whereby the amplitude ratio input/output is plotted instead of output/input. The inverse Nyquist technique was introduced independently by H.T. Marcy in the USA and by A.L. Whiteley in the UK.[26] The advantage of the inverse plot is that response of components in the feedback loop, or in inner feedback loops, can be added vectorily to the basic plot of the forward loop transfer function. The example which Marcy gives is shown in Figure 8.4

$$\frac{\theta_0}{\theta_i}(j\omega) = \frac{K_d G_d(j\omega)}{1 + K_d G_d(j\omega) K_f G_f(j\omega) - K_d G_d(j\omega)}$$

If an alteration to either the direct or forward loop function $K_d G_d(j\omega)$ or the feedback components $K_f G_f(j\omega)$ is made the effect becomes very difficult to unscramble but the inverse function gives

$$\frac{\theta_i}{\theta_0}(j\omega) = \frac{1}{K_d G_d(j\omega)} + K_f G_f(j\omega) - 1$$

which enables the separation of frequency characteristics of the direct and feedback components.

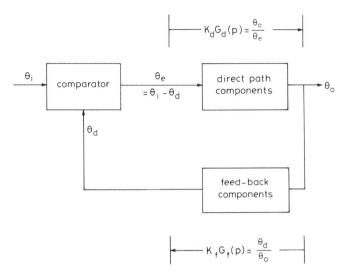

Figure 8.4 *Block diagram of a servomechanism, incorporating components in the feedback circuit*

Reproduced (with partial redrawing) by permission of H.T. Marcy, from *AIEE Trans.*, 1946, **65**, p. 521

The two papers published by Whiteley in 1946 and 1947 made a substantial contribution to the literature and they were discussed in some detail in Chapter 6.[27] In the earlier paper Whiteley gave a detailed discussion of errors in servomechanisms and devised a classification system based on steady-state errors. He also gave tables of typical filters for stabilisation, both cascade and feedback. The overall approach in this paper is based on transient response techniques, and to ease the problem of repeated calculation to determine roots of the characteristic equation, Whiteley proposed a set of standard values for the coefficients of the characteristic equations of differing orders. These became known as Whiteley's standard forms. The design technique was to modify the component parameters to get the coefficients of the characteristic equation to match the appropriate standard form. In the second paper he attempted a unified theoretical approach across servomechanisms, process control and industrial regulators. The formal classification of systems as type 0, 1, 2, 3 based on the number of integrators appeared in print for the first time. This paper also presented the inverse Nyquist approach.

Whiteley's combination of extensive practical experience and deep theoretical understanding of feedback control systems can be discerned from reading his last major contribution to the literature of control systems, a review of progress published in 1951.[28] In this review he emphasised the increasing use of continuous controllers in preference to relay or on-off controllers, the great increase in both the services and in industry of servomechanisms and the widespread interest in, and study of, servomechanism theory, as the outstanding changes of the decade. In his opinion, the current practice of designing mechanical assemblies specifically for servomechanisms and the acceptance that gearing must be of the highest quality was a major advance compared with the practice at the beginning of the decade when the lack of mechanical rigidity in the drives, use of irreversible worm drives, gear backlash and the radial float of the large mountings that were originally designed for manual operation were a serious limitation to the performance of gun control servomechanisms.

He observed that the current practice was to take position feedback from close to the drive motor as this gives maximum stiffness by avoiding resilience and backlash in the main gear train. It does not, however, compensate for errors in the power train. And he went on to commend the 'divided reset' technique of Belsey and Broadbent whereby part of the position feedback is taken from the motor shaft and part from the output shaft, claiming that a two to threefold reduction in steady-state and transient errors can be achieved by this method.

In discussing methods of modifying the response of a system, he strongly supported Harold Chestnut in recommending the use of feedback stabilisation and his comment 'stabilisation by feedback alone is common, although there are few adherents to the principle of modifying the error signal to obtain approximations to its time derivatives' suggest that it was the normal practice in the UK. Performance is limited, he suggested, by instrument accuracies, signal/noise ratios (particularly on error signals) and mechanical rigidity.

In reviewing component developments he drew attention to the reduction in size of selsyn units and their increased accuracy, the improvements in tachogenerators, and in rotating amplifiers, and he noted a growing interest in magnetic amplifiers (these had been used by the German navy and were also used in the control of the V2 rockets).[29] He noted that as yet there has been no industrial use of pulsed data systems but that they may find use in the chemical industry (in fact pulsed data systems in the form of what Hazen termed 'definite correction systems' were in widespread use in the process industries but they were not recognised as such, an interest in pulsed or sampled data systems in these industries did not occur until the move to digital computer control in the 1960s).

In surveying theoretical developments, Whiteley observed that 'some theoretical treatments which have appeared in recent years are more in the nature of mathematical exercises than aids to good engineering'.[30] He particularly criticised many of the proposals which had been made for design criteria, for example the integrated square error put forward by Phillips in *Theory of Servomechanisms* which he said gives too oscillatory response for industrial regulator work. He argued in favour of the minimum moment of integrated square error proposed by John Westcott.

8.2 The root locus technique

The principal advantage of the Nyquist and Bode approach to feedback system design is that measured values of gain and phase can be used and it is not necessary to have an analytical model of the blocks forming the system. An alternative approach based on generating a frequency domain locus $p(j\omega)$ from the characteristic polynomial $p(s)$ of a system described by a linear differential equation was introduced by the Russian A.V. Mikhailov in 1938. Similar techniques were developed independently by Cremer and by Leonhard.[31]

The final weapon in the armoury of classical techniques was the root-locus method proposed by Evans in 1948 and extended by him in 1950. This method enables a rapid assessment to be made of the way in which the closed-loop characteristics vary with changes in gain.

Walter R. Evans, an assistant professor in the Department of Electrical Engineering, Washington University, St. Louis, Missouri, conceived the idea that is now known as the root locus method as a simple graphical means of finding the transient response of a control system. He gave credit to P. Profos, who in the Sulzer Technical Review in 1945 pointed out that the error/output is a function of a complex variable with frequency as the imaginary part and damping as the real part. Evans argued that 'the Nyquist plot is...one line of a conformal map with the root of the equation being the value of the variable which makes the function equal to -1'.

Using the example of adding a compensating network as shown in Figure 8.5, Evans refers to the tediousness of determining the roots of such a system and argues that obtaining a frequency response plot in the form of e/θ_0, that is the inverse open loop response, either by calculation or by direct measurement, is much simpler. Obtaining such a plot for a particular value of gain K gives the curves shown in Figure 8.6. He notes that as the value of K is changed and the curve brought closer to the -1 point then the system damping is reduced. An experienced designer can estimate from a series of plots the gain required to give a desired transient performance. The key idea put forward by Profos is to consider this vector plot as the base line for the determination of the complex roots of the

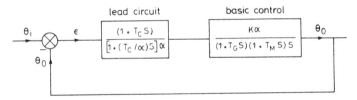

Figure 8.5 *Block diagram of a system considered by Evans*

Reproduced (with partial redrawing) by permission of W.R. Evans, from *AIEE Trans.*, 1948, **67**, p. 549

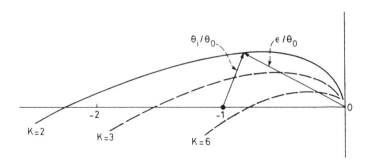

Figure 8.6 *Vector plot of error/output ratio with sinusoidal signal impressed*

Reproduced (with partial redrawing) by permission of W.R. Evans, from *AIEE trans.*, 1948, **67**, p. 549

main damped sinusoid term in the system response. To do this the function in s has to be plotted in terms of its real $(-\sigma)$ and complex $(i\omega)$ parts. Using the analogy of the plotting of an electrostatic field in terms of a complex variable of flux and voltage, he argues that a pattern of curvilinear squares can be sketched on the top of the vector of e/θ_0 which is a function of s. This is shown by Figure 8.7. To ease the problem of sketching the curves and in particular the problem of determining numerical values, Evans suggested using a protractor as shown in Figure 8.8.

In the 1948 paper there is only the sketchiest outline of the method and it was only with the publication of the paper, 'Control System Synthesis by Root Locus Method' in 1950, that the full implications of the method were recognised. Evans had some difficulty in getting this second paper published: it was first submitted in November 1948 and not accepted for publication until November 1949. 'The root locus method', he explained, 'is the result of an effort to determine the roots of the differential equation of a control system by using the concepts now associated with frequency response methods'. In this second paper the technique is explained in a manner similar to that typically used in present day text books, although the rules for sketching loci are now greatly extended. Evans also gives details for the construction of a spirule, a specially constructed protractor that eases the computation of angles, etc. when constructing the loci. He later arranged for the manufacture of spirules and for many years he himself sold them.

8.3 Analogue simulation

The linear design techniques attracted considerable attention but as the 'refinement of design' proceeded with increasing emphasis on accuracy, speed of response and general quality of performance, 'one becomes', C. Concordia of the General Electric

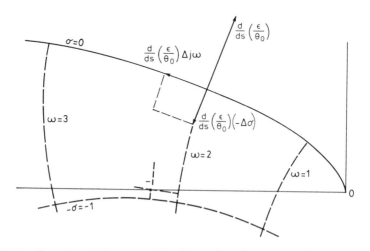

Figure 8.7 *Complex plot for system showing conformal map properties*
Reproduced (with partial redrawing) by permission of W.R. Evans, from *AIEE Trans.*, 1948, **67**, p. 549

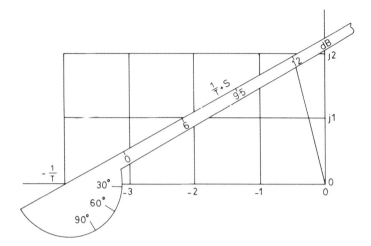

Figure 8.8 *Vector determination of $1/T + s$*

Reproduced (with partial redrawing) by permission of W.R. Evans, from *AIEE Trans.*, 1948, **67**, p. 551

Co. observed during the discussion of Brown's and Hall's 1946 paper, 'primarily interested in studying more and more the effects of those nonlinear elements such as friction, backlash, saturation'.[32] In order to do this Concordia recommended using a differential equation approach with a stability check by means of Routh for the linear analysis with simulation using the differential analyser to allow for the consideration of the nonlinear effects. In their reply Brown and Hall noted that not everyone had access to a differential analyser but that such devices were becoming more readily available. Just before the outbreak and during the war, George A. Philbrick had created considerable interest in simulators through his demonstrations of POLYTHEMUS, the electronic process simulator. With this he had shown that much less complex and cumbersome devices than the mechanical differential analyser could provide useful and informative results. During the war other instrument manufacturing companies began to make use of simulated processes, for example D.P. Eckman and W.H. Wannamaker of the Brown Instrument Co. devised a simple adjustable simulator based on a resistor capacitor network which they used in conjunction with a real controller and valve actuator to investigate the effect of dead-time on the performance of P + I controllers.[33]

Engineers working for the Westinghouse Company investigated much more general methods of representing a variety of systems using electrical analogies and G.D. McCann, S.W. Herwald and W.O. Osbon used such methods to investigate servo problems. Westinghouse developed an electronic analogue computer. Two of them were built initially: one for themselves and one for California College of Technology, Pasadena, where G.D. McCann was now teaching. McCann and colleagues used the analogue computer for a range of investigations including the performance of a servomechanisms with coulomb friction.[34]

8.4 The transition to modern control

The close of the classical era and the beginnings of the transition period leading to modern control theory is marked by two conferences, the Conference on Automatic Control held in July 1951 at Cranfield in the UK, and the other, Frequency Response Symposium, in December 1953 in New York. The first of these, organised by the Department of Scientific and Industrial Research, with the assistance of the IEE and the IMechE, was the first major international conference on automatic control.[35] Arnold Tustin chaired the organising committee, and 33 papers were presented, 16 of which dealt with problems of noise, non-linearity or sampling systems. There were also sessions on analogue computing and the analysis of the behaviour of economic systems (this latter reflecting both the particular interest of Arnold Tustin and the way in which interest in applications of feedback theory was growing).[36]

An important contribution to understanding the behaviour of systems containing relays was the publication in 1953 of the book *Discontinuous Automatic Control* written by I. Flugge-Lotz.[37] This work made more widely available the extensive work on the problem which had been done in Germany.

Work on relay control systems done after about 1950 divides into two categories: that concerned with stability and the determination of limit cycles or system performance for specific inputs (including random signal inputs), and that concerned with the design of optimum relay systems.

Many people realised that by exploiting the nonlinear effects of a relay it should be possible to get a better system performance than with a linear amplifier. Theoretically, it should be possible to improve performance by selecting the appropriate point at which to switch from one relay state to the other. Flugge-Lotz and Klotter[38] investigated the performance of relay systems with a simple linear switching function. They studied a system described by the equation

$$d^2y/dt^2 + 2\zeta dy/dt + y = \text{sgn}(ay + bdy/dt) \quad \text{for} \quad 0 < \zeta < 1 \tag{8.1}$$

That is, the relay forcing function is unity with the same sign as $ay + bdy/dt$. They studied behaviour of the system in the four quadrants of the phase plane and showed that only in the quadrant with both a and b negative did the system seek the desired equilibrium state, but even then it chattered (i.e. oscillation at high frequency and small amplitude) as it approached the equilibrium state. Thus linear switching is not optimal.

For a second order system described by the equations

$$dy/dt = y'$$
$$dy'/dt + g(y, y') = \varphi(y, y') \tag{8.2}$$

where $\varphi(y, y')$ is a discontinuous function with two possible values $+1$ and -1. The general optimum switching problem is to find a function $\varphi(y, y')$ such that from an arbitrary point α in the phase plane the solution will pass through the origin O and the length of time necessary to move along the path α to O is a minimum. D.W. Bushaw investigated this problem for the case when $g(y, y') = 2\zeta y' + y$ and solved it for all real values of ζ. The technique used by Bushaw is complicated and not directly able to be extended to other systems.[39] An attempt to solve the problem for higher order systems was made by Kang and Fett when they tried to determine the switching surfaces in phase space.

The classical control techniques developed during the 1940s were concerned with the servomechanism problem as illustrated in Figure 8.9. The design methodologies were for linear single input systems — that is systems which can be described by linear differential equations with constant coefficients and have a single control input. The design tools were necessarily limited by the computational tools and simulation facilities available. The tools are essentially analytical and graphical.

One set of techniques, the frequency response techniques, based on the use of Nyquist, Bode or Nichols charts, assessed performance in terms of bandwidth, resonances, and gain and phase margins. An alternative set based on use of the Laplace transform expressed performance in terms of rise time, percentage overshoot, steady-state error and damping. The root locus technique provides a method of assessing the time domain performance from construction, that is graphical techniques, based on frequency response ideas.

Experience during the war showed that to obtain high performance — fast response, high dynamic and static accuracy, good rejection of noise and external disturbances — required more than simple linear analysis could provide. In particular non-linearities and noisy signals must be taken into account. The work also showed that if the 'best' performance was to be obtained then some form of model of the plant was needed. The model can be mathematical — equations — or in the form of frequency response data.

It was gradually recognised that the classic representation was not an adequate model of the general control system. The recognition of the more general nature of the problem arose from the immediate post war work on aircraft and missile control systems. In aircraft and missile control the motion can be modified by several different available controls and hence the system is what is now termed multivariable. In general there are several inputs and several outputs and there are interactions between the inputs and outputs.

A second realisation — but one which was implicit in the work of Black and began

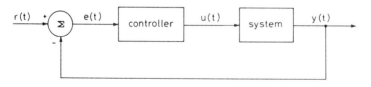

Figure 8.9 *Traditional servomechanism problem*

r(t) = reference input
y(t) = actual output
e(t) = error signal [e(t) = r(t) − y(t)]
u(t) = control input
Redrawn from Athans: 'Perspectives in modern control theory' in Science Technology and the Modern Navy, Thirtieth Anniversary 1946-76, edited by E.I. Solkovitz, Department of the Navy, Office of Naval Research, Arlington, VA., 1976, figure 3

to emerge explicitly during the war — is that systems possess uncertainty. Michael Athans[40] listed these uncertainties as

- errors inherent in modelling a physical system by means of mathematical equations,
- errors in the parameters that appear in differential equations of motion (e.g. the submarine hydrodynamic derivatives),
- exogenous stochastic disturbances that influence the time evolution of the system state variables in a random manner (e.g. the effects of surface waves on submarine depth), and
- sensor errors and related noise in measurements.

Such uncertainties are modelled as random variables and/or random processes and, he notes, the 'complete description of any real physical system requires the use of *stochastic differential equations*'. A typical system is thus represented as shown in Figure 8.10.

As A.J.G. MacFarlane[41] has pointed out, the development of modern control theory was strongly influenced by two factors: first, the nature of the problem that society saw as important — the launching, manoeuvring, guidance and tracking of missiles and space vehicles; and secondly by the advent of the digital computer. The problem was essentially one of the control of ballistic objects and hence detailed physical models could be constructed in terms of differential equations, both linear and non-linear; also measuring instruments and other components of great accuracy and precision could be developed and used. Performance requirements as well as involving positional accuracy also involved constraints that were expressed in the form of optimisation requirements, typically requirements to reach a specified position in minimum time, or to carry out a particular set of manoeuvres with minimum fuel consumption. A consequence was to focus attention once again on the differential equation approach to the analysis and design of control systems. In considering dynamical problems that involve minimising or maximising some performance index there is, MacFarlane points out, 'an obvious and strong analogy

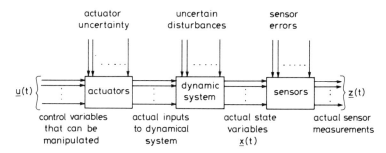

Figure 8.10   *Realistic stochastic dynamic system. From a pragmatic point of view, the only variables available for real-time measurement are control inputs u(t) and sensor measurements z(t)*

> Redrawn from Athans, 'Perspectives in modern control theory' in Science Technology and the Modern Navy, Thirtieth Anniversary 1946–76, edited by E.I. Solkovitz, Department of the Navy, Office of Naval Research, Arlington, VA., 1976, figure 1

with the classical variational formulations of analytical mechanics given by Lagrange and Hamilton'.

To deal with the multivariable nature of the problems, engineers working in the aerospace industries turned to formulating the general differential equations in terms of a set of first order equations. This was a technique of which Poincaré was first to see the significance and first to exploit. The whole approach became known as the 'state-space' approach. The method makes explicit a fundamental characteristic of a dynamical system: the present behaviour is influenced by the past behaviour and thus instantaneous relationships between the present input and present output variables do not specify its behaviour. An additional set of variables, the so-called state variables describe the past behaviour of the system and the present output variables are thus determined by both the state variables and the present input variables. In mathematical notation this is generally now expressed in the form

$$x = Ax + Bu$$
$$y = Cx$$

where x represents the set of state variables, u the set of inputs and y the set of outputs.

The formulation of the control problem in state space led to extensive and deep studies of mathematical problems of automatic control. The growing availability of the digital computer, during the late 1950s, made an algorithmic recursive algorithm solution possible, as opposed to the search for a closed form solution in the classical approach.

Michael Athans has placed the date of the origins of what is now referred to as *modern control theory* as being 1956 and in September of that year an international conference on automatic control, organised by the joint control committee of the VDI and VDE was held in Heidelberg, Germany, 1956. During the conference a group of delegates met and agreed to form an international organisation to promote progress in the field of automatic control. An organising group of seven people — Broida (France), Chairman, Grebe (Germany), Letov (USSR), Nowacki (Poland), Oldenburger (USA), Welbourn (UK), with Ruppel (Germany) as Secretary — was charged with drawing up plans for an international federation. The organisation, The International Federation of Automatic Control, abbreviated to IFAC, was officially formed by the adoption of a constitution at a meeting held in Paris on 11th and 12th September 1957. Attendees at the meeting elected Harold Chestnut as the first president, with A.M. Letov and V. Broida elected as vice presidents, G. Ruppel was chosen as secretary and G. Lehmann as treasurer. At this meeting the Russian delegate extended an invitation to hold the first conference in Moscow in 1960.[42]

The constitution of IFAC restricted membership to one organisation from each country and as a result many countries formed joint councils or committees of existing scientific and technical bodies to represent them as members of IFAC. For example in the USA the AIEE, SME, IRE, ISA and AIChemE joined together to form the American Automatic Control Council.

The Moscow Conference was an important and highly visible symbol of the change in direction that had been slowly developing during the 1950s and it is fitting that at the conference Rudi Kalman presented a paper, 'On the general theory of control systems'[43] that clearly showed 'that a deep and exact duality existed between the problems of multivariable feedback control and multivariable feedback filtering' and hence ushered in a new treatment of the optimal control problem.[44]

The conferences of 1951 and 1953, together with the publication of numerous text books; articles such as Tustin's in *Engineering* in 1950, Brown's in the *Scientific American* in 1951, and numerous articles on control topics in *Mechanical Engineering* during the early 1950s brought automatic control to the attention of engineers. The publication of Wiener's book, *The Human Use of Human Beings* and a series of articles published in *Scientific American* in 1952 attracted the attention of a wider technical community. By the mid-1950s there was a growing general awareness of the potential of automatic control. Many books on the subject intended for the general reader were published and the British Government quietly encouraged a debate on the subject. The emphasis in these popular and semi-popular works was on automation in the sense of mechanisation and remote control of production lines and other assembly processes. There was also great interest in the possibilities of numerical control of machine tools.

Central to this debate were issues that many of the engineers and administrators involved in control system work during the war had intuited — control systems had moved beyond feedback amplifiers and single loop servomechanisms and had become concerned with large scale, complex systems. Gordon Brown and Duncan Campbell, in 1949, laid out clearly what they saw as the areas of application of control in the future:

> Improved automatic control...is the co-ordinated design of plant, instruments, and control equipment. We have in mind more a philosophic evaluation of systems which might lead to the improvement of product quality, to better co-ordination of plant operation, to a clarification of the economics related to new plant design, and to the safe operation of plants in our composite social-industrial community. These general remarks are illustrated by mention that certain industries operating at large production might show appreciable increase in economy and quality on standard production items by improved automatic control. The conservation of raw materials used in a process often prompts reconsideration of control. The expenditure of power or energy in product manufacture is another important factor related to control. The protection of health of the population adjacent to large industrial areas against atmospheric poisoning and water-stream pollution is a sufficiently serious problem to keep us constantly alert for advances in the study and technique of automatic control, not only because of the human aspect but because of the economy aspect.[45]

This they viewed as a long term programme with many technical and human problems that 'may take a decade or more to resolve'. It carried with it the hopes of many but there were also fears. Wiener, writing in 1947, saw the changes brought about by our deepening understanding of feedback mechanisms — cybernetics — as the second industrial revolution and he expressed the fears of many when he wrote 'the modern industrial revolution is...bound to devalue the human brain at least in its simpler and more routine decisions. Of course, just as the skilled carpenter, the skilled mechanic, the skilled dressmaker have in some degree survived the first industrial revolution, so the skilled scientist and the skilled administrator may survive the second. However, taking the second revolution as accomplished, the average human of mediocre attainments or less has nothing to sell that it is worth anyone's money to buy'. Wiener's answer was a 'society based on human values other than buying and selling' but he was not optimistic about the creation of such a society and thought that the best that any of us could do to fulfil our moral responsibility was to work to create the wide and general public understanding.[46]

In this book, although I have not tackled the difficult task of relating the technical changes brought about by the development of feedback theory and devices to social and economic change, I hope that what I have done provides a starting point that will help others with the task. It is an important task, for even though we are nearing the end of this century, we have not solved and do not understand many of the problems that emerged with the first use of automatic control mechanisms at the beginning of the century. As control technology and theory continues to change it is vitally important that we do begin to understand more deeply its social consequences and learn to adapt the technology to society's needs rather than society to the technology.

8.5 Notes and references

1. Quoted from Penick, James L., Pursell, Carroll W., Jr., Sherwood, Morgan B., and Swain, Donald C.: (eds) *The Politics of American Science, 1939 to the Present* (Rand McNally, Chicago, 1965), p. 87. The attitude of the Administration to Bush's proposals are made abundantly clear in a letter from Harold D. Smith, Director of the Bureau of the Budget to Rear Admiral J.A. Furer dated 12th June 1945. He wrote, 'the physical scientists are worried about government and governmental controls largely because most of them — as they make rather clear to me by what they say and do not say — do not know even the first thing about the basic philosophy of democracy... However, most of them have learned to accept governmental funds with ease, and I think they can adapt themselves to governmental organisation with equal ease'. NARS Folder Weaver, Office Files of Harold Hazen.
2. Porter, Arthur: *Memorandum on the 'Application of control systems in industry'*, prepared for the Interdepartmental Technical Committee on Servo Mechanisms, Ministry of Supply, October 1945.
3. PRO AVIA 22/2427 minutes of meeting on 28th November 1946.
4. Porter, A.: 'The design of automatic and manually operated control systems', paper read 28th May 1945, published in *Trans. Society of Instrument Technology*, **1** (1947), pp. 23-42; also published in 1947 was a paper by Farrington, G.H.: 'Automatic temperature control of jacketed pans', *Trans. Society of Instrument Technology*, **1** (1947), pp. 2-22 which had been read on 13th December 1945.
5. Hayes, K.A.: 'Servo-mechanisms: recent history and basic theory', *Trans. Society of Instrument Technology*, **2** (1950), pp. 2-13; Uttley, A.M.: 'Stabilisation of closed loop control systems', *Trans. Society of Instrument Technology*, **2** (1950), pp. 14-18; Tustin, A.: 'Problems to be solved in the development of control systems', *Trans. Society of Instrument Technology*, **2** (1950), pp. 19-27.
6. IEE: Convention on automatic regulators and servo mechanisms, papers published in *J. IEE*, **94** part IIa (1947).
7. IEE Annual report of Council 1950-1, *Proc. IEE*, **98** (1951), p. 242.
8. These included Werey, R.B.: *Instrumentation and Automatic Control in the Oil Refining Industry* (The Brown Instrument Company, Philadelphia, 1941); Weber, R.L.: *Temperature Measurement and Control* (Philadelphia, 1941); Griffiths, R.: *Thermostats and Temperature-Regulating Instruments* (Charles Griffin, London, 1943).

9 Rhodes, T.J.: *Industrial Instruments for Measurement and Control* (McGraw Hill, New York, 1941). Rhodes was Head of Mechanical and Process Development in the Research and Development Division of US Rubber Company.
10 Smith, Ed.S.: *Automatic Control Engineering* (McGraw Hill, New York, 1944).
11 Eckmann, Donald P., 1945.
12 Ziegler, J.G., Nichols, N.B.: 'Optimum settings for automatic controllers', *Trans. American Society of Mechanical Engineers*, **64**, 1942, pp. 759-768. Ziegler, J.G.: ' 'On-the-job' adjustments of air operated recorder-controllers', *Instruments*, **16**, (1941), pp. 394-7, 594-6, 635. Ziegler, J.G., Nichols, N.B.: 'Process lags in automatic control circuits', *Trans. American Society of Mechanical Engineers*, **65**, 1943, pp. 433-444.
13 Herwald, S.W.: 'Considerations in servomechanism design', *Trans. American Institute of Electrical Engineers*, **63**, 1944, pp. 871-876.
14 *Servomechanisms*, Selected Government Research Reports, **5** (HMSO, London, 1951).
15 James, H.M., Nichols, N.B. and Phillips, R.S.: *Theory of Servomechanisms*, **25**, Radiation Laboratory Series (McGraw Hill, New York, 1946).
16 Ferrell, E.B.: 'The servo problem as a transmission problem', *Proc. Institute of Radio Engineers*, **33**, 1945, pp. 763-767.
17 Bode, H.W.: 'Relations between attenuation and phase in feedback amplifier design', *Bell System Technical J.*, **19**, 1940, pp. 421-454.
18 Bode, H.W.: *Network Analysis and Feedback Amplifier Design* (Van Nostrand, Princeton, NJ, 1945).
19 MacColl, L.A.: *Fundamental Theory of Servomechanisms* (Van Nostrand, Princeton, NJ, 1945), reprinted (with new Preface by R.W. Hamming), Dover Books, 1968.
20 Frey, W.: 'A generalisation of the Nyquist and Leonhard stability criteria', *Brown Boveri Review*, **33**(3), 1946, pp. 59-65.
21 Brown, G.S., Hall, A.L.: 'Dynamic behaviour and design of servomechanisms', *Trans. American Society of Mechanical Engineers*, **68**, 1946, p. 517.
22 Hall, A.C.: 'Application of circuit theory to the design of servomechanisms', *J. Franklin Institute*, **242**, 1946, pp. 279-307.
23 Brown, G.S. and Campbell, D.P.: *Principles of Servomechanisms* (Wiley, New York, 1948), p. 235.
24 Harris, H.: 'The frequency response of automatic control systems', *Trans. American Institute of Electrical Engineers*, **65**, 1946, pp. 539-546.
25 Chestnut, Harold, Mayer, R.W.: *Servomechanisms and Regulating System Design*, **1** (Wiley, New York, 1951).
26 Marcy, H.T.: 'Parallel circuits in servomechanisms', *Trans. American Institute of Electrical Engineers*, **65**, 1946, pp. 521-529, discussion p. 1128. Whiteley, A.L.: 'Theory of servo systems with particular reference to stabilisation', *J. Institution of Electrical Engineers*, **93** (Pt. II), 1946, pp. 353-367, discussion pp. 368-372.
27 Thaler, G.J.: *Automatic Control and Classical Linear Theory*, Benchmark Papers in Electrical Engineering and Computer Science (Dowden, Hutchinson and Ross, Stroudsburg, 1974).
28 Whiteley, A.L.: 'Servomechanisms: a review of progress', *Proc. IEE*, **98** (Pt. 1), 1951, p. 290.
29 The magnetic amplifier has a long history, patents appeared during the early years of the century and use gradually grew. It was widely used during the 1950s and early 1960s but with the development of high power semi-conductor devices its use has declined.
30 Whiteley, A.L. *op. cit.*, (n.28), p. 296.
31 Cremer, L.: 'Ein neues Verfahren zur Beurteilung der Stabilitat Linearer Regelungssysteme', *Z. Angew. Math. Mech.*, **26**-7, 1947, pp. 161-163.
32 C. Concordia, contribution to discussion of paper by Brown, *op. cit.*, (n.21), p. 522.
33 Eckmann, Donald P., Wannamaker, W.H.: 'Electrical analogy method for fundamental investigations in automatic control', *Trans. American Society of Mechanical Engineers*, **67**, 1945, p. 81-86.
34 McCann, G.D., Herwald, S.W., Kirschbaum, H.S.: 'Electrical analogy methods applied to servomechanisms', Part I, *Trans. American Institute of Electrical Engineers*, **65**, 1946, pp. 91-96. Herwald, S.W., McCann, G.D.: 'Dimensional analysis of servomechanisms by electrical analogy', Part II, *Trans. American Institute of Electrical Engineers*, **65**, 1946, pp. 636-640. McCann, G.D., Osbon, W.O., Kirschbaum, H.S.: 'General analysis of

speed regulators under impact loads', *Trans. American Institute of Electrical Engineers,* **66**, 1947, pp. 1243-1252. Harder, E.L., McCann, G.D.: 'A large-scale general purpose electric analog computer', *Trans. American Institute of Electrical Engineers,* **67**, 1948, pp. 664-673. McCann, G.D., Lindvall, F.C., Wilts, C.H.: 'The effect of Coulomb friction on the performance of servomechanisms', *Trans. American Institute of Electrical Engineers,* **67**, 1948, pp. 540-546.
35 The proceedings of the conference were published as Tustin, Arnold, (ed): *Automatic and Manual Control* (Butterworth, London, 1952).
36 In his presentation to the session on economic systems Tustin pointed out that the relationships given by J.M. Keynes in *The general Theory of Employment, Interest, and Money* (1935), could be modelled as closed loop feedback systems. This led to an interesting exchange with J.H. Westcott who pointed out that Tustin's model omitted an important loop, the influence of 'news, public opinion or gossip on the behaviour of consumers, and hence on the behaviour of the economic system itself'. Tustin conceded the point that the introduction of such a loop might be necessary. *ibid.* pp. 559-560.
37 Flugge-Lotz, I., *Discontinuous Automatic Control* (Princeton University Press, 1953).
38 See Tsien, 1954, p. 145 for full details.
39 *Ibid.*, pp. 151-158, for an outline of the solution.
40 Athans, Michael: 'Perspectives in modern control theory', in *Science, Technology, and the Modern Navy, Thirtieth Anniversary 1946-1976,* ed. Edward I. Solkovitz (Department of the Navy, Office of Naval Research, Arlington, Va, 1976), **ONR-37**, p. 143.
41 MacFarlane, A.G.J.: 'The development of frequency-response methods in automatic control', *IEEE Trans. Automatic Control,* **AC-24**(2), 1979, pp. 250-265.
42 Bennett, S.: 'The emergence of a discipline: automatic control 194?-1960', *Automatica,* **12**, 1976, pp. 113-121.
43 Kalman, R.E.: 'On the general theory of control systems', *Proc. First IFAC Congress in Moscow,* **1** (Butterworth, London, 1960), pp. 481-492.
44 MacFarlane, A.G.J.: *op. cit.*, (n.41), 1979, p. 258.
45 Brown, G.S., Campbell, D.P.: 'Instrument Engineering: Its Growth and Promise in Process-Control Problems', Fall Meeting of Industrial Instruments and Regulators Division of ASME, September 1949. Published in *Mechanical Engineering,* **72**, 1950, pp. 124-7 and 136, discussion pp. 587-589.
46 Wiener, Norbert: *Cybernetics: or Control and Communication in the Animal and the Machine* (Wiley, New York, 1948), pp. 37-8 (Wiener dated the Introduction from which the quotation is taken, November 1947.)

Bibliography

The bibliography is divided into three sections: primary sources and public records; general works and technical papers. The section containing technical papers is subdivided into years and covers the period 1920 to 1960. It is not a complete bibliography of technical papers that refer to feedback control and associated subjects, but in each year I have selected papers which are either important or which give a flavour of the issues being considered during the period.

Primary sources and public records

The major collections of primary material and public records which have been used are to be found in the following depositories:
Hagley Museum & Library, PO Box 3630, Wilmington, Delaware 19807 Major collections of relevance are the Leeds & Northrup Company collection and the Brown Instrument Company Collection; the library also has holdings of Trade Catalogues.
Massachusetts Institute of Technology, Archives Center, MIT Holds some of Hazen's papers and the records of the Servomechanism Laboratory. Also of value are the papers of the Office of the President of MIT.
Museum of American History, Smithsonian Institution, Washington, DC 2056 The library hold an extensive collection of Trade Catalogues
National Archives, Washington DC 20408 The major collection of relevance is Record Group 227, the records of the Office for Scientific Research and Development. These records cover the period 1941 to 1946.
Navy Yard, Washington DC The Naval Operational Archives.
Public Record Office, London This is the major source of information on government supported activities and is particularly important for the work between 1939 and 1945. Relevant files are to be found in the papers of the Ministry of Supply, Ministry of Aviation, War Office, Admiralty and the Cabinet Office.

General works

AIEE Committee Report, Bibliography on feedback control, *Applications and Industry*, 1954, **10**, pp. 430-462
ALFORD, L.P.: 'Ten years progress in management 1923-1932', *Trans. Amer. Soc. Mech. Eng.*, 1933, **55**, pp. 7-21

AITKEN, H.G.J.: 'The continuous wave: technology and American radio, 1900-1932' (Princeton Press, Princeton, 1985)
ANDERSON, F.W.: 'Orders of magnitude: a history of NACA and NASA 1915-1976'
ANDREW, Alex: 'Norbert Wiener (1894-1964): a discussion and review of a biography by P.R. Mesani', *Kybernetes*, 1991, **20**(1), pp. 11-16
ANDRONOV, A.A.: 'I.A. Vyshnegradskii and his role in the creation of automatic control theory', *Automation and Remote Control*, 1978, Pt 1, **39**(4), pp. 469-478
ANON: 'Growth of automation in the USSR (1917-1957)', *Automation and Remote Control*, 1957, **18**, pp. 569-617
ANON: 'Historical museum on paper of American instruments in science and industry', *Instruments*, 1948, **21**, pp. 35-42, 128-148, 248-250, 334-348
APPLEYARD, R.: 'The History of IEE (1871-1931)' (Institution of Electrical Engineers, London, 1939)
ARBIB, M.A.: 'Cybernetics after 25 years', *IEEE Trans. Systems, Man. and Cybernetics*, 1975, **5**, pp. 359-365
ARTOBOLEVSKII, I.I.: 'Some aspects of the development of the theory of automatic machines by Soviet scientists', *Proc. Inst. Mech. Eng.*, 1967, **182**(1), pp. 825-34
ARTZT, M: 'Survey of d-c amplifiers', *Electronics*, 1945, **18**(8), pp. 112-118
ASTROM, K.J.: 'Process control — past, present and future', *IEEE Control Systems Magazine*, 1985, **5**(3), pp. 3-10
ATHERTON, D.P.: 'Nonlinear systems: history', *in* SINGH, M.G.(Ed.): Systems and Control Encyclopedia (Pergamon, Oxford, 1987), pp. 3383-90
ATHERTON, W.A.: 'From compass to computer' (Macmillan, London, 1984)
AUTOMATIC CONTROL, articles from *Scientific American* (Bell, London, 1955)
BABB, M.: 'Pneumatic instruments gave birth to automatic control', *Control Engineering*, Oct. 1990, **37**(12), pp. 20-22, 24, 26
BAGRIT, L.: 'The age of automation', BBC Reith Lectures 1964 (Weidenfeld & Nicolson, London, 1965)
BAKER, W.J.: 'A history of the Marconi company' (Methuen, London, 1970)
BAXTER, J.P., III: 'Scientists against time' (G.P.O. Washington, 1947)
BELEVITCH, V.: 'Summary of the history of circuit theory', *Proc. Inst. Radio Eng.*, 1962, **50**, pp. 848-855
BELLMAN, R.: 'Eye of the hurricane, an autobiography' (World Scientific, Singapore, 1984)
BELLMAN, R., and KALABA, R. (Eds): 'Selected papers on mathematical trends in control theory' (Dover, New York, 1964)
BELLMAN, R.E., and LEE, E.S.: 'History and development of dynamic programming', *IEEE Control Systems Magazine*, 1984, **4**(4), pp. 24-28. A more detailed account is in BELLMAN, R.: 'Some vistas of modern mathematics' (University of Kentucky Press, Lexington, 1968)
BENNETT, S.: 'The search for "uniform and equable motion": a study of the early methods of control of the steam engine', *Int. J. Control*, 1975, **21**(1), pp. 113-147
BENNETT, S.: 'The emergence of a discipline: automatic control 1940-1960', *Automatica*, 1976, **12**, pp. 113-121
BENNETT, S.: 'A history of control engineering 1800-1930' (Peter Peregrinus, Stevenage, 1979)
BENNETT, S.: 'F.C. Williams, his contribution to the development of automatic control', *Electronics and Power*, Nov.-Dec. 1979, **25**(11), pp. 800-803
BENNETT, S.: 'Nicolas Minorsky and the automatic steering of ships', *IEEE Control Systems Magazine*, 1984, **4**(4), pp. 10-15
BENNETT, S.: 'Harold Hazen and the theory and design of servomechanisms', *Int. J. Control*, 1985, **42**, pp. 989-1012
BENNETT, S.: 'The industrial instrument — master of industry, servant of management: automatic control in the process industries, 1900-1940', *Technology & Culture*, 1991, **32**(1), pp. 69-81
BENNETT, S.: 'Industrial instruments — a brief history', *Measurement & Control*, May 1992, **25**, pp. 111-114
BENNETT, S.: 'The development of process control instruments: the early years', *Transactions of the Newcomen Society*, 1992, **63**, pp. 133-164
BENNETT, W.R.: 'Reminiscences', *IEEE Trans. Circuits & Syst.*, 1977, **24**, pp. 667-669

BERKOVICH, D.M.: 'Twenty-fifth anniversary of the Institute of Automation and Remote Control', *Automation and Remote Control*, 1964, Pt 2, **25**(11), pp. 1473-1479
BERKOVICH, D.: 'From the history of control science', *Automation and Remote Control*, 1971, Pt 1, **32**(6), pp. 1004-1011
BERKELEY, E.C.: 'Giant brains or machines that think' (Chapman and Hall, New York, 1949)
BIRKENHEAD, LORD: 'The Prof. in Two Worlds' (Collins, London, 1961)
BISSELL, C.C.: 'Modelling sampled-data systems: a historical outline', *Trans. Inst. Measurement and Control*, 1985, **7**(3), pp. 159-64
BISSELL, C.C.: 'Karl Kupfmuller: a German contributor to the early development of linear systems theory', *Int. J. Control*, 1986, **44**(4), pp. 977-989
BISSELL, C.C.: 'Control engineer and much more: aspects of the work of Aurel Stodola', *Measurement and Control*, 1989, **22**(4), pp. 117-122
BISSELL, C.C.: 'Stodola, Hurwitz and the genesis of the stability criterion', *Int. J. Control*, 1989, **50**(6), pp. 2313-332
BISSELL, C.C.: 'Secondary sources for the history of control engineering: an annotated bibliography', *Int. J. Control*, 1991, **54**(3), pp. 517-28
BISSELL, C.C.: 'Six decades in control: an interview with Winifried Oppelt', *IEE Review*, January 1992, pp. 17-21
BISSELL, C.C.: 'Pioneers of control: an interview with Arnold Tustin', *IEE Review*, June 1992, pp. 223-226
BLACK, H.S.: 'Inventing the negative feedback amplifier', *IEEE Spectrum*, 1977, **14**, p. 54
BLICKLEY, G.J.: 'Modern control started with Ziegler-Nichols tuning', *Control Eng.*, Oct. 1990, **37**(12), pp. 11, 13, 15, 17
BODE, H.: 'Feedback—the history of an idea', Symposium on Active Networks and Filters, Polytechnic Institute of Brooklyn, 19-21 April 1960 (Polytechnic Press, Brooklyn, NY), pp. 1-17. Reprinted in 'Selected Papers on Mathematical Trends in Control Theory' (see Bellman, R., and Kabala, R.), pp. 106-123
BOLLAY, W.: 'Aerodynamic stability and automatic control', *J. Aeronautical Sci.*, 1951, **18**(9), pp. 569-617, discussion pp. 617-23, 640
BOYCE, J.C.: 'New weapons for air warfare: fire-control equipment, proximity fuzes and guided missiles' (Little Brown, Boston, 1947)
BRAVERMAN, H.: 'Labour and monopoly capital: the degradation of work in the twentieth century' (Monthly Review Press, New York, 1974)
BRITTAIN, J.E.: 'C.P. Steinmetz and E.F.W. Alexanderson: creative engineering in a corporate setting', *Proc. IEEE*, 1976, **64**, pp. 1413-1417
BRITTAIN, J.E.: 'Turning points in American electrical engineering' (IEEE Press, New York, 1977)
BROWN, G.S.: 'Eloge: Harold Locke Hazen 1901-1980', *Annals of the History of Computing*, 1981, **3**(1)
BURCHARD, J.E.: 'QED: MIT in World War II' (J. Wiley, New York, 1948)
BUSH, V.: 'Science the endless frontier' (GPO, Washington, 1945)
BUSH, V.: 'Modern arms and free men' (MIT Press, Cambridge, MA, 1968)
BUSH, V.: 'Pieces of the action' (Morrow, New York, 1970)
CATHERON, A.R.: 'Clesson E. Mason, Oldenburger Medallist, 1973', *Transactions ASME, J. of Dynamic Systems, Measurements, and Control*, 1974, **96**, p. 12
CATHERON, A.R.: 'On "those magnificent men and their controlling machines"', *Trans. Amer. Soc. Mech. Eng.*, Ser. G, *J. of Dynamic Systems, Measurement and Control*, 1975, **97**(4), p. 456
CHESTNUT, H.: 'Automatic control and electronics', *Proc. Inst. Radio Eng.*, 1962, **50**, pp. 787-792
CHRISTMAN, A.B.: 'Sailors, scientists and rockets: history of Naval Weapons Centre China Lake, California (GPO, Washington, 1971)
CLARK, J. (Ed.): 'Technological trends and employment Volume 2 Basic Process Industries (Gower, Aldershot, 1985)
CLARK, R.W.: 'Tizard' (Methuen, London, 1965)
COALES, J.F.: 'Historical and scientific background of automation', *Engineering*, 1956, **182**, pp. 363-370
COCKCROFT, J.D.: 'Memories of radar research' *Proc. IEE*, 1985, **132**, Pt A, pp. 327-339

COHN, N.: 'Recollections of the evolution of real-time control applications to power systems', *Automatica*, 1984, **20**(2), pp. 145-162
CONWAY, H.G.: 'Some notes on the origins of mechanical servo-mechanisms', *Transactions of the Newcomen Society*, 1953-55, **29**, pp. 55-75
CSAKI, F.: 'Survey of previous and recent trends in control engineering', *Periodical of the Polytechnic of Electrical Engineers (Hungary)*, **17**(2), pp. 99-119
DAVIES, I.L.: 'Electronics and the aeroplane', *Proc. IEE*, 1976, **123**, pp. 13-15
DRAPER, C.S.: 'Flight control', 43rd Wilbur Wright Memorial Lecture, *J. Royal Aeronautical Soc.*, 1955, **59**, pp. 451-477
DRAPER, C.S.: 'The control of flight: automation in the air', *Engineering*, 1955, **179**, p. 652
DUFFY, M.C.: 'Technomorphology and the Stephenson traction system', *Transactions of the Newcomen Society*, 1983, **54**, pp. 55-78
DUMMER, G.W.A.: 'Electronic inventions 1745-1976' (Pergamon Press, Oxford, 1977)
DUPREE, A.H.: 'Science in the federal government' (Bellknap Press, Cambridge, 1957)
EAMES, C., and EAMES, R.: 'A computer perspective' (Harvard University Press, 1973)
FAGEN, M.S. (Ed.): 'A history of engineering and science in the Bell System: the early years (1875-1925)' (Bell Telephone Laboratories, Murray Hill, NJ, 1975)
FAGEN, M.S. (Ed.): 'A history of engineering and science in the Bell System: national service in war and peace (1925-1975)' (Bell Telephone Laboratories, Murray Hill, NJ, 1979)
FISCHER, C.S.: 'Touch somebody: the telephone industry discovers sociability', *Technology & Culture*, 1988, **29**, 37
FLUGGE-LOTZ, I.: 'Memorial to N. Minorsky', *IEEE Trans. Automatic Control*, 1971, **16**, pp. 289-91
FOXBORO COMPANY, THE: 'A little of ourselves' (Foxboro Instrument Company, Foxboro)
FUCHS, A.M.: 'A bibliography of the frequency-response method as applied to automatic-feedback-control systems', *Trans. Amer. Soc. Mech. Eng.*, 1954, **76**(8), pp. 1185-1194
FULLER, A.T.: 'Directions of research in control', *Automatica*, 1963, **1**, pp. 289-96
FULLER, A.T. (Ed.): 'Stability of Motion' (Taylor & Francis, London, 1975)
FULLER, A.T.: 'Linear control of nonlinear systems', *Int. J. Control*, 1967, **5**(3), pp. 197-243
FULLER, A.T.: 'The early development of control theory', *Trans. Amer. Soc. Mech. Eng., J. of Dynamic Systems, Measurement, and Control*, 1976, **98**, pp. 109-118, 224-235
FULLER, A.T.: 'Edward John Routh', *Int. J. Control*, 1977, **26**(2), pp. 169-173
GALE, F.W.J.: 'The evolution of the automatic helmsman', *Automation Progress*, 1958, **10**, pp. 374-377
GARCIA Y ROBERTSON, R.: 'Failure of the heavy gun at sea', *Technology & Culture*, 1987, **28**(3), pp. 539-557
GHELFI, G.: 'A classified bibliography of major books and papers on the history of the development of automatic control systems theory and practice', M.S.E.E. Thesis, University of Wisconsin, Madison, USA, 1978
GILBERT, K.R.: 'The control of machine tools — a historical survey', *Transactions of the Newcomen Society*, 1971-2, **44**, pp. 119-127
GODWIN, G.S.: 'Marconi 1939-45: a war record' (Chatto & Windus, London, 1946)
GOLDSTINE, H.H.: 'The computer from Pascal to von Neumann' (Princeton University Press, Princeton, 1972)
GORZ, A.: 'Farewell to the working class' (Pluto Press, London, 1982)
GOSSICK, B.: 'Heaviside and Kelvin — study and contrast', *Annals of Science*, 1976, **33**(3), pp. 275-87
GREEN, C.M., THOMSON, H.C., and ROOTS, P.C.: 'The Ordnance Department: planning munitions for war, US Army in World War II' (GPO, Washington, 1955)
GROSS, D.: 'Historical perspective [of operations research]', *J. Washington Academy of Sciences*, 1989, **79**(2), pp. 47-60
GUERLAC, H.: 'Radar in World War II'
HACKMANN, W.: 'Seek & strike: sonar, anti-submarine warfare and the Royal Navy 1914-54' (HMSO, London, 1984)
HALL, A.C.: 'Early history of the frequency-response field', *Trans. Amer. Soc. Mech. Eng.*, 1954, **76**(8), pp. 1153-4

HARTCUP, G.: 'The challenge of war: scientific and engineering contributions to World War II' (David & Charles, Newton Abbot, 1970)
HAWKINS, L.A.: 'The story of General Electric research' (General Electric Company, Schenectady, 1950)
HEIMS, S.J.: 'John von Neumann and Norbert Wiener' (MIT Press, Cambridge, MA, 1980)
HERWALD, S.W.: 'Recollections of the early development of servomechanism and control systems', *IEEE Control Systems Magazine*, 1984, **4**(4), pp. 29-32
HIGGINS, W.H.C., HOLBROOK, B.D., and EMLING, J.W.: 'Defense research at Bell Laboratories', *Annals of the History of Computing*, 1983, **4**(3), pp. 218-245
HOGG, I.V.: 'Anti-aircraft: a history of air defence' (McDonald & Jane's, London, 1979)
HOGG, I.V.: 'British and American artillery of World War 2' (Hippocrene, New York, 1979)
HOROWITZ, I.: 'History of personal involvement in feedback control history', *IEEE Control Systems Magazine*, 1984, **4**(4), pp. 22-23
HUGHES, T.P.: 'Elmer Sperry: inventor and engineer' (Johns Hopkins Press, Baltimore, 1971)
HUGHES, T.P.: 'Networks of power: electrification in western society 1880-1930' (Johns Hopkins Press, Baltimore, 1983)
HUGHES, T.P.: 'Model builders and instrument makers', *Science in Context*, 1988, **2**(1), pp. 59-75
JONES, R.V.: 'Instruments and advancement of learning', *Trans. Soc. of Instrument Technology*, 1967, **19**, pp. 3-11
JONES, R.V.: 'More and more about less and less', *Proc. Royal Institute*, **43**, pp. 323-345
JONES, R.V.: 'The pursuit of measurement', *Proc. IEEE*, 1970, **117**, pp. 1185-1191
JONES, R.V.: 'Influence of two world wars on the organization of science', *Proc. Royal Society of London*, 1975, **342**, Series A
JURY, E.I.: 'Stability tests for one, two and multi-dimensional linear systems', *Proc. IEEE*, 1977, **124**(12), pp. 1237-40. Reprinted in *Trans. Amer. Soc. Mech. Eng.*, Ser. G, *J. of Dynamic Systems, Measurement and Control*, 1978, **100**, pp. 105-109
JURY, E.I.: 'Sampled-data systems revisited: reflections, recollections and reassessments', *Trans. Amer. Soc. Mech. Eng.*, Ser. G, *J. of Dynamic Systems, Measurement and Control*, 1980, **102**(4), pp. 208-216
JURY, E.I.: 'On the history and progress of sampled-data systems', *IEEE Control Systems Magazine*, 1987, **7**(1), pp. 16-21
JURY, E.I., and TSYPKIN, Y.Z.: 'On the theory of discrete systems', *Automatica*, 1971, **7**(1), pp. 89-107
KHRAMOI, A.V.: *'History of automation in Russia before 1917'* (English translation: Jerusalem (1969) (1956))
KINSEY, G.: 'Orfordness — secret site: a history of the establishment 1915-1980' (Terence Dalton, London, 1981)
KITAMORI, T. et al.: 'Control engineering in Japan: past and present', *IEEE Control Systems Magazine*, 1984, **4**(4), pp. 4-10
KLEMM, F.: 'A history of western technology' (Allen & Unwin, London, 1959)
KNAPP, C.H.: 'In memoriam Kochenburger, R.J., 1919-1980' *IEEE Trans. Automatic Control*, 1982, **27**(6)
LAYTON, E.T.: 'Scientists and engineers, the evolution of the IRE', *Proc. IEEE*, 1976, **64**, pp. 1390-1392
LEFKOWITZ, I.: 'Tribute: Don Eckman and his impact on process control', *IEEE Control Systems Magazine*, 1984, **4**(4), pp. 32-34
LEVIDOW, L., and YOUNG, R.: 'Science and technology and the labour process, Marxist Studies Volume 1' (CSE Books, London, 1981)
LILIENFELD, R.: 'The rise of systems theory: an ideological analysis' (J. Wiley, New York, 1978)
LUTZEN, J.: 'Heaviside's operational calculus and the attempts to rigorise it', *Archive for History of the Exact Sciences*, 1979-80, **21**, pp. 161-200
MABON, P.C.: 'Mission communications — the story of Bell Laboratories' (Bell Laboratories, Murray Hill, NJ, 1975)
MacFARLANE, A.G.J.: 'The development of frequency-response methods in automatic control', *IEEE Trans. Automatic Control*, 1979, **24**, pp. 250-265

MACLAURIN, W.R.: 'Invention and innovation in the radio industry' (Macmillan, New York, 1949)
McMAHON, A.M.: 'The making of a profession: a century of electrical engineering in America (IEEE Press, New York, 1984)
MACMILLAN, R.H.: 'The literature of control engineering', *Engineering*, June 1955, **179**, p. 687
MASANI, P.R.: 'Norbert Wiener 1894-1964' (Birkhauser Verlag, Basel, 1990)
McNEILL, W.H.: 'The Pursuit of Power: technology armed force and society since AD 1900' (Blackwell, Oxford, 1983)
McRUER, GRAHAM, D.: 'A historical perspective for advances in flight control systems', *AGARD—Advances in Control Systems*, Conference Report No. 137, 1974, pp. 2.1-2.7
McRUER, GRAHAM, D.: 'Eighty years of flight control: triumphs and pitfalls of the systems approach', *Amer. Inst. Aeronautics and Astronautics, J. of Guidance and Control*, 1981, **4**, pp. 353-362
MASKREY, R.H., and THAYER, W.J.: 'Brief history of electro-hydraulic servomechanisms', *Trans. Amer. Soc. Mech. Eng., J. of Dynamic Systems, Measurement, and Control*, 1978, **100**(2), p. 110
MAYR, O.: 'The origins of feedback control' (MIT Press, Cambridge, MA, 1970)
MAYR, O.: 'Feedback mechanisms in the historical collections of the National Museum of History and Technology' (Smithsonian Institution Press, Washington DC, 1971)
MAYR, O.: 'Adam Smith and the concept of the feedback system', *Technology and Culture*, 1971, **12**, pp. 1-22
MAYR, O.: 'Victorian physicists and speed regulation: an encounter between science and technology', *Notes and Records of the Royal Society of London*, 1971, **26**, pp. 205-228
MAYR, O.: 'James Clerk Maxwell and the origins of cybernetics', *Isis*, 1971, **62**, pp. 425-444
MAYR, O.: 'Yankee practice and engineering theory: Charles T. Porter and the dynamics of the high speed steam engine', *Technology and Culture*, 1975, **16**, pp. 570-602
MAYR, O. (Ed.): 'Philosophers and machines' (Science History Publications, New York, 1976)
MAYR, O.: '*Authority, liberty and automatic machinery in early modern Europe*' (Johns Hopkins, Baltimore, MD, 1987)
MESCH, F.: 'The contribution of systems theory and control engineering to measurement science', VIIth IMEKO Congress, London 1976
MILLER, J.A.: 'Men and volts at war: the story of General Electric in World War II' (McGraw-Hill, New York, 1947)
MILLER, J.A.: 'Workshop of engineers: story of the General Electric Company 1895-1952' (General Electric, Schenectady, 1953)
MILLMAN, S.: 'A history of engineering and science in the Bell System: Communication Sciences (1925-1980)' (AT & T Bell Laboratories, Murray Hill, NJ, 1984)
NAHIN, P.J.: 'Oliver Heaviside: genius and curmudgeon', *IEEE Spectrum*, 1983, **20**(7), pp. 63-69
NAHIN, P.J.: 'Oliver Heaviside: sage in solitude' (IEEE Press, New York, 1988)
NOBLE, D.: 'America by Design' (A.A. Knopf, New York, 1977)
NOBLE, D.: '*Forces of production: a social history of industrial automation*' (Alfred A. Knopf (Ed.), New York, 1984)
NYQUIST, H.: 'The regeneration theory', *Trans. Amer. Soc. Mech. Eng.*, 1954, **76**(8), p. 1151
OLDENBURGER, R.F.: 'IRD frequency-response symposium: foreword', *Trans. Amer. Soc. Mech. Eng.*, 1954, **76**(8), pp. 1145-1149
OGBURN, W.F., MERRIAM, J.C., and ELLIOTT, E.C.: 'Technological trends and national policy: including the social implications of new inventions' Report of Subcommittee on Technology to the National Resources Committee (GPO, Washington, DC, 1937)
O'NEILL, E.F.: 'A history of engineering and science in the Bell System: transmission technology (1925-1975)' (AT & T Bell Laboratories, Murray Hill, NJ, 1985)
OPPELT, W.: 'A historical review of autopilot development, research and theory in Germany', *Trans. Amer. Soc. Mech. Eng., J. of Dynamic Systems, Measurement, and Control*, 1976, **98**(3), pp. 213-223

OPPELT, W.: 'On the early growth of conceptual thinking in control system theory—the German role up to 1945', *IEEE Control Systems Magazine*, 1984, **4**(4), pp. 16-22

OWENS, LARRY: 'Vannevar Bush and the differential analyzer: the text and context of an early computer', *Technology and Culture*, 1986, **27**, pp. 63-95

PADFIELD, P.: 'Guns at sea' (St. Martin's Press, New York, 1974)

PAYNTER, H.M.: 'In memoriam, George A. Philbrick (1913-1974)', *Trans. Amer. Soc. Mech. Eng., J. of Dynamic Systems, Measurement, and Control*, 1975, **97** pp. 213-215

PENICK, JAMES L., PURSELL, CARROLL W., Jr., SHERWOOD, MORGAN B., and SWAIN, DONALD C. (Eds): 'The politics of American science, 1939 to the present' (Rand McNally, Chicago, 1965)

PERAZICH, G., SCHIMMEL, H., and ROSENBERG, B.: 'Industrial instruments and changing technology', Work Projects Administration, National Research Project, Report M-1 (Philadelphia, 1938). Reprinted in COHEN, I. BERNARD (Ed.): 'Research and Technology' (Arno Press, New York, 1980)

PERAZICH, G., and FIELD, P.M.: 'Industrial research and changing technology', Work Projects Administration, National Research Project, Report M-4 (Philadelphia, 1940). Reprinted in 'Research and Technology', *ibid.*

PIERCE, J.R.: 'The early days of information theory', *IEEE Trans. Information Theory*, 1973, **19**(1), pp. 3-8

POLAK, P.: 'An historical survey of computational methods in optimal control', *SIAM Review*, 1973, **15**(2), p. 553

PORTER, A.: 'The servo-panel—a unique contribution to control systems engineering', *Electronics and Power*, Oct. 1965, pp. 330-333

POSTAN, M.M., HAY, D., and SCOTT, J.D.: 'Design and development of weapons: studies in government and industrial organization' (HMSO, London, 1964)

PROFOS, P.: 'Professor Stodola's contribution to control theory', *Trans. Amer. Soc. Mech. Eng.*, Ser. G, *J. of Dynamic Systems, Measurement, and Control*, 1976, **98**(2), pp. 119-20

RAMSEY, A.R.J.: 'The thermostat or heat governor: an outline of its history', *Transactions of the Newcomen Society*, 1945-1947, **25**, pp. 53-72

RANT, B.: 'Technical change and British Naval Policy' (Hodder & Stoughton, London, 1977)

READER, W.J.: 'A history of the Institution of Electrical Engineers 1871-1971' (Peter Peregrinus, Stevenage, UK, 1987)

REINTJES, J.F.: 'Numerical control: making a new technology' (Oxford University Press, New York, 1991)

REICH, L.S.: 'The making of American industrial research: science and business at GE and Bell, 1876-1926' (Cambridge University Press, 1985)

ROBINSON, E.A.: 'A historical perspective of spectrum estimation', *Proc. IEEE*, 1982, **70**(9), pp. 885-907

RORENTROP, K.: 'Entwicklung der modernen Regelungstechnik' (Oldenbourg, Munich, 1971)

ROSENBERG, N.: 'Perspectives on technology' (Cambridge University Press, 1976)

ROSENBERG, N.: 'Inside the Black Box: technology and economics' (Cambridge University Press, 1982)

ROSENBROCK, H.H.: 'On the history of system zeros and the work of Kronecker', Proceedings of the 27th Institute of Electrical and Electronics Engineers Conference on Decision Control, 1988, pp. 887-889

ROTHSCHILD, J. (Ed.): 'Machina ex dea: feminist perspectives on technology' (Elmsford, New York, 1983)

ROWE, A.P.: 'One story of radar' (Cambridge University Press, 1948)

ROWLAND, B., and BOYD, W.B.: 'US Navy Bureau of Ordnance in World War II' (Bureau of Ordnance, Department of Navy, Washington, DC, 1953)

SADLER, P.: 'Social implications of automation', *Proc. Inst. Mech. Eng.*, 1972, **186**, pp. 141-147

SAHAL, D.: 'Patterns of technological innovation' (Addison Wesley, New York, 1981)

SAIN, M.K., and SCHRADER, C.B.: 'The role of zeros in the performance of multiinput, multioutput feedback systems', *IEEE Transactions on Education*, 1990, **33**(3), pp. 244-257

SCOTT, J.D.: 'Vickers: a history' (Weidenfeld & Nicolson, London, 1962)

SHARLIN, H.I.: 'The making of the electrical age from the telegraph to automation' (Abelard Schaum, London, 1963)

SHAW, T.: 'The conquest of distance by wire telephony', *Bell System Technical Journal*, 1944, **2**, pp. 337-42
SIEMASZKO, Z.A.: 'Control before the 20th century — Part 1', *Control and Instrumentation*, July 1969, **1**(3), pp. 65-67
SIEMASZKO, Z.A.: 'Control before the 20th century — Part 2', *ibid.* Aug. 1969, **1**(4), pp. 45-47
SILJAK, D.D.: 'Alexander Michailovich Liapunov (1857-1918)', *Trans. Amer. Soc. Mech. Eng.*, Ser. G, *J. of Dynamic Systems, Measurement, and Control*, 1976, **98**(2), pp. 121-22
SIMPSON, R.J., and POWER, H.M.: 'Electric arc light, an episode in the history of feedback and control', *Electronics and Power*, 1979, **25**(9), pp. 645-650
SINCLAIR, B.: 'Notions on engineering professionalism in America', *Technology & Culture*, 1986, **27**(4)
SOBEL, R.: 'The age of giant corporations: a microeconomic history of American business 1914-1970' (Greenwood Press, Westport, CT, 1972)
SOULE, G.: 'What automation does to human beings' (Sidgwick and Jackson, London, 1956)
STEVENS, R.A.: 'Control engineering: historical background', *Proc. IEE*, 1974, **121**(5), p. 396
STEWART, C.A.: 'P.J. Daniell', *J. of the London Mathematical Society*, 1947, **22**, pp. 75-80
STEWART, I.: 'Organizing scientific research for war: the administrative history of the Office of Scientific Research and Development' (Little Brown, Boston, 1948)
STOCK, J.T.: 'Pneumatic process controllers: the early history of some basic components', *Transactions of the Newcomen Society*, 1984-5, **56**, pp. 169-77
STOCK, J.T.: 'Pneumatic process controllers: the ancestry of the proportional-integral-derivative controller', *Transactions of the Newcomen Society*, 1987-8, **59**, pp. 15-29
SUMIDA, JON TESTURO: 'British capital ship design and fire control in the Dreadnought Dra: Sir John Fisher, Arthur Hungerford Pollen and the battle cruiser,' *J. Modern History*, 1979, **5**(2), pp. 205-230
SWORDS, S.S.: 'Technical history of the beginnings of RADAR' (Peter Peregrinus, London, 1986)
SYDENHAM, P.H.: *'Measuring instruments: tools of knowledge and control'* (Peter Peregrinus, Stevenage, 1979)
TERRELL, EDWARD: 'Admiralty brief: the story of inventions that contributed to victory in the Battle of the Atlantic' (London, 1958)
THALER, G.J.: 'Automatic control and classical linear theory', Benchmark Papers in Electrical Engineering and Computer Science (Dowden, Hutchinson and Ross, Stroudsburg, 1974)
THIESMEYER, L., and BURCHARD, J.E.: 'Combat scientists: science in World War II' (OSRD, Boston, 1947)
THOMSON, H.C., and MAYO, L.: 'The technical services: The Ordnance Department: Procurement and Supply' (GPO, Washington, DC, 1960)
TUCKER, D.G.: 'The history of positive feedback: the oscillating audion, the regenerative receiver, and other applications up to around 1923', *Radio and Electronic Engineering*, 1972, **42**, pp. 69-80
URE, A.: 'The philosophy of manufacturers: or an exposition of the scientific, moral and commercial economy of the factory system of Great Britain' (2nd edn.) (London, 1835)
USELDING, P.: 'Studies of technology in economic history', in GOLLMAN, ROBERT E. (Ed.): Research in economic history, Supplement 1, Recent developments in the study of business, pp. 159-219
VAN VALKENBURG, M.E. (Ed.): 'Circuit theory: foundations and classical contributions' (Dowden, Hutchinson & Ross, Stroudsburg, 1981)
WADDINGTON, C.H.: 'O R in World War II' (London, 1973)
WARD, J.E.: 'Predecessors of the IEEE Control Systems Society', *IEEE Control Systems Magazine*, 1987, **7**(1), pp. 76-77
WARREN, C.A.: 'In defence of the nation', *Bell Laboratory Record*, 1975, **53**, pp. 96-107
WATSON-WATT, R.A.: 'Three steps to victory' (Odhams, London, 1957)
WEAVER, WARREN: 'Scene of change: a lifetime in American science' (New York, 1970)
WEST, J.C.: 'Forty years in control', *Proc. IEE*, 1985, Pt A, **132**(1), pp. 1-8
WILDES, K., and LINDGRAN, N.: 'The history of electrical engineering and computer science at MIT 1882-1982'

WILKES, M.: 'Memoirs of a computer pioneer' (MIT Press, Cambridge, MA, 1985)
WILKINSON, B.: 'The shopfloor politics of new technology' (Heinemann, London, 1983)
WILLIAMS, A.J.: 'Bits of recorder history', *Trans. Amer. Soc. Mech. Eng., J. of Dynamic Systems, Measurement, and Control*, 1973, **1**, pp. 6-16
WILLIAMS, M.A.: 'A history of computing technology' (Prentice Hall, Englewood Cliffs, NJ, 1985)
WILLIAMS, T.J.: 'Two decades of change — review of the 20 year history of computer control', *Control Engineering*, 1977, **24**(9), pp. 71-6; *Oil and Gas Journal*, 1977, **75**(12), pp. 83-9
WOOD, A.B.: 'From board of invention and research to Royal Naval Scientific Service', *J. Royal Naval Scientific Service*, 1965, **20**, p. 59
WOODBURY, D.O.: 'Battlefronts of industry: Westinghouse in World War II' (Wiley, NY, 1948)
WOOLF, ARTHUR G.: 'Electricity, productivity, and labor saving: American manufacturing, 1900-1929', *Explorations in Economic History*, 1984, **21**(2), pp. 176-191
ZIEGLER, J.G.: 'Those magnificent men and their controlling machines', *Trans. Amer. Soc. Mech. Eng., J. of Dynamic Systems, Measurement, and Control*, Sept. 1975, **97**, pp. 279-80

Selected technical publications

Abbreviations

AIEE	Journal of the American Institute of Electrical Engineers
AIEE	Transactions of the American Institute of Electrical Engineers
ASME	Transactions of the American Society of Mechanical Engineers
BSTJ	Bell System Technical Journal
C.S.Memo	Metropolitan-Vickers Electrical Company Ltd., Control Section Memorandum
GER	General Electric Review
IEE	Journal of the Institution of Electrical Engineers
IRE	Proceedings Institute of Radio Engineers
IMechE	Proceedings of the Institution of Mechanical Engineers
INA	Transactions of the Institute of Naval Architects
JFI	Journal of the Franklin Institute
JSI	Journal of Scientific Instruments
RSI	Review of Scientific Instruments
SIT	Transactions of the Society of Instrument Technology

1920

BUSH, V.: 'A simple harmonic analyzer', *AIEE*, 1920, **39**, p. 903

CAMPBELL, G.A., and FOSTER, R.M.: 'Maximum output networks for telephone substation and repeater circuits', *AIEE*, 1920, **39**, pp. 231-280

DANIELL, P.J.: 'Further properties of the general integral', *Ann. of Math.*, 1920, **21**, pp. 203-220

DOHERTY, R.E.: 'Oscillating frequency of two dissimilar synchronous machines', *GER*, 1920, **23**, pp. 125-129

ELLIS, A.L., and ST. CLAIR, B.W.: 'Inherent regulation of continuous current circuits', *AIEE*, 1920, **39**, pp. 309-330, 331-336

JACKSON, P.R.: 'The stabilization of ships by means of gyroscopes', *INA*, 1920, **62**, pp. 83-88, 88-92

LEWIS, W.W.: 'A new short circuiting calculating table', *GER*, 1920, **23**, pp. 669-71

TAYLOR, G.I.: 'Diffusion by continuous movements', *Proc. London Math. Soc.*, 1920, **20**, pp. 279-289

VAN DER POL, B.: 'A theory of the amplitude of free and forced triode vibrations', *Radio Review*, 1920, **1**, p. 701

1921

APPLETON, E.V., and VAN DER POL, B.: 'On the form of free triode vibrations', *Phil. Mag.*, 1921, **42**, pp. 201-220

BAUM, F.G.: 'Voltage regulation and insulation for large power, long distance transmission systems', *AIEE*, 1921, **40**, pp. 1017-1077

COLPITTS, E.H., and BLACKWELL, O.B.: 'Carrier current telephony and telegraphy', *ibid.*, pp. 205-300

DEVENDORF, N.L.: 'Speed regulation in the hydraulic plant', *Power*, 1921, **54**, pp. 764-767

DUNCAN, R.D.: 'Stability conditions in vacuum tube circuits', *Physical Review*, 1921, **77**, p. 302

PAULY, R.: 'Some methods of obtaining adjustable speed with electrically driven rolling mills', *GER*, 1921, p. 422

POCOCK, L.C.: 'Distortion in thermionic tube circuits', *Electrician*, 1921, **86**, p. 246

SCHEYER, E.: 'Control of machines by perforated records', *American Machinist*, 1921, **55**, pp. 743-747

SCHUR, J.: 'On algebraic equations which only contain roots with negative real parts', *Z. fur Angewandte Mathematik und Mechanik*, 1921, **1**(4), pp. 307-311

TOLLE, M.: 'Regelung der Kraftmaschinen' (Springer, Berlin, 1921, 3rd edn.)

1922

'The latest Sperry gyro stabilizer installation', *J. Amer. Soc. Naval Eng.*, 1922, **34**, pp. 487-489

GRADENWITZ, A.: 'Self-steering vessels: recent German developments in automatic gyro practice', *Scientific American*, 1922, **137**, p. 96

GRIFFIN, R.S.: 'Radio ship control', *Mech. Eng.*, 1922, **44**, pp. 43-44, 70

HOWARD, H.S.: 'Hydraulic steering gears', *J. Amer. Soc. Naval Eng.*, 1922, **34**, pp. 259-279

ISSERTELL, H.G.: 'The control of blower motors', *GER*, 1922, **XXV**, pp. 288-295

MINORSKY, N.: 'Directional stability of automatically steered bodies', *J. Amer. Soc. Naval Eng.*, 1922, **34**(2), pp. 280-309

SPERRY, E.A.: 'Automatic Steering', *Trans. Soc. Naval Arch. & Marine Eng.*, 1922, **XXX**, pp. 53-57

1923

BETHENOD, J.F.J.: 'Distortion free telephone receivers', *IRE*, 1923, **11**, p. 163

BUSH, V.: 'Transmission line transients', *AIEE*, 1923, **42**, pp. 878-893

BUSHMAN, V.: 'Adjustable speed main roll drives', *GER*, 1923, p. 681

CLARK, A.B.: 'Telephone transmission over long cable circuits', *AIEE*, 1923, **42**, pp. 86-97

DULA, H.: 'Sur les cycles limites', *Bull. Soc. Mat. de France*, 1923, **51**

KERSHAW, J.B.E.: 'The control of boilers by temperature and draught measurements', *The Engineer*, April 1923, **135**, pp. 437-441

SCHURING, O.R.: 'A miniature A-C transmission system for the practical solution of network and transmission problems', *AIEE*, 1923, **XLII**, pp. 831-840

SPERRY, E.A.: 'The gyro ship-stabilizer', *J. Soc. Naval Arch.*, 1923, **XXXII**, pp. 232-248

1924

BALEK, D.K.: 'The application of the saturated core reactor and regulator', *AIEE*, 1924, **43**, pp. 937-940

BOYAJIAN, A.: 'Theory of DC excited iron-core reactors and regulators', *ibid.*, pp. 919-936

CLINKER, R.C.: 'A dynamic model of a valve and oscillating circuit', *JIEE*, 1924, **62**, pp. 125-128

EVANS, R.D., and BERGVALL, R.C.: 'Experimental analysis of stability and power limitations', *AIEE*, 1924, **43**, pp. 39–58, discussion pp. 71–103
FORTESCUE, C.L., and WAGNER, C.F.: 'Some theoretical considerations of power transmission', *ibid.*, pp. 16–23, discussion pp. 71–103
FOSTER, C.J.: *BSTJ*, 1924, **3**
FRIIS, H.T., and JENSEN, A.G.: 'High frequency amplifiers', *ibid.*, pp. 181–205
GATES, S.B.: 'Notes on the aerodynamics of automatic directional control', *RAE Rep. No. BA 487*, 19 Feb. 1924
GATES, S.B.: 'Notes on the aerodynamics of an attitude elevator control', *RAE Rep. No. BA 494*, 19 March 1924
HORTON, J.W.: 'Vacuum tube oscillators', *BSTJ*, 1924, **3**, pp. 508–24
NYQUIST, H.: 'Certain factors affecting telegraph speed', *AIEE*, 1924, **43**
POCOCK, L.C.: 'Faithful reproduction in radio-telephony', *JIEE*, 1924, **62**, pp. 791–815
TURNER, L.B.: 'The relations between damping and speed in wireless reception', *JIEE*, 1924, **62**, pp. 192–201, discussion pp. 202–207

1925

'Automatic regulators', *Engineer*, March 1925, **139**, pp. 351–52
BILES, H.J.R.: 'Model experiments with anti-rolling tanks', *INA*, 1925, **67**, pp. 179–188, discussion pp. 189–206
BUSH, V., and BOOTH, R.D.: 'Power system transients', *AIEE*, 1925, **XLIV**, pp. 80–97, discussion pp. 97–103
CARR, L.H.A.: 'The use of induction regulators in feeder circuits', *JIEE*, 1925, **63**, pp. 864–873, discussion pp. 874–876
CRISSON, G.E.: 'Irregularities in loaded telephone lines', *BSTJ*, 1925, **4**
FORTESCUE, C.L.: 'Transmission stability, analytical discussion of some factors entering into the problem', *AIEE*, 1925, **44**, pp. 984–1003
HARVEY, H.F., and THAN, W.E.: 'Electric propulsion of ships', *AIEE*, 1925, **XLIV**, pp. 497–522
KELLOGG, E.W.: 'Design of non-distorting power amplifiers', *AIEE*, 1925, **XLIV**, pp. 302–17
SPENCER, H.H., and HAZEN, H.L.: 'Artificial representation of power systems', *AIEE*, 1925, **44**, p. 72
TUSTIN, A.: 'Economics and industrial electrification', *JIEE*, 1925, **62**, pp. 1141–1146

1926

CLARK, E.: 'Steady state stability in transmission systems, calculation by means of equivalent circuits and circle diagrams', *AIEE*, 1926, **XLV**, pp. 22–41
EVANS, R.D., and WAGNER, C.F.: 'Studies of transmission stability', *ibid.*, pp. 51–94
FLETCHER, H.: 'The theory and the operation of the howling telephone with experimental confirmation', *BSTJ*, 1926, **5**
GARNER, H.M.: 'Lateral stability with special reference to controlled motion', *ARC R&M 1077*, Oct. 1926
PUTMAN, H.V.: 'Synchronizing power in synchronous machines under steady and transient conditions', *AIEE*, 1926, **XLV**, pp. 1116–1130
SHIRLEY, O.E.: 'Stability characteristics of alternators', *ibid.*, pp. 1108–1115
STEIN, T.: 'Regelung und Ausgleich in Dampfanlagen' (Springer, Berlin, 1926)
VAN DER POL, B.: 'On relaxation oscillations', *Phil. Mag.*, 1926, **7**, p. 978

1927

BUSH, V., GAGE, F.D., and STEWART, H.R.: 'A continuous integraph', *JFI*, 1927, **211**, pp. 63–84
BUSH, V., and GOULD, K.E.: 'Temperature distribution along a filament', *Phys. Review*, 1927, **29**, pp. 337–345

BUSH, V., and HAZEN, H.L.: 'Integraph solution of differential equations', *JFI*, 1927, **211**, pp. 575-615
CLOUGH, F.H.: 'Stability of large power systems', *JIEE*, 1927, **65**, p. 653
GATES, S.B.: 'A survey of longitudinal stability below the stall with an abstract for designer's use', *R&M No. 1118*, July 1927
GLAUERT, H.: 'A non-dimensional form of the stability equations of an aeroplane', *R&M No. 1093, ARC*, March 1927
HALL, J.H.: 'Automatic control of synchronous motors', *Iron & Steel Eng.*, 1927, **4**, pp. 427-435
HONES, B.W.: 'Time-limit control of loads having large inertia', *Ind. Engng.*, 1927, **85**, pp. 453-456
MEREDITH, F.W.: 'The stability and accuracy of the Larynx pilotless aircraft', *RAE Rep. No. H 1195*, Nov. 1927
NIEMAN, C.W.: 'Bethlehem torque amplifier', *American Machinist*, 1927, pp. 895-897
TURNER, F.C.: 'Thermionic valve type close voltage regulator', *Engng.*, 1927, **124**, pp. 537-538
VAN DER POL, B: 'Forced oscillations in a circuit with non-linear resistance', *The London, Edinburgh & Dublin Philosophical Magazine & J. Science*, 1927, **3**, pp. 65-80

1928

BEATTY, R.T.: 'The stability of the tuned-grid, tuned-plate high frequency amplifier', *Exp. Wireless Eng.*, 1928, **5**, p. 3
BEATTY, R.T.: 'The stability of a valve amplifier with tuned circuits and internal reaction', *Proc. Phys. Soc. (London)*, 1928, **40**, p. 261
COWLEY, W.L.: 'On the stability of controlled motion', *ARC R&M 1235*, 1928
GOULD, K.E.: 'A new machine for integrating a functional product', *MIT J. Mat. & Physics*, 1928, **3**, p. 309
HARTLEY, R.V.L.: 'Theory of information', *BSTJ*, 1928, **7**
JONES, B.M.: 'Research on the control of airplanes', *NACA TIM No. 485*, 1928, *Nature*, **121**, pp. 755-762
LIENARD, A.: 'Etude des oscillations entretenues', *Revue Gen. Elect.*, 1928, **23**, pp. 901-946
MORRISON, W.A.: 'Thermostat design for frequency standards', *IRE*, 1928, **16**(7), p. 976
NICKLE, C.A., and CAROTHERS, R.M.: 'Automatic voltage regulators, applications to power transmission systems', *AIEE*, 1928, **47**, pp. 957-974
NYQUIST, H.: 'Certain topics in telegraph transmission theory', *AIEE*, 1928, **47**, pp. 617-44
POINCARE, H.: 'Oeuvres Volume 1' (Gauthier-Villars, Paris, 1928)
SCHMIDT, H.F.: 'Fluid governors for prime movers', *Elec. J.*, 1928, **25**, pp. 168-171
STEIN, T.: 'Selbstregelung ein neues Gesetz der Regeltcnik', *Zeitschrift des Vereins deutscher Ingenieure*, 1928, **72**
TEPLOW, L.: 'Stability of synchronous motors under variable torque loads as determined by the recording product integraph', *GER*, 1928, **31**, pp. 356-365

1929

'New compass and path indicator', *Shipbldg. and Shipping Record*, Dec. 1929, p. 684
BUSH, V.: 'Operational circuit analysis' (Wiley, New York, 1929)
FRAZER, R.A., and DUNCAN, W.J.: 'On the criteria for the stability of small motions', *Proc. Royal Soc.*, 1929, **A124**, p. 642
HAZEN, H.L.: 'Solving power system problems by means of the power network analyzer', *Power Plant Eng.*, Nov. 1929, p. 1220
HULL, A.W.: 'Hot-cathode thyratrons', *GER*, 1929, **32**, pp. 213-223, 390-394
KNAPP, O.: 'Die Selbstregler und ihre Verwendung in der Glasindustrie', (Automatic regulators and their utilization in the glass industry), *Glas-Industrie*, Sept. 1929, pp. 167-170, 179-180
McREA, H.A.: 'Automatic control of frequency and load', *GER*, June 1929, **32**(6), pp. 309-313

MASON, C.E.: 'Control in continuous distillation', *Oil & Gas J.*, March 1929
MOELLER, M.: 'Electric control of steam boilers', *Elec. World*, Sept. 1929, **94**(3), p. 580
NORDEN, C.L.: 'Ship stabilizer', US Patent 1708679, April 1929
PARK, R.H., and BANCKER, E.H.: 'System stability as a design problem', *AIEE*, 1929, **48**, pp. 170-194
ROBERTSON, T.F.: 'Automatic frequency control', *Elec. World*, Aug. 1929, **94**, pp. 267-268
STOLLER, H.M., and POWER, J.R.: 'A precision regulator for alternating voltage', *AIEE*, 1929, **48**, pp. 808-811
TUPHOLME, C.H.S.: 'Automatic combustion control', *Electrical Review*, Nov. 1929, **29**, pp. 942-943

1930

'Progress in the petroleum industry', Report of the Petroleum Division, *ASME*, 1930, pp. 1-7
'New device uses light to control machinery', *GER*, 1930, **33**(7), p. 398
ANDRONOV, A.A., and VITT, A.A.: 'Contribution to Van der Pol's theory of entrainment', *Archiv fur Elektrotechnik*, 1930, **24**, p. 99
BOOTH, R.D., and DAHL, O.G.C.: 'Power system stability: a non-mathematical review', *GER*, 1930, **33**(12), pp. 677-681
BRYANT, L.W., and WILLIAMS, A.H.: 'The application of the method of operators to the calculation of the disturbed motion of an aeroplane', *ARC R&M 1346*, July 1930
BUTTERWORTH, S.: 'On the theory of filter amplifiers', *Exp. Wireless*, 1930, **7**, pp. 536-541
CARWILE, P.B., and SCOTT, F.A.: 'Automatic neutralization of the variable grid bias in a direct current feed-back amplifier', *RSI*, 1930, **1**, p. 203
CLAPP, J.K.: 'Temperature control for frequency standards', *IRE*, 1930, **18**(2), pp. 2003-2010
CORBY, R.A.: 'The versatility of application of Selsyn equipment', *GER*, 1930, **32**, p. 706
DOW, H.H.: 'Economic trends in the chemical industry', *Ind. & Eng. Chem.*, 1930, **22**, pp. 113-116
DE FLOREZ, L.: 'Applications of Selsyn remote control in the oil refining industry', *GER*, 1930, **33**(7), pp. 378-83
FREEMAN, N.L.: 'Control systems for oil and gasoline electric engines', *AIEE*, 1930, **49**, pp. 1262-70
GUNN, R.: 'A new frequency-stabilized oscillation system', *IRE*, 1930, **18**(9), pp. 1560-1574
HAZEN, H.L., SCHURIG, O.R., and GARDENER, M.F.: 'The MIT network analyzer: design and application to power-system problems', *AIEE*, 1930, **49**, pp. 1102-1113
HEATON, V.E., and BRATTAIN, W.H.: 'Design of a portable temperature controlled piezo-electric oscillator', *IRE*, 1930, **18**(7), pp. 1239-46
HOLDER, L.F.: 'Principles of Selsyn equipments and their operation', *GER*, 1930, **33**, pp. 500-504
HUGGINS, M.: 'Gyropilot goes cross-country', *Aero. Digest*, 1930, **17**(1), pp. 51-52
JONES, D.M.: 'Controlling load, maintaining frequency', *Elec. World*, 1930, **95**, pp. 1072-76
KHAIKIN, S.E.: 'Continuous and discontinuous oscillations', *ZH Fiz*, 1930, **7**(6), p. 21
KEARSLEY, W.R.: 'Thyratron stabilizer for X-ray tubes', *GER*, 1930, **33**, pp. 571-572
LEWIS, W.A.: 'Motor control for wind tunnel: precision speed regulation for the wind tunnel motor at California Institute of Technology', *AIEE*, 1930, **49**, pp. 99-104
McMASTER, A.J.: 'Photoelectric cells in chemical technology', *Ind. & Eng. Chem.*, 1930, **22**, pp. 1070-1073
MINORSKY, N.: 'Automatic steering tests', *J. Amer. Soc. Naval Eng.*, 1930, **42**, pp. 285-310
MOORE, F.: 'A new type of self-balancing potentiometer', *RSI*, 1930, **1**(3), pp. 125-139
NELSON, J.R.: 'Stability of balanced high frequency amplifiers', *IRE*, 1930, **18**, p. 88

PESTARINI, J.M.: 'The theory of the dynamic operation of the metadyne', *Revue Générale de l'Electricité*, 1930, **27**, pp. 355-395
PESTARINI, J.M.: 'Metadynes and their derivatives', *ibid.*, 1930, **28**, pp. 813, 851-890
PURCELL, T.E., and HAYWARD, A.P.: 'Operating characteristics of turbine governors', *AIEE*, 1930, **49**, pp. 715-722
SUMMERS, I.H., and McCLURE, J.B.: 'Progress in the study of system stability', *AIEE*, 1930, **49**, pp. 132-161
STEIN, I.M.: 'Precision industrial recorders and controllers', *JFI*, 1930, **209**, pp. 201-228
STYER, C.A., and VEDDER, E.H.: 'Process control with the electric eye', *Ind. Eng. Chem.*, 1930, **22**, pp. 1062-1069
VERMAN, L.C., and RICHARDS, L.A.: 'A vacuum-tube regulator for alternators', *RSI*, 1930, **1**, pp. 581-591
VON HANDEL, P., KRUGER, K., and PLENDL, H.: 'Quartz control for frequency stabilization in short-wave receivers', *IRE*, 1930, **18**(2), pp. 307-320
WAGNER, C.F.: 'Effect of armature resistance upon hunting of synchronous machines', *AIEE*, 1930, **49**, pp. 1011-1026
WATANABE, Y.: 'The piezo-electric resonator in high frequency oscillation circuits', *Proc. IRE*, 1930, **1**(5), pp. 862-893
WUNSCH, G.: 'Regler für Druck und Menge' (Oldenbourg, Munich, 1930)

1931

'Automatic adjustment for steel mill rolls', *Electrical Review*, Jan. 1931, **CVIII**, p. 38
'Boykow automatic pilot', *Aero. Rev.*, 1931, **VI**(8), p. 116
'Combustion control equipment', *Electrical Review*, Aug. 1931, **CIX**, pp. 283-284
'Hydraulic turbine governors and frequency control', *Proc. NELA*, 1931, **88**, pp.542-572
'Kent recording and controlling apparatus', *Engineering*, 1931, **132**, pp. 407-408
'Kirkstall: the new Leeds power station', *Electrical Review*, Nov. 1931, **CVII**, pp. 539-543
'Mechanization of industry', *ibid.*, Jan. 1931, **CVIII**, pp. 166-167
'A large rolling-mill drive', *ibid.*, March 1931, p. 464
BEHAR, M.F.: 'Timing and scheduling instruments', *Instruments*, 1931, **4**, pp. 309-352
BENIOFF, H.: 'The operating frequency of regenerative oscillatory systems', *IRE*, 1931, **19**(7), pp. 1274-1277
BERGVALL, R.C.: 'Series resistance method of increasing transient stability limit', *AIEE*, 1931, **50**, pp. 490-497
BUSH, V.: 'The differential analyzer', *JFI*, 1931, **212**(4), pp. 447-488
BUSH, V., and CALDWELL, S.H.: 'Thomas-Fermi equation solution by the differential analyzer', *Physical Review*, 1931, **38**, pp. 1898-1902
CAMPBELL, G.A., and FOSTER, R.M.: 'Fourier Integrals for Practical Applications', Bell System Monograph B584, Sept. 1931
CARNEGIE, H.S.: 'The electrical driving of rolling mills', *JIEE*, 1931, **69**, pp. 1279-1291
CHALMERS, T.W.: 'The automatic stabilization of ships' (Chapman & Hall, London, 1931)
COOK, W.R.: 'On curve fitting by means of least squares', *Philosophical Magazine*, 7th series, 1931, **12**, pp. 1025-1039
CONKLIN, FINCH, HANSELL, C.W.: 'New methods of frequency control employing long lines', *IRE*, 1931, **19**, pp. 1918-30
CRISSON, G.: 'Negative impedance and the twin 21-type repeater', *BSTJ*, 1931, *10*, pp. 485-513
FRISCH, E.: 'Controlling a gyro stabilizer', *Electrical J.*, 1931, **28**(7), pp. 408-410
GRAY, T.S.: 'A photo-electric integraph', *JFI*, 1931, **212**, pp. 77-102
GUYOMAR: 'Fire-control from aeroplanes against targets and from the ground against aeroplanes', *Rev. F. Aer.*, Aug. 1931, **25**, pp. 935-945 (French; for abstract see *Aeron. J.*, 1932, p. 78)
HARRISON, T.R.: 'The new Brown potentiometric recorder', *RSI*, 1931, **2**, pp. 618-625
HINMAN, W.S.: 'Automatic volume control for aircraft radio receivers', *Bur. Stds. J. Res.*, 1931, **7**(1), pp. 37-46

LE CORBELLIER, P.: 'Les systemes auto entretenues et les oscillations de relaxation' (Herman, Paris, 1931)
LLEWELLYN, F.B.: 'Constant frequency oscillators', *IRE*, 1931, **19**, pp. 2063-2094
SLOTTMAN, G.V.: 'Fuel control in the iron and steel industry', *J. Inst. Fuel*, 1931, **4**, pp. 275-280, discussion. pp 280-284
SUITS, C.G.: 'Non-linear circuits for relay applications', *Electrical Engineering*, 1931, **50**, pp. 963-964
VOLTERRA, V.: 'Theorie mathematique de la lutte pour la vie' (Gauthier-Villars, Paris, 1931)
WARD, R.H.: 'Anti-aircraft gun control', *Army Ordnance*, 1931, **XI**, pp. 452-457
YOUNG, J.M.: 'Automatic remote control of boilers', *J. Inst. Fuel*, 1931, **5**, pp. 217-223

1932

'The gyroscopic stabilizing equipment of the Lloyd Sabado liner *Conte de Savoia*', *The Engineer*, 1932, **153**, pp. 32-35, 62-65
'Motora ship stabilizer', *Shipbdg. & Shipping Record*, Oct. 1932, **40**(17), pp. 399-400
BEHAR, M.F.: 'The manual of instrumentation' (Instrument Publishing Co., Pittsburgh, 1932)
CALDWELL, S., OLER, C.B., and PETERS, J.C.: 'An improved form of electrocardiograph', *RSI*, 1932, **3**, pp. 277-286
HAUS, F.: 'Automatic stability of airplanes', *NACA Tech. Memo. 695*, Dec. 1932, translated from 'Stabilite automatique des avions', *L'aeronautique*, 1932, **14**, pp. 243-251
HEATON, V.E., and LAPHAM, E.G.: 'Quartz plate mountings and temperature control for piezo-oscillators', *IRE*, 1932, **20**(2), pp. 261-271
ISRAEL, D.D.: 'Sensitivity controls—manual and automatic', *IRE*, 1932, **20**(3), pp. 461-477
KUSONOSE, Y., and ISHIKAWA, S.: 'Frequency stabilization of radio transmitters', *IRE*, 1932, **20**, pp. 310-339
LA PIERRE, C.W.: 'A precision photoelectric controller', *GER*, 1932, **35**(7), p. 403
LEE, Y.W.: *J. for Maths. & Phys.* (1932)
MITCHELL, G.S.: 'Ship's gyro stabilizer', *Electrical Review*, 110 n 2829, Feb. 1932, p. 225
NYQUIST, H.: 'Regeneration theory', *BSTJ*, 1932, **11**, pp. 126-147
POTTER, A.A., SOLBERG, H.L., and HAWKINS, G.A.: 'Characteristics of a high pressure series steam generator', *ASME*, 1932, **54**, pp. 9-27
SCHILOVSKY, P.: 'The gyroscopic stabilization of ships', *Engineering*, 1932, **134**, pp. 689-690
SPORN, M.: 'Frequency, time and load control on interconnected systems', *Electrical World*, 1932, **99**, pp. 618-624
SPERRY, E.A., Jr.: 'Description of the Sperry automatic pilot', *Aviation Eng.*, Jan. 1932, pp. 16-18
VAN DER POL, B., and NIESSEN, K.F.: 'Symbolic calculus', *Philosophical Magazine*, March 1932, **85**, pp. 537-577
WEINLAND, C.E.: 'Thyratron voltage regulator for an alternator', *RSI*, 1932, **3**, pp. 9-19

1933

'Progress in petroleum mechanical engineering', *ASME*, 1933, **55**
'Automatic yacht steerer', *Scientific American*, April 1933, p. 238
ALFRIEND, J.V.: 'Light sensitive process control', *AIEE* paper no. 33-45, June 1933, pp. 512-515
ARGUIMBAU, L.B.: 'An oscillator having a linear operating characteristic', *IRE*, 1933, **21**(1), pp. 14-28
BEST, E.H.: 'A recording transmission measuring system for telephone circuit testing', *BSTJ*, 1933, **12**, pp. 22-34
BYRD, H.L., and PRITCHARD, S.R.: 'Solution of the two-machine stability problem', *GER*, 1933, **36**(2), pp. 81-93
CLARK, A.B., and KENDALL, B.W.: 'Carrier in cable', *BSTJ*, 1933, **12**, pp. 251-263

COLEBROOK, F.M.: 'Valve oscillators of stable frequency: a critical survey of present knowledge', DSIR Radio Research Special Report No. 13, 1933
CURTIS, A.M.: 'An oscillograph for ten thousand cycles', *BSTJ*, 1933, **12**, pp. 76-90
GREBE, J.J., BOUNDY, R.F., and CERMAK, R.W.: 'The control of chemical processes', *Trans. Amer. Inst. Chem. Eng.*, 1933, **29**, pp. 211-255
GULLIKSEN, F.H.: 'Recent developments in electronic devices for industrial control', *AIEE*, 1933, **52**, pp. 33-42
HOYT, R.S.: 'Probability theory and telephone transmission engineering', *BSTJ*, 1933, **12**, pp. 35-75
HULL, A.W.: 'Characteristics and functions of thyratrons', *Physics*, 1933, **4**, pp. 66-75
LISTON, J.: 'Developments in the electrical industry during 1932', *GER*, 1933, **36**, p. 7
LLEWELLYN, F.B.: 'Vacuum tube electronics of ultra high frequencies', *IRE*, 1933, **21**, pp. 1532-1573
La PIERE, C.W.: 'An improved photoelectric recorder', *GER*, 1933, **36**, pp. 271-274
MIDWORTH, C., and TAGG, G.F.: 'Some electrical methods of remote indication', *JIEE*, 1933, **73**, p. 33
SCHILOVSKY, P.: 'Preliminary calculations of the sizes of gyroscopes required to stabilize a ship', *Trans. Inst. Naval Arch.*, 1933, **75**, pp. 153-162, 163-170
SOHON, H.: 'Supervisory and control equipment for audio frequency amplifiers', *IRE*, 1933, **21**(2), pp. 228-237
THOMPSEN, J.W.: 'Automatic control applications increase as refineries find operation economies', *National Petroleum News*, March 1933, **27**, p. 27

1934

ALGRAIN, P.: 'Calculating machine for directing anti-aircraft fire', *Rev. de l'armee de l'air*, 1934, **58**, pp. 568-613
ALLEN, J.E.: 'Oscillograph analyses governor performance', *Power*, 1934, **78**, pp. 610-12
BAGALLY, W.: 'Stability of resistance coupled amplifiers', *Wireless Engineer*, 1934, **11**, p. 179
BEDFORD, L.H., and PUCKLE, O.S.: 'A velocity-modulation television system', *IEE*, 1934, **75**, pp. 63-92
BLACK, H.S.: 'Stabilized feedback amplifiers', *BSTJ*, 1934, **13**, pp. 1-18
CLARK, P.H.: 'Temperature-control systems and equipment for electric heating', *GER*, 1934, **37**(5), pp. 208-17
DYCHES, H.E., and HELLMUND, R.E.: 'The Pitt-Westinghouse graduate program', *AIEE*, 1934, **53**, pp. 103-4
EPENSCHIED, L., and STRIEBY, M.E.: 'Wide band transmission over coaxial cables', *ibid.*, pp. 1371-80
FOOTE, E.B.: 'The "Fulscope" temperature regulator', *Instruments*, 1934, **7**, p. 81
FOPPL, O.: 'Theory of anti-rolling tanks', *Ing. Arch.*, 1934, **5**(1), pp. 35-42
GARRATT, G.R.M.: 'Description of RAE Mk 1/Smith autopilot', *Aircraft Engineer*, 1934-35
GRIFFITHS, R.: 'Thermostats and temperature regulating instruments' (Griffin, 1934)
GROSZKOWSKI, J.: 'Oscillators with automatic control of the threshold of regeneration', *IRE*, 1934, **22**(2), pp. 145-51
GULLIKSEN, F.H.: 'Electronic regulator for AC generators', *AIEE*, 1934, **53**, pp. 877-81, discussion pp. 1530-31
HAZEN, H.L.: 'Theory of Servomechanisms', *JFI*, 1934, **218**, pp. 283-331
HAZEN, H.L.: 'Design and test of a high performance servo-mechanism', *ibid.*, pp. 543-80
HENDERSON, J.B.: 'The automatic control of the steering of ships and suggestions for its improvement', *Trans. INA*, 1934, **v?**, pp. 20-31
HODGKINSON, R.P.: 'Stabilization of "Conte di Savoia" ', *Engineer*, Oct. 1934, **158**(4109), p. 369
HODGSON, J.L., and ROBINSON, L.L.: 'Development of automatic combustion control systems for industrial and power station boilers', *Proc. IMechE*, 1934, **120**, p. 59
HORT, H.: 'Pneumatic stabilizing system', *Shipbldg. & Shipping Record*, Dec. 1934, **44**(20), p. 647
IVANOFF, A.: 'Theoretical foundations of the automatic regulation of temperature', *J. Inst. Fuel*, 1934, **7**, pp. 117-30, discussion pp. 130-38

KHINTCHINE, A.: 'Korrelationstheorie der stationaren stochastichen Prozesse', *Mathematischen Annalen*, 1934, **109**, pp. 604-615
KJOLSETH, K.E.: 'An automatic electrode regulator for three-phase arc furnaces', *GER*, 1934, **37**(6), pp. 301-303
KOHRS, W.: 'Flamm system of stabilizing ships', *Engineering*, April 1934, **137**(3563), p. 502
LINVILLE, T.M., and WOODWARD, J.S.: 'Selsyn instruments for position systems', *Electrical Engineering*, 1934, **53**, p. 953
MEAHL, H.R.: 'Quartz crystal controlled oscillator circuits', *IRE*, 1934, **22**(6), pp. 732-37
MINORSKY, N.: 'Ship stabilization by activated tanks', *The Engineer*, 1934, pp. 154-157
NYQUIST, H.: 'Regeneration theory', *BSTJ*, 1934, **11**, pp. 126-147
PETERSON, E., KREER, J.G., and WARE, L.A.: 'Regeneration theory and experiment', *BSTJ*, 1934, **13**, pp. 680-700; see also *IRE*, 1934, **22**, pp. 1191-1210
RELLSTAB, L.: 'Activated anti-rolling tank system', *Engineer*, June 1934, **157**(4094), pp. 648-650
STEVENSON, A.R., and HOWARD, A.: 'An advanced course in engineering', *AIEE*, 1934, **54**, pp. 265-268
VAN DER POL, B.: 'The non-linear theory of electric oscillations', *IRE*, 1934, **22**(8), pp. 1051-1086
WHITELEY, A.L.: 'Industrial applications of thyratrons', *The Engineer*, 1934, **158**, p. 141
ZABEL, R.M., and HANCOX, R.R.: *RSI*, 1934, **5**, pp. 28-29

1935

'Twenty-four foot wind tunnel at Farnborough', *The Engineer*, 1935, **159**, p. 534
BASSETT, P.R., and HODGKINSON, F.P.: 'New studies of ship motion', *Trans. Soc. Naval Arch. & Marine Eng.*, 1935, **43**, pp. 286-95, discussion pp. 295-306
CALLENDER, A., HARTREE, D.R., and PORTER, A.: 'Time-lag in a control system', *Philosophical Trans. Royal Soc. London*, 1935, **235**, pp. 415-444
GARDNER, G.W.H., and SUDWORTH, J.: 'Trials of catapulting and controlling a Queen Bee from land', *RAE Rep. H1396*, 1935
HARTREE, D.R.: 'The differential analyser', *The Engineer*, 1935, **160**, pp. 56-82; also in *Nature*, 1935, **135**, p. 940
HOWARTH, O.: 'The control of voltage and power factor on interconnected systems', *JIEE*, 1935, **76**, pp. 353-68
MINORSKY, N.: 'Problems of anti-rolling stabilization of ships by the activated tank method', *J. Amer. Soc. Naval Eng.*, 1935, **47**, pp. 87-119
MITEREFF, S.G.: 'Principles underlying the rational solution of automatic-control problems', *ASME*, 1935, **57**, pp. 159-63
OTT, P.W.: 'The usefulness of mathematics to engineers', *GER*, 1935, **38**, p. 138
PENICK, D.B.: 'Direct-current amplifier circuits for use with the electrometer tube', *RSI*, 1935, **6**, pp. 115-120
TENER, R.S., and DIBBLE, E.S.: 'Constant voltage variable-speed DC generator-exciter set', *GER*, 1935, **38**(60), pp. 263-67
TRAVIS, C.: 'Automatic frequency control', *IRE*, 1935, **23**(10), pp. 1125-1141
WHITELEY, A.L.: 'Thyratron control of resistance welding', Iron & Steel Welding Symposium, 1935

1936

ANON: *Engineering*, Aug. 1936, **142**(3682), pp. 175-78
ANON: *Engineer*, Sept. 1936, **162**, pp. 312-14, 319
BORDEN, P.A., and BEHAR, M.F.: 'Automatic control of voltage and current', *Instruments*, 1936, **9**, pp. 201-210
CALLENDER, A., HARTREE, D.R., and PORTER, A.: 'Time-lag in a control system', *Philosophical Trans. Royal Soc. London*, 1936, **235A**, pp. 415-444
CALLENDER, A., and STEVENSON, A.B.: 'The application of automatic control to a typical problem in chemical industry', *Soc. Chemical Industry: Proc. Chemical Engineering Group*, 1936, **18**, pp. 108-116

CLARK, W.J.: 'The automatic control of chemical processes', *ibid.*, pp. 125-137
GILBERT, R.W.: 'A potentiometric direct-current amplifier and its applications', *IRE*, 1936, **24**(9), pp. 1239-1246
GROVE-WHITE, *RUSI* (Feb. 1936)
HANSELL, C.W., and CARTER, P.S.: 'Frequency control by low power factor line circuits', *IRE*, 1936, **24**(4), pp. 597-619
HAZEN, H.L., JAEGER, J.J., and BROWN, G.S.: 'An automatic curve follower', *RSI*, 1936, **7**, pp. 353-357
IVANOFF, A.: 'The influence of the characteristics of a plant on the performance of an automatic regulator', *Soc. Chemical Industry: Proc. Chemical Engineering Group*, 1936, **18**, pp. 138-151
JONES, R.T.: 'A simplified application of the method of operators to the calculation of disturbed motions of an aeroplane', *NACA Rep. No. 560*, 1936
JONES, R.T.: 'Calculations of the motion of an airplane under the influence of irregular disturbances', *J. Aer. Sci.*, 1936, **3**, pp. 419-425
SALZBERG, B.: 'Notes on theory of the single stage amplifier', *Proc. IRE*, 1936, **24**(6), pp. 879-897
SANTIS, R.DE, RUSSO, M.: 'Rolling of the *S.S. Conte di Savoia* in tank experiments and at sea' *Trans. Soc. Naval Architects & Marine Engineers*, 1936, **44**, pp. 169-194
SMITH, E.S.: 'Automatic regulators, their theory and application', *ASME*, 1936, **58**, pp. 291-303, discussion 1937 **59**, p. 125
THOMAS, H.A.: 'A method of stabilizing the frequency of a radio transmitter by means of an automatic monitor', *JIEE*, 1936, **78**, pp. 717-722
WEST, C.P., and APPLEGATE, T.N.: 'A generator-voltage regulator without moving parts', *Electric J.*, 1936, **33**, pp. 181-183
WHITELEY, A.L.: 'Application of the hot-cathode grid-controlled rectifier or thyratron', *JIEE*, 1936, **78**, pp. 516-539

1937

'Mechanized mathematics', *Engineer*, 1937, **164**, pp. 6-7, 36-9
'Problems of anti-aircraft fire control', *Revue de l'armee de l'air*, April 1937, pp. 461-471 (French); abstract in *J. Roy. Aer. Soc.*, 1937, p. 625
BLACK, H.S.: 'Wave translation system', US Patent 2102671, Dec. 1937
DIETZOLD, R.L.: 'The isograph—a mechanical root finder', *Bell Telephone Laboratories Record*, 1937, **16**, pp. 130-134
GARMAN, G.W.: 'Thyratron D-C motor control', *Electronics*, 1937, **10**, pp. 20-21
GREBE, J.J.: 'Elements of automatic control', *Ind. Eng. Chem.*, 1937, **29**(11), pp. 1225-1228
HARTREE, D.R., *et al.*: 'Time-lag in a control system—II', *Proc. Roy. Soc. London*, 1937, **161**(A), pp. 460-476
KELLOGG, E.W.: 'A review of the quest for constant speed', *RCA Review* II, 1937, pp. 220-239; see also *J. Soc. Motion Picture Eng.*, April 1937
McMAHON, J.B.: 'Control of liquid level in vessels under pressure', *Ind. & Eng. Chem.*, 1937, **29**, pp. 1219-1224
MEREDITH, F.W., and COOKE, P.A.: 'Aeroplane stability and the automatic pilot', *J. Roy. Aero. Soc.*, 1937, **61**, pp. 415-436
MINORSKY, N.: 'Principles and practice of automatic control', *Engineer*, Jan.-April 1937, **163**, pp. 94-97, 122-124, 150-151, 176-177, 204-205, 236-237, 268-269, 294-295, 322-323, 352-3, 380-382, 408-409, 438-439, 467-469
REID, D.G.: 'Necessary conditions for stability (or self-oscillation) of electrical circuits', *Wireless Engineer*, 1937, **14**, pp. 588-596
RICH, T.A.: 'The ballistic use of instruments', *GER*, 1937, **40**, pp. 583-589
SCHMITT, H.M.: 'Continuous control systems with variable characteristics', *Ind. & Eng. Chem.*, 1937, **29**, pp. 1229-1231
SMITH, E.S., and FAIRCHILD, C.O.: 'Industrial instruments, their theory and application', *ASME*, 1937, **59**, pp. 595-607
SYKES, B.: 'Generator regulating systems', *Aircraft Eng.*, 1937, **9**, pp. 37-40

1938

ALEXANDERSON, E.F.W., EDWARDS, M.A., and WILLIS, C.H.: 'Electronic speed control of motors', *Electrical Engineering*, 1938, **57**, pp. 343-352
BEDEAU, F., and MARE, J. DE: 'Etude des pricipes de la retroaction', *Onde Elec.*, 1938, **17**, pp. 153-73, 247-70
BERGMAN, D.J.: 'Application of automatic control in the oil industry', *ASME*, 1938, **60**, pp. 651-656, discussion 1938, **61** pp. 350-354
BODE, H.W.: 'Variable equalizers', *BSTJ*, 1938, **17**, pp. 229-244
BRISTOL, E.S., and PETERS, J.C.: 'Some fundamental considerations in the application of automatic control to continuous processes', *ASME*, 1938, **60**, pp. 641-50
BUTLER, O.I.: 'Metadyne control', *The Railway Gazette*, Sept. 1938, pp. 472-477
BUTLER, O.I.: 'Basic theory of the Metadyne', *The Electrical Times*, Pt. 1 Jan. 1938, pp. 51-52, Pt. 2 April 1938, pp. 591-601
CONCORDIA, C., CRARY, S.B., and LYONS, F.M.: 'Stability characteristics of turbine generators', *AIEE*, 1938, **57**, p. 732
FISCHEL, E.: 'The principle and constituent elements of automatic steering apparatus for aircraft', Collected papers of the Lilenthal Soc., 1938, pp. 231-236
FOERSTER, E.: 'Ship stabilization in theory and practice', *The Shipbuilder*, 1938, **45**, p. 247
HAIGLER, E.D.: 'Application of temperature controllers', *ASME*, 1938, **60**, pp. 633-640
HAUS, F.: 'Aerodynamic principles of automatic stabilizers', *Proc. of the Lilenthal Soc.*, Oct. 1938, pp. 273-306
HAYES, T.J.: *'Elements of ordnance'* (Macmillan, New York, 1938)
HENNING, W.: 'Remote control with Siemens' equipment employing the selector process', *Siemens Zietschrift*, Aug. 1938, pp. 402-06
KLEMIN, A., PEPPER, P.A., and WITTNER, H.A.: 'Longitudinal stability in relation to the use of an automatic pilot', *NACA Tech. No. 666*, 1938
MASON, C.E.: 'Quantitative analysis of process lags', *ASME*, 1938, **60**, pp. 327-334
MIKHAILOV, A.V.: 'Methods for harmonic analysis in automatic control systems', *Avtomat Telemekh*, 1938, **3**, pp. 27-81 (Russian)
REICHEL, W.A., and SYLVANDER, R.C.: 'Autosyn application for remote indication of aircraft instruments', *J. Aero. Sci.*, 1938, **6**, pp. 464-67
PERAZICH, G., SCHIMMEL, H., and ROSENBURG, B.: *'Industrial instruments and changing technology'* (W.P.A., Philadelphia, 1938)
SPITZGLASS, A.F.: 'Quantitative analysis of single-capacity processes', *ASME*, 1938, **60**, pp. 665-74
STEIN, S.: *'Technical memorandum on automatic control'* (Scientific Apparatus Makers of America, Chicago, 1938)
WEINBLUM, G.: 'Theory of active anti-rolling tanks', *ZAMM*, 1938, **18**, pp. 122-27 (German)

1939

'Galileo system for the gunnery control and torpedo firing in Italian ships "Duca degli Abruzzi" and "Garibaldi" ', BIOS/Gp. 2/HEC 13,580
BEDFORD, A.V., and FREDENDALL, G.L.: 'Transient response of amplifiers', *IRE*, 1939, **27**, p. 277
BLACKETT, P.M.S., and WILLIAMS, F.C.: 'An automatic curve follower for use with the differential analyser', *Proc. Cambridge Philosophical Soc.*, 1939, **35**, pp. 494-505
DRAPER, C.S., and SCHLIESTETT, G.V.: 'General principles of instrument analysis', *Instruments*, 1939, **12**, pp. 137-142
FLETCHER, G.H., and TUSTIN, A.: 'The Metadyne, and its application to electric traction', *JIEE*, 1939, **85**, pp. 370-399
HANNA, C.R., OPLINGER, K.A., and VALENTINE, C.E.: 'Recent developments in generator voltage regulation', *AIEE*, 1939, **58**, p. 838
JOFEH, L.: 'An operational treatment of the design of electro-magnetic time-base amplifier', *JIEE*, 1939, **85**, p. 400
REICHEL, W.A., and SYLVANDER, R.C.: 'Autosyn application for remote indication of aircraft instruments', *J. Aero. Sci.*, 1939, **6**, p. 464

SUDWORTH, J.: 'Mathematical note on the stability of controlled longitudinal motion', *RAE Dept. Note No. Inst. 332*, April 1939

THOMAS, H.A.: *'Theory and design of valve oscillators'* (Chapman & Hall, London, 1944, 3rd impression)

WEISS, H.K.: 'Constant speed control theory', *J. Aero. Sci.*, 1939, **6**, pp. 147–152

WILLIAMS, F.C.: 'A reversible head for the automatic curve following device', *Proc. Cambridge Philosophical Society*, 1939, **35**, pp. 506–511

1940

'The stability of certain linear systems', Metro-Vick Report 4616, Nov.1940

ALEXANDERSON, E.F.W., EDWARDS, M.A., and BOWMAN, K.K.: 'The amplidyne generator – a dynamoelectric amplifier for power control', *AIEE*, 1940, **59**, pp. 937–939, discussion p. 1136

ALLAN, G.R.: 'An investigation into centrifugal governor vibration', *J. IMechE*, March 1940

BIOT, M.A.: 'Coupled oscillations of aircraft engine-propeller systems', *J. Aero. Sci.*, 1940, **7**(9)

BODE, H.W.: 'Relations between attenuation and phase in feedback amplifier design', *BSTJ*, July 1940, **19**, pp. 421–454

BOICE, W.K., *et al.*: 'The direct-acting generator voltage regulator', *AIEE*, 1940, **59**, pp. 149–157

BROWN, G.S.: 'Behaviour and design of servomechanisms', Fire Control Committee (D-2) of the NDRC, Nov. 1940

BUTLER, O.I.: 'Amplidyne generator system of power control', *The Electrical Times*, 1940, **98**(2543-4), pp. 43-4, 62-7

BUTLER, O.I.: 'A new form of constant current generator', *The Electrical Times*, March 1940, pp. 217–218

DAVIS, A.: 'The field system of governing', *ASME*, 1940, **62**, pp. 692–700

DEN HARTOG, J.P.: *'Mechanical vibrations'* (McGraw Hill, New York, 1940)

DRAPER, C.S., and BENTLEY, G.P.: 'Design factors controlling the dynamic performance of instruments', *ASME*, 1940, **62**, pp. 421–432

FAIRCHILD, C.O.: 'Elementary theory of automatic control', *Instruments*, 1940, **13**, pp. 334–339

FISHER, A.: 'Design characteristics of amplidyne generators', *AIEE*, 1940, **59**, pp. 939–944, discussion p. 1135

HANNA, C.R., OPLINGER, K.A., and MIKINA, S.J.: 'Recent developments in speed regulation', *AIEE*, 1940, **56**, pp. 692–700

HAZEN, H.L., and BROWN, G.S.: 'The Cinema Integraph: a machine for evaluating a parametric product integral', *JFI*, 1940, **230**, pp. 19–44, 183–205

IMLAY, F.H.: 'A theoretical study of lateral stability with an automatic pilot', *NACA Report No. 693*, 1940

McDONAUGH, E.W.: 'Servo operation of the control surfaces of large aircraft', *Aero. Digest*, 1940, **37**(2), pp. 76–80

MASON, C.E., and PHILBRICK, G.A.: 'Automatic control in the presence of process lags', *ASME*, 1940, **62**, pp. 295–308

MOHLER, F.: 'Amplidyne: a new tool of many uses', *Iron & Steel Eng.*, Sept. 1940

MOLLOY, E.: *'Aeroplane instruments Part I'* (Chemical Publishing Co., New York, 1940)

O'BREEN, J.E.: 'Automatic control for land and marine boilers', *Trans. Inst. Engrs. & Shipbldr. in Scotland*, 1940, p. 80

POMEROY, L. Jr.: 'Hydraulic servomechanisms', *Aeroplane*, 1940, **59**, pp. 577–80

SHOULTS, D.R., EDWARDS, M.A., and CREVER, F.E.: 'Industrial applications of amplidyne generators', *AIEE*, 1940, **59**, pp. 944–949, discussion p. 1136

SPENCER, J.: 'Report on a trial flight with the Smith type V automatic control', *RAE Report No. Inst. 1434*, June 1940

SPITZGLASS, A.F.: 'Quantitative analysis of single-capacity processes', *ASME*, 1940, **62**, pp. 51–62

TUSTIN, A.: 'Metadyne generators and exciters as applied to the electric drive of guns', *C.S.Memo 108*, 8 April 1940

TUSTIN, A.: 'The metadyne system of remote power control as applied to a mark VI Pom-Pom mounting for trials during August-September 1940', *C.S.Memo 110*, 14 Oct. 1940
VON KARMAN, T.: 'The engineer grapples with non-linear problems', *Bull. Amer. Math. Soc.*, 1940, **46**, pp. 615–83
ZIEBOLZ, H.: *'Relay devices and their application to the solution of mathematical equations'* (Askania Regulators, Chicago, 1940)

1941

'Rapidity of follow-up of dead-beat system', *C.S.Memo 3*, circa July 1941
'Design of correction unit', *C.S.Memo 6*, circa July 1941
'Stability of position control of a mass', *C.S.Memo 17*, 22 Aug. 1941
'General stability criterion', *C.S.Memo 18*, 23 Aug. 1941
'Performance of Ward Leonard system', *C.S.Memo 20*, Aug. 1941
'Performance of cyclic controls', *C.S.Memo 22*, 4 Sept. 1941
BOMBERGER, D.C., and WEBER, B.T.: 'Stabilization of Servomechanisms', BTL, MM-41-110-52, 10 Dec. 1941
CHURCH, A.M.: 'Characteristics of centrifugal governors', *J. Amer. Soc. Naval Eng.*, 1941, **53**, pp. 831–839
CONCORDIA, C., CRARY, S.B., and PARKER, E.F.: 'Effect of prime-mover speed governor characteristics on power system frequency variations and tie-line power swings', *AIEE*, 1941, **60**, p. 559
DOUGHERTY, J.J., et al.: 'Power system governing', *AIEE*, 1941, **60**, p. 731
DRAPER, C.S.: 'Instrument instruction at MIT', *Instruments*, Oct. 1941, p. 248
FAIRCHILD, C.O., et al.: *'Temperature — its measurement and control in science and industry'*, Temperature Symposium of the American Institute of Physics, Reinhold, New York, 1939
GARTSIDE, M.J.: 'The principles of automatic temperature control', *The Engineer*, May 1941, pp. 324–326
GODET, S.: 'Analysis of amplitude follow-up systems using armature voltage for stabilizing', General Electric Co. Schenectady Report, circa 1941
HARRIS, H.: 'The analysis and design of servomechanisms', *OSRD* Report No. 454, 1941
HIGGS-WALKER, G.W.: 'Some problems connected with steam turbine governing', *IMechE*, 1941, **146**, pp. 117–125
HOLLIDAY, T.B.: 'Application of electric power in aircraft', *AIEE*, 1941, **60**, pp. 218–225
IMLAY, F.H.: 'The theoretical lateral motions of automatically controlled aeroplane subjected to a yawing moment disturbance', NACA Tech. Note 809, 1941
KOTELNIKOV, V.A.: 'Longitudinal dynamic stability of an aeroplane with an automatic pilot', *Aeron. Eng. (USSR)*, 1941, **15**, pp. 27–31; also in *RTP Trans.* (1241)
LIU, Y.J.: 'Servomechanisms: charts for verifying their stability and for finding the roots of their third and fourth degree characteristics equations', *Dept. of Electrical Engineering MIT*, 1941
McCLURE, J.B., and CAUGHEY, R.J.: 'Prime-mover speed governors for interconnected systems', *AIEE*, 1941, **60**, p. 147
MASON, C.E.: 'Mathematics of surge vessels and automatic averaging control', *ASME*, 1941, **63**, pp. 589–601
MINORSKY, N.: 'Note on the angular motion of ships', *ibid.*, pp. 111–120
MINORSKY, N.: 'Control Problems', *JFI*, Nov.–Dec. 1941, **232**, pp. 451–487, 519–551
MORRIS, D.G.O.: 'Response and stability of correction unit', *C.S.Memo 2*, 12 June 1941
MORRIS, D.G.O.: 'Impedance characteristics of loaded motor', *C.S.Memo 16*, 20 Aug. 1941
MORRIS, D.G.O.: 'Stability of position control of mass', *C.S.Memo 17*, 22 Aug. 1941
MORRIS, D.G.O.: 'General stability criterion', *C.S.Memo 18*, 23 Aug. 1941
MORRIS, D.G.O.: 'Performance of Ward-Leonard system', *C.S.Memo 20*, 25 Aug. 1941
MORRIS, D.G.O.: 'Performance of cyclic controls', *C.S.Memo 22*, 4 Sept. 1941
MORRIS, D.G.O.: 'Analysis of correction unit performance', *C.S.Memo 26*, 4 Sept. 1941
MORRIS, D.G.O.: 'Equations of control for G.L. Mark III aerial', *C.S.Memo 27*, 22 Sept. 1941

OPPELT, W.: 'Comparison of automatic control system', NACA Tech. Memo No. 966 Feb. 1941; translated from *Luftfahrt Forschung*, 1939, **16**(8)
OSBON, W.O.: 'A turbine governor performance analyzer', *AIEE*, 1941, **60**, pp. 963-967
PETERS, J.C.: 'Getting the most from automatic control', *Ind. & Eng. Chem.*, Sept. 1941, **33**, pp. 1095-1103
RHODES, T.J.: *'Industrial instruments for measurement and control'* (McGraw Hill, New York, 1941)
RICHARDS, W.J.: 'Note on the development of automatic pilots, with particular reference to the proposed Mk VII pilot', RAE Tech. Note No. Inst. 593, 1941
SUDWORTH, J.: 'Note on a new form of gyroscopic pitch control for aeroplanes', RAE Dept. Note No. Inst., 1941 (485)
WEBER, R.L.: *'Temperature measurement and control'* (Philadelphia, 1941)
WEREY, R.B.: *'Instrumentation and automatic control in the oil refining industry'* (The Brown Instrument Company, Philadelphia, 1941)

1942

'Description of electric drive with automatic control applied to 4.5 mounting', *C.S.Memo 41*, 1942
'The choice of optimum motor dimensions and gear ratio for acceleration duty', *C.S.Memo 135*, 20 Feb. 1942
'A note on gear ratio value in servo systems', ADRDE Report, Sept. 1942
'Testing machine for aeroplane automatic pilot', *Engineering*, 1942, **154**, pp. 385-86
BEDFORD, A.V., and FREDENDALL, G.L.: 'Analysis, synthesis and evaluation of the transient response of television apparatus', *IRE*, 1942, **3**, p. 440
BROWN, W.S.: 'A simple method of constructing stability diagrams', *ARC R & M 1905*, Oct. 1942
CALDWELL, G.A., and FORMAHLS, W.H.: 'Electrical drives for wide speed ranges', *AIEE*, 1942, **61**, pp. 54-56
CLYMER, C.C.: 'Large adjustable speed wind tunnel drives', *AIEE*, 1942, **61**, pp. 156-58
DANIELL, P.J.: 'Analogy between the interdependence of phase-shift and gain in a network and the interdependence of potential and current flow in a conducting sheet', Report in Servo Panel Library B. 39 (see Tustin 1945)
DUNN, R.W.: 'Automatic pilots', *RAE Tech. Notes*, 1942 (643, 644, 666)
FORMAHLS, W.H.: 'Rototrol — a versatile electric regulator', *Westinghouse Engineer*, 1942, **2**, p. 51
FORRESTER, J.W., *et al.*: 'Fundamental studies in servomechanisms rated approximately 100 watts', *Servomechanisms Laboratory Project No. 6097*, 1942
GARDNER, M.A., and BARNES, J.L.: *'Transients in linear systems'* (Wiley, New York, 1942)
GRAYBEAL, T.D.: 'Steady-state theory of the amplidyne generator', *AIEE*, 1942, **61**, pp. 750-56, discussion p. 1049
HARRIS, H.J.: 'The analysis and design of servomechanisms', OSRD Report No. 454, 1942
HIGGS-WALKER, C.W.: 'Some problems connected with steam turbine governors', *IMechE*, 1942, **146**, pp. 117-25
HOLM, J.G.: 'Stability study of A-C power transmission systems', *AIEE*, 1942, **61**, pp. 893-905, discussion pp. 1046-48
HUREWICZ, W., and NICHOLS, N.B.: 'Servos with torque saturation', part 1, *MIT Radiation Lab. Report No. 555*, 1942; part 2, *MIT Radiation Lab. Report No. 592*, 1943
JONES, R.C., EXNER, D.W., and WRIGHT, S.H.: 'Aircraft voltage regulator and cutout', *AIEE*, 1942, **61**, pp. 334-339
KAUFMANN, P.G.: 'Distribution of boiler load', *IMechE*, 1942, **148**, pp. 42-52; also discussion (1943), **150**, pp. 68-73
LIWSCHITZ, M.M., and KILGORE, L.A.: 'A study of the modified Kramer or asynchronous-synchronous cascade variable speed drive', *AIEE*, 1942, **61**, pp. 255-60
LAWRY, C.: 'An analysis of an amplidyne servomechanism', *Group B Radiation Laboratory*, Feb. 1942
MINORSKY, N.: 'Self-excited oscillations in dynamical systems possessing retarded actions', *J. Applied Mechanics*, 1942, **9** (A65-71)

MORRIS, D.G.O.: 'Control of G.L. aerial Mark III', *C.S.Memo 11/12*, 6 June 1942
MORRIS, D.G.O.: 'Low pass filter for H.S.F.', *C.S.Memo 38*, 27 Feb. 1942
PETERS, J.C.: 'Experimental study of automatic control systems', *ASME*, 1942, **64**, pp. 247-55
PRESCOTT, J.C.: 'A clock controlled governor for close speed regulation', *IEE*, 1942, **89**(II), pp. 210-16, discussion p. 520
SCHWENDNER, A.F.: 'Analysing governor system performance by means of special instruments', *ASME*, 1942, **64**, pp. 43-47
SHANNON, C.: 'The theory and design of linear differential equation machines', OSRD Report 411, Jan. 1942
SMITH, E.S.: *'Automatic control engineering'* (McGraw-Hill, New York, 1942)
THOMPSON, L.W., and CREVER, F.E.: 'The application of voltage regulators to aircraft generators', *AIEE*, 1942, **61**, pp. 363-365
TUSTIN, A.: 'Scheme for automatic acceleration in following targets', *C.S.Memo 47*, 29 Sept. 1942
TUSTIN, A.: 'The performance of cyclic control systems with special reference to systems incorporating metadynes', *C.S.Memo 48*, 3 Oct. 1942
TUSTIN, A.: 'Proposed power drive and control for type G.L. Mark III aerial', *C.S.Memo 1004*, 10 Sept. 1942
VAN ANTWERPEN, F.J.: 'The automatic control problem', *Ind. & Eng. Chem.*, 1942, **34**(4), pp. 387-91
VARNEY, R.N.: 'An all-electric integrator for solving differential equations', *RSI*, 1942, **13**, p. 10
WASS, C.A.A.: 'Feedback amplifiers', *Nature*, 1942, **150**, pp. 381-382
WHITELEY, A.L.: 'Theory of servomechanisms', BTH Report L315-S, June 1942
WHITELEY, A.L.: 'An introductory analysis of position control servos', BTH Research Report L184-s, Sept. 1942
WHITELEY, A.L.: 'Servo systems—new form of phase-amplitude diagram', BTH Research Laboratory Report 1288-5, Oct. 1942
WHITELEY, A.L.: 'Techniques of electrical position control servo-systems', BTH Report L193-s, Nov. 1942
WILLIAMS, F.C.: 'Automatic following mirror systems', TRE Report No. T1505, 1942
ZIEGLER, J.G., and NICHOLS, N.B.: 'Optimum settings for automatic controllers', *ASME*, 1942, **64**, pp.759-768

1943

'Electric motor control regulates speed through use of electronic tubes', *Scientific American*, Aug. 1943, **169**, pp. 82-83
'Electrical gun director demonstrated', *Bell Lab. Record*, 1943, **22**, pp. 157-67
BROWN, W.S.: 'A method for the solution of biquadratic equations', ARC 6775 S & C 1544, June 1943
BULGAKOV, B.V.: 'Periodic processes in free pseudo-linear oscillatory system', *JFI*, 1943, **235**, pp. 591-616
CREVER, F.E.: 'Fundamental principles of amplidyne applications', *AIEE*, 1943, **62**, pp. 603-6
DANIELL, P.J.: 'Interpretation and use of harmonic response diagrams (Nyquist diagrams) with particular reference to servomechanisms', *Servo Lib. S2*, June 1943; reprinted in *Servomechanisms*, 1951
EVANS, L.W.: 'Solution of the cubic equation and the cubic charts', *Department of Electrical Engineering, MIT*, 1943
EWALD, P.R.: 'Forced oscillation approach to the problem of the automatically controlled continuous process', *Instruments*, 1943, **16**, pp. 474-476, 506
FIELD, H.: 'An introduction to aircraft hydraulic systems', *ASME*, 1943, **66**, pp. 569-76
GRIFFITHS, R.: *'Thermostats and temperature regulating instruments'* (Charles Griffin, London, 1943)
HALL, A.C.: *'The analysis and synthesis of linear servomechanisms'* (The Technology Press, Cambridge, MA. (Restricted Circulation) 1943)

JOFEH, L.: 'The effect of stiction on cyclic control systems', A.C. Cossor Report MR 110, Oct. 1943
JOFEH, L.: 'The effect of wire wound potential dividers on the behaviour of displacement-displacement servomechanisms', A.C. Cossor Report MR 133, circa 1943-45
KRON, G.: 'Steady-state and hunting equivalent circuits of long-distance transmission systems'. *GER.* 1943, **46**. pp. 337-42
KRYLOV, N., and BOGOLIUBOV. N.: *'Introduction to nonlinear mechanics'* (Princeton University Press, 1943)
MELVILL JONES, B.: *'Dynamics of the aeroplane'*, in DURAND, W.F. (Ed.): *Aerodynamic Theory* V, (Durand Reprinting Committee, Pasadena, 1943)
MORAN. R.F.: 'Determination of the precision of analytical control methods', *Ind. & Eng. Chem. (Anal. Ed.)*, 1943, **15**(6), pp. 361-64
NEUMARK, S.: 'The disturbed longitudinal motion of an uncontrolled aeroplane and of an aeroplane with automatic control', *ARC R & M No. 2078*, 1943; also in *RAE Report No. Aero. 1793*, 1943
PETERS, J.C., and OLIVE, T.R.: 'Fundamental principles of automatic control', *Chem. Metall. Engng.*, 1943, **50**(5), pp. 98-107
PHILLIPS, R.S.: 'Servomechanisms', Radiation Laboratory Report No. 372, 11 May 1943
PORTER, A.: 'Some servo-mechanisms used in the position control of A.A. equipment', ADRDE Minor Report No. 8, PRO(AVIA 1235), 14 April 1943
SUDWORTH, J., and HOPKIN, H.R.: 'Influence of automatic pilots on stabilisation and dynamic stability in pitch', RAE Tech. Note. No. Inst. 775, July 1943
TUSTIN, A.: 'A comparison of the characteristics and performance of Ward-Leonard generators, Amplidynes and Metadynes', *C.S.Memo 54A*, 18 Feb. 1943
TUSTIN, A.: 'The possibility of using the thermal time constant of a temperature sensitive resistor to obtain phase advance of an electrical signal', *C.S.Memo 68*, 8 Feb. 1943
TUSTIN, A.: 'A basis for the analysis of servo control systems', *C.S.Memo 80*, 12 April 1943
TUSTIN, A.: 'Uni-directional operative relations', *C.S.Memo 95*, 14 Sept. 1943
ZIEGLER, J.G.: ' "On-the-job" adjustments of air operated recorder-controllers', *Instruments*, 1943, **16**, pp. 394-397, 594-596, 635
ZIEGLER, J.G., and NICHOLS, N.B.: 'Process lags in automatic control circuits', *ASME*, 1943, **65**, pp. 433-444

1944

'Aerodynamic marine engine governor', *Engineering*, 1944, **157**, pp. 447-48
'Development of the electric director', *Bell Lab. Record*, 1944, **22**, pp. 225-30
'The metadyne', *The Engineer*, 1944, **178**(4628), pp. 227-28
ALEXANDERSON, E.W.F., EDWARDS, M.A., and BOWMAN, K.K.: 'The amplidyne system of control', *IRE*, Sept. 1944
BERRY, T.M.: 'Polarized light servo system', *AIEE*, 1944, **63**, pp. 195-7
CONCORDIA, C.: 'Steady-state stability of synchronous machines as affected by voltage regulator characteristics', *ibid.*, pp. 215-20, discussion p. 490
DANIELL, P.J.: 'Operational methods for servo systems', Servo Panel Report S1, July 1944
DANIELL, P.J.: 'Digest of manual on the extrapolation, interpolation and smoothing of stationary time series with engineering applications, by Norbert Wiener, OSRD Report 370', Servo Panel Lib. p. 47, circa 1944
EMERY, J.N., *et al.*: 'A general account of the control system of the German A4 rocket', RAE Report No. Inst. 1447, Dec. 1944
GINZTON, E.L.: 'D-C amplifier design technique', *Electronics*, 1944, **17**(3), pp. 98-102
HARDER, E.L., and CHEEK, R.C.: 'Regulation of AC generators with suddenly applied loads', *AIEE*, 1944, **63**, p. 310
HERWALD, S.W.: 'Considerations in servomechanism design', *AIEE*, 1944, **63**, pp. 871-876
HOPKIN, H.R.: 'A theoretical analysis of the behaviour of an automatic control applying elevator as a linear function of aircraft pitch deviation from a gyro vertical, and the time integral of that deviation', RAE Tech. Note No. Inst. 842, March 1944
HOPKIN, H.R.: 'Stability and stabilisation during automatic landing approach', RAE Tech. Note No. Inst. 114, Oct. 1944

LOURIE, A.I.: 'Concerning the theory of stability of regulating systems', *Prikl. Mat. Mekh.*, 1944, **8**, pp. 256-258 (Russian)
MINORSKY, N.: 'On non-linear phenomenon of self-rolling', *Proc. Nat. Acad. Sci. USA*, Oct. 1944, **30**
MITCHELL, K., COAD-PRYOR, A., and PARNELL, M.V.: 'The application of automatic control to the rudder as a possible means of reducing sideslip errors in dive bombing', RAE Tech. Note No. Aero. 1551, 1944
OLDENBURG, R.C., and SARTORIUS, H.: *'Dynamik Selbsttatiges Regelungen'* (R. Oldenburg, Berlin, 1944); translation in *The Dynamics of Automatic control'* (ASME, New York, 1948)
OSBORNE, S.P.: 'Flight tests of the G.E.C. three axes automatic pilot (model no. 2CJIAI) in a high performance twin engined bomber', RAE Tech. Note No. Inst. 874, 1944
PHILLIPS, R.S., and WEISS, P.R.: 'Theoretical calculation on best smoothing of position data for gunnery prediction', NDRC Report 532, Feb. 1944
PRINZ, D.G.: 'Contributions to the theory of automatic controllers and followers', *JSI*, 1944, **21**, pp. 53-64, 110-111
PROFOS, P.: *'Vektorielle Regeltheorie'* (Zurich, 1944)
ROBSON, J.M.: 'The design of servomechanisms which are required to smooth as well as retransmit data', RRDE Res. Report 264, Nos. 1944
SMITH, Ed S.: *'Automatic control engineering'* (McGraw-Hill, New York, 1944)
STRECKER, F.: *'Regelungstechnik: Begriffe und Beziechungen'* (VDI, Berlin, 1944)
SUDWORTH, J.: 'German flying bomb: examination of automatic pilot', RAE Tech. Note No. Inst. 869, 1944
THOMAS, H.A.: *'Theory and design of valve oscillators'* (Chapman & Hall, London, 1944)
THOMPSON, J.D.: 'Electric gun turrets for aircraft', *AIEE*, 1944, **63**, pp. 799-802
THORPE, J.B.J., and HOPKIN, H.R.: 'Description and flight tests of the Smith lateral electrical autopilot in a four engined bomber', RAE Tech. Note No. Inst. 867, 1944
TUSTIN, A.: 'The anti-stabilising effect of back-lash in positional servo controls', *C.S.Memo 168*, 22 June 1944
TUSTIN, A.: 'An investigation of the operator's response in a manual control power driven gun', *C.S.Memo 169*, 22 Aug. 1944
TUSTIN, A.: 'The choice of response characteristics for controllers for power driven guns', *C.S.Memo 184*, 10 Oct. 1944
TUSTIN, A.: 'Methods of phase advancing harmonic components of amplitude modulation of an A.C. carrier', *C.S.Memo 186*, 2 Nov. 1944
TUSTIN, A.: 'The effect of backlash in positional control systems with phase advance', *C.S.Memo 193*, 19 Dec. 1944
WHITELEY, A.L.: 'Servo systems—new form of phase-amplitude diagram', BTH Research Laboratory Report 1288-5, 9 Oct. 1944
WINN, R.B.: 'Cooling and governing problems on a water-cooled auxiliary generating plant', *J. Roy. Aero. Soc.*, 1944, **XLVIII**, pp. 516-537

1945

ALLAN, J.F.: 'The stabilization of ships by activated fins', *Trans. I. Naval Arch.*, 1945, **87**, pp. 123-159; see also *Engineering*, 1945, **159**(1), pp. 434-35
BATEMAN, H.: 'The control of an elastic fluid', *Bulletin of the American Mathematical Society*, 1945, **51**, pp. 601-646
BELSEY, F.H.: 'Effect of backlash and resilience between motor and load on the stability of a servo system', Metro-Vick Report C. 565, June 1945
BETHELL, P.: 'The development of the torpedo', *Engineering*, 1945, **159**, pp. 403, 442; **160**, pp. 4, 41, 201, 341, 365, 529; 1946, **160**, pp. 73, 121, 169, 242
BODE, H.W.: *'Network analysis and feedback amplifier design'* (Van Nostrand, Princeton, NJ, 1945)
BOICE, W.K., and LEVOY, L.G.: 'Governor requirements for aircraft alternator drives', *AIEE*, 1945, **64**, pp. 534-40
BOWER, J.L.: 'Fundamentals of the amplidyne generator', *AIEE*, 1945, **64**, pp. 873-81
BREITWIESER, C.J.: 'Constant-speed drives for aircraft alternators', *AIEE*, 1945, **64**, pp. 763-68, discussion p. 991
BUJAK, J.Z.: 'Variable speed hydraulic governing', *IMechE*, 1945, **153**, pp. 193-205

BUNCE, H.B., and CUNNINGHAM, J.C.: 'Instability in DC aircraft systems', *AIEE*, 1945, **64**, pp. 786-91

CAGE, J.M.: 'Amplifier theory applied to regulators', *Electronics*, 1945, **18**, p. 140

CAMPBELL, D.P.: 'Theory of automatic control systems', *Industrial Aviation*, 1945, pp. 62-64, 94-95

CAMPBELL, W.W.: 'Stability from the Nyquist diagram', *Servo Lib. G. 47, 14 Feb. 1945*

CHIN, P.T. and WALTER, G.E.: 'Transient response of controlled rectifier circuits', *AIEE*, 1945, **64**, pp. 208-14

CREVER, F.E., and JACKSON, R.L.: 'Automatic load control for turbine generators', *ibid.* pp. 656-60

DANIELL, P.J.: 'Backlash in reset mechanisms', *C.S.Memo 199*, 16 March 1945

DANIELL, P.J.: 'An explanatory note on H.W. BODE's paper on the relation between phase-lag and attenuation (Bell Tel. Journal 19 (1940) p. 421)', *C.S.Memo 201*, 21 March 1945

EBERHARDT, W.W., *et al.*: 'Automatic regulation of turbine generators to relieve power supply systems of steel-mill load swings', *AIEE*, 1945, **64**, pp. 751-58

ECKMAN, D.P.: *'The principles of industrial process control'* (John Wiley, New York, 1945)

ECKMAN, D.P., and WANNAMAKER, W.H.: 'Electrical analogy method for fundamental investigations in automatic control', *ASME*, 1945, **67**, pp. 81-86

EYRES, N.R., *et al.*: 'A theoretical investigation of the effect of stiction and Coulomb friction on the minimum smooth running speed of a metadyne controlled motor as incorporated in A.F.1.', ADRDE Report No. 180, circa 1945

FERRELL, E.B.: 'The servo problem as a transmission problem', *IRE*, 1945, **33**, pp. 763-67

HANNA, G.R., OSBON, W.O., and HARTLEY, R.A.: 'Tracer-controlled position regulator for propeller milling machine', *AIEE*, 1945, **64**, pp. 201-205, discussion p. 452

HARDER, E.L., and VALENTINE, C.E.: 'Static voltage regulator for Rototrol exciter', *ibid.*, pp. 601-06

HAYES, K.A.: 'Automatic controllers', *Proc. I. Chem. Engrs.*, 1945, **27**, pp. 173-79

HAYES, K.A.: 'Some fundamental characteristics of control systems', *Proc. Joint Conf. Inst. Chem. Engrs., Inst. Physics. Chem. Engng. Group*, 1945, pp. 27-33

HICK, W.E.: 'Basic principles of servos', *Aeronautics*, 1945, **13**(5), pp. 34-39

HOPKIN, H.R.: 'Calculations on the lateral stability of an aircraft controlled by an autopilot using a rate-rate system or a displacement-displacement system', RAE Tech. Note No. Inst. 132, 1945

HUREWICZ, W.: 'On servos with pulsed error data', Radiation Lab. Report 721, April 1945

JOFEH, L.: 'Study of the effects of saturation on the behaviour of a type of servomechanism', A.C. Cossor Report MR165, 1945

JOHNSON, T.C.: 'Selsyn design and application', *AIEE*, 1945, **64**, pp. 703-08

KAUFMANN, P.G.: 'On the theory of automatic control regulator equations', *IMechE*, 1945, **153**, pp. 249-59

KAUFMANN, P.G.: 'Quantitative analysis of single and multiple capacity plants', *ibid.*, pp. 237-48

MACCOLL, L.A.: *'Fundamental theory of servomechanisms'* (Van Nostrand, Princeton, NJ, 1945); reprinted by R.W. Hamming, Dover Books, 1968

MIDDEL, H.D.: 'Electronic tools for process control applications', *Electrical World*, April 1945, **123**, p. 103

PRINZ, D.G.: 'Automatic control: a survey of criteria for stability', *Proc. Inst. Chem. Eng.*, 1945, **27**, pp. 34-38

PROFOS, P.: 'A new method for the treatment of regulation problems', *Sulzer Technical Review*, 1945, **2**, p. 1

RESTEMEYER, W.E.: 'Operational methods in servo design', *J. Aero Sci.*, 1945, **12**(3), pp. 313-19

RICE, S.O.: 'Mathematical analysis of random noise', *BSTJ*, 1945, **25**, pp. 46-156

ROGERS, R.H.: 'Servomechanisms—how they work', *Machine Design*, 1945, **17**(4), pp. 119-24

SMITH, E.S.: 'Stabilizing a suction relief valve', *ASME*, 1945, **67**, P. 87

SUDWORTH, J.: 'Note on the use of an automatic pilot with reference to stability and control', RAE Tech. Note No. Inst. 167, 1945
TUSTIN, A.: 'The effects of mechanical elasticity in a servo system, especially when resetting from the load', *C.S.Memo 195*, 20 Feb. 1945
TUSTIN, A.: 'An analogy suggested by Prof. P.J. DANIELL between the interdependence of phase-shift and gain in a network and the interdependence of potential and current flow in a conducting sheet', *C.S.Memo 200*, 20 March 1945
TUSTIN, A.: 'A method proposed by W.W. CAMPBELL (C.E.A.D.) to obtain information relating to the response to "unit signals" from the Nyquist diagram', *C.S.Memo 202*, 22 March 1945
TUSTIN, A.: 'Notes on manual control of guns further to C.S. 169 and C.S. 184', *C.S.Memo 203*, 4 April 1945
TUSTIN, A.: 'Obtaining harmonic response loci from tests of the response to suddenly applied constant inputs and the reverse', *C.S.Memo 207*, 30 April 1945
WEISS, H.K.: 'Dynamics of constant speed propellers', *J. Aero. Sci.*, 1945, **10**(2), pp. 58–69
WHITELEY, A.L.: *'Theory of servo systems with particular reference to stabilisation'*, D.Sc. dissertation University of Leeds, 1945
ZAHORSKY, L.A.: 'The amplidyne generator applied to speed-controlled electric gun turrets', *AIEE*, 1945, **64**, pp. 221–24, discussion p. 436

1946

ADKINS, B.: 'The analysis of hunting by means of vector diagrams', *JIEE*, 1946, **93**(II), pp. 541–548
AHRENDT, W.R., and TAPLIN, J.F.: *'Automatic regulation'* (published by the authors, 1946)
BEDFORD, L.H.: 'The development of gun-laying radar receivers type G.L. Mk. 1 and G.L. Mk. 1* and GL/E.F.; Radiolocation Convention Proceedings', *JIEE*, 1946, **93**(IIIA), pp. 287–88
BROWN, G.S., and HALL, A.L.: 'Dynamic behaviour and design of servomechanisms', *ASME*, 1946, **68**, pp. 503–524
CALDWELL, G.A.: 'Wide-speed-range drive with servomechanisms', *Westinghouse Engineer*, 1946, **5**, p. 157
CHESTNUT, H.: 'Electrical accuracy of Selsyn generator control transformer system', *AIEE*, 1946, **65**, pp. 570–76
DAVIDSON, M.: *'The gyroscope and its applications'* (Hutchinson, London, 1946)
ECKMAN, D.P.: 'Automatic control terms', *Chemical Industry*, 1946, **58**, pp. 832, 1020; 1946, **59**, pp. 528, 710
ECKMAN, D.P.: 'Automatic control terminology', *Instrumentation*, 1946, **2**, pp. 9–14
ECKMAN, D.P.: 'The effect of measurement of dead time in the control of certain processes', *ASME*, 1946, **68**, p. 707
FREY, W.: 'A generalisation of the Nyquist and Leonhard stability criteria', *Brown Boveri Review*, 1946, **33**(3), pp. 59–65
GRAHAM, R.E.: 'Linear Servo Theory', *BSTJ*, 1946, **25**, pp. 616–51
HALL, A.C.: 'Application of circuit theory to the design of servomechanisms', *JFI*, 1946, **242**(4), pp. 279–307
HARRIS, H.: 'The frequency response of automatic control systems', *AIEE*, 1946, **65**, pp. 539–46
HERWALD, S.W., and McCANN, G.D.: 'Dimensional analysis of servomechanisms by electrical analogy', *AIEE*, 1946, **65**, pp. 636–40
HOPKIN, H.R.: 'Routine computing methods for stability and response investigation on linear systems', RAE Tech. Note No. Inst. 954, 1946
JELONEK, Z.: 'Analysis of operation of a thermostat with contact thermo-regulator', *Proc. Camb. Phil. Soc.*, 1946, **42**(1), pp. 62–72
MARCY, H.T.: 'Parallel circuits in servomechanisms', *AIEE*, 1946, **65**, pp. 521–29, discussion p. 1128
McCANN, G.D., HERWALD, S.W., and KIRCHSBAUM, H.S.: 'Electrical analogy methods applied to servomechanisms', *ibid.*, pp. 91–96

MOXON, L.A., et al.: 'Some automatic control circuits or radar receivers', *JIEE*, 1946, **93**(IIIA), pp. 1143-58
OSBORNE, S.P.: 'Automatic controls Mk. V system as applied to the Queen Wasp target aircraft', RAE Tech Note Inst. 676, 1946
PERKINS, G.S.: 'Analysis of servomechanisms for instrument and power drives', *Product Engineering*, 1946, **17**, pp. 332-34
SCHREINER, K.E.: 'High performance demodulators for servomechanisms', *Proc. Natl. Electronics Conf.*, 1946, **2**, pp. 393-403
SHERMAN, S.: 'Generalized Routh-hurwitz discriminant', *Phil. Mag.*, 1946, **37**(271), pp. 537-51
WEISS, H.K.: 'Analysis of relay servomechanisms', *J. Aero. Sci.*, 1946, **13**(7), pp. 364-76
WHITELEY, A.L.: 'Theory of servo systems with particular reference to stabilization', *JIEE*, 1946, **93**(II), pp. 353-67, discussion pp. 368-72
WILLIAMS, F.C.: 'Introduction to circuit techniques for radiolocation', *JIEE*, 1946, **93**, pp. 289-308
WILLIAMS, F.C., RITSON, F.J.U., and KILBURN, T.: 'Automatic strobes and recurrence-frequency selectors', *JIEE*, 1946, **93**(IIIA), pp. 1275-1300
WILLIAMS, F.C., and UTTLEY, A.M.: 'The Velodyne', *ibid.*, pp. 1256-74
ZIEBOLZ, H.: *'Analysis and design of translator chains'* (Askania Regulator Co., Chicago, 1946)

1947

'Preliminary observations on flywheel energy storage as applied to metadyne sets for RPC', *C.S.Memo 228*, 1947
'Phase advance of A.C. modulation', *C.S.Memo*, 1947
'The effect of friction at creep speeds in servos having velocity reset', *C.S.Memo 233A*, 1947
ADKINS, B.: 'Amplidyne voltage control of generators', *JIEE*, 1947, **94**(IIA), pp. 49-60
BARWICK, E.C.: 'The amplidyne: a review of its basic principles and applications', *Elect. Times*, 1947, **111**, pp. 449-53, 485-88
BATES, J.A.V.: 'Some characteristics of the human operator', *JIEE*, 1947, **94**(IIA), pp. 298-304
BELL, J.: 'Data-transmission systems', *ibid.*, pp. 222-35
BROADHURST, J.W., et al.: 'Automatic control in the chemical industry', *ibid.*, pp. 79-95
CHESTNUT, H.: 'Obtaining attenuation frequency characteristics for servomechanisms', *GER*, 1947, **50**, pp. 38-44
COOMBES, J.E.M.: 'Hydraulic remote position controllers', *JIEE*, 1947, **94**(IIA), pp. 270-82
CREMER, L.: 'Ein neues Verfahren zur Beurteilung der Stabilitat linearer Regelungssysteme', *Z. Angew. Math. Mech.*, 1947, **26**, pp. 161-163
CRISS, G.B.: 'The inverse Nyquist plane in servomechanism theory', *IRE*, 1947, **35**(12)
DOUCH, E.J.H.: 'The use of servos in the army during the past war', *JIEE*, 1947, **94**(IIA), pp. 177-89
FARRINGTON, G.H.: 'Automatic temperature control of jacketed pans', *Trans. Soc. Instrument Technology*, 1947, **1**, pp. 2-22
FARRINGTON, G.H.: 'Theoretical foundations of process control', *JIEE*, 1947, **94**(IIA), pp. 23-38
FORSTER, E.W., and LUDBROOK, L.C.: 'Some industrial electronic servo and regulator systems', *ibid.*, pp. 100-111, discussion p. 125
GAIRDNER, J.O.H.: 'Some servo mechanisms used by the Royal Navy', *ibid.*, pp. 209-13
GARVEY, R.J.: 'Electrical remote position-indicating systems as applied to aircraft', *ibid.*, pp. 283-91
GOLDFARB, L.C.: 'On some nonlinear phenomena in regulatory systems', *Automatika i Telmekhanika*, 1947, **8**, pp. 349-83 (Russian); translated in OLDENBURGER, R. (Ed.): *Frequency Response* (Macmillan, New York, 1956)
GREENBERG, H.: 'Frequency-response method for determination of dynamic stability characteristics of airplanes with automatic controls', NACA Tech. Note No. 1229, 1947
GRIFFITHS, E.: *'Methods of measuring temperature'* (Charles Griffin, London, 1947)
HALL, A.A.: 'The use of servomechanisms in aircraft', *JIEE*, 1947, **94**(IIA), pp. 256-82

HARRIS, H. Jr.: 'The analysis and design of servomechanisms', *ASME*, 1947, **69**(3), pp. 267-80
HARRIS, H. Jr.: 'A comparison of two basic servomechanism types', *AIEE*, 1947, **66**, pp. 83-92
HAYES, K.A.: 'Elements of positions control', *JIEE*, 1947, **94**(IIA), pp. 161-76
HERR, D., and GERST, I.: 'The analysis and an optimum synthesis of linear servomechanisms', *AIEE*, 1947, **66**, pp. 959-70
HEUMANN, G.W.: *Magnetic control of industrial motors* (Wiley, New York, 1947)
HOPKIN, H.R., and DUNN, R.W.: *Theory and development of automatic pilots 1937-47* (RAE Monograph No. 2.5.03, Farnborough, 1947)
JAMES, H.J., NICHOLS, N.B., and PHILLIPS, R.S.: *Theory of servomechanisms*, Radiation Laboratory Series Vol. 25 (McGraw-Hill, New York, 1947)
LAMBERT, W.E.C.: 'Naval applications of electrical remote-positional controllers', *JIEE*, 1947, **94**(IIA), pp. 236-50
LAUER, H., LESNIK, R., and MATSON, L.: *Servomechanism fundamentals*, (McGraw-Hill, New York, 1947)
LEONHARD, A.: 'New techniques for investigating stability', RAE Library Translation No. 195, Sept. 1947
MARCHANT, E.W., and ROBB, A.C.: 'Methods of testing small servo mechanisms and data-transmission systems', *JIEE*, 1947, **94**(IIA), pp. 292-97
McCANN, G.D., OSBON, W.O., and KIRSCHBAUM, H.S.: 'General analysis of speed regulators under impact loads', *AIEE*, 1947, **66**, pp. 1243-52
MINORSKY, N.: 'Experiments with activated tanks', *ASME*, 1947, **69**, pp. 735-47
MINORSKY, N.: *Introduction to non-linear mechanics* (Edwards, Ann Arbor, 1947)
NEWTON, G.C.: 'Hydraulic variable speed transmissions as servomotors', *JFI*, 1947, **243**(6), pp. 439-69
NEWTON, G.C., and WHITE, W.T.: 'Laboratory aids for electromechanical system development', *AIEE*, 1947, **66**, pp. 315-19
PORTER, A.: 'The design of automatic and manually operated control systems', *SIT*, 1947, **1**, pp. 23-42
RAGAZZINI, J.R., RANDALL, R.H., and RUSSELL, F.A.: 'Analysis of problems in dynamics by electronic circuits', *IRE*, 1947, **35**, pp. 444-52
SCHWARTZ, G.J.: 'The application of lead networks and sinusoidal analysis to automatic control systems', *AIEE*, 1947, **66**, pp. 69-77
STEWART, C.: 'Automatic control of generators', *JIEE*, 1947, **94**(IIA), pp. 39-48
TARPLEY, H.I.: 'Instrument to measure servomechanism performance', *RSI*, 1947, **18**, pp. 39-43
TUSTIN, A.: 'A method of analysing the behaviour of linear systems in terms of time series', *JIEE*, 1947, **94**(IIA), pp. 130-42
TUSTIN, A.: 'The effects of backlash and of speed-dependent friction on the stability of closed-cycle control systems', *JIEE*, 1947, **94**(IIA), pp. 143-51
TUSTIN, A.: 'A method of analysing the effect of certain kinds of non-linearity in closed-cycle control systems', *ibid.*, pp. 152-60
TUSTIN, A.: 'The nature of the operator's response in manual control and its implications for controller design', *ibid.*, pp. 190-207
WATSON WATT, R.: *JIEE*, 1947, **93**(IIIA), p. 14
WHITELEY, A.L.: 'Fundamental principles of automatic regulators and servomechanisms', *ibid.*, pp. 5-19
WILLIAMS, F.C., and RITSON, F.J.U.: 'Electronic servo simulators', *ibid.*, pp. 112-24
YOUNG, L.: 'Automatic control applied to modern high-pressure boilers', *ibid.*, pp. 66-78
ZIEGLER, J.G., and NICHOLS, N.B.: 'Industrial process control', *Chem. Engng. Progress*, 1947, **49**, p. 309

1948

BROWN, G.S., and CAMPBELL, D.P.: *Principles of servomechanisms* (Wiley, New York, 1948)
CADY, W.M., KARELITZ, M.B., and TURNER, L.A. (Eds): *Radar scanners and radomes*, Radiation Laboratory Series 26 (McGraw-Hill, New York, 1948)

EVANS, W.R.: 'Graphical analysis of control systems', *AIEE*, 1948, **67**, pp. 547–51

GREENWOOD, I.A., HOLDAM, J.V., and MACRAE, D.: *'Electronic instruments'* (McGraw-Hill, New York, 1948)

HANNAH, M.R.: 'Frequency response measurement of hydraulic power units', *ASME*, 1948, **70**, p. 363

HARDER, E.L., and McCANN, G.D.: 'A large-scale general purpose electric analog computer', *AIEE*, 1948, **67**, pp. 664–73

McCANN, G.D., LINDVALL, F.C., and WILTS, C.H.: 'The effect of Coulomb friction on the performance of servomechanisms', *AIEE*, 1948, **67**, pp. 540–46

OLDENBURG, R.C., SARTORIUS, H.: *'The dynamics of automatic control'* (ASME, New York, 1948)

OPPELT, W.: 'Locus curve method for regulators with friction', *Z. Deut Ingr. Berlin*, 1948, **90**, pp. 179–83 (German); translation in Report 1691, National Bureau Standards, Washington, DC, 1952

PORTER, A.: 'Basic principles of automatic control systems', *IMechE*, 1948, **159**, pp. 25–34

SEAMANS, R.C. Jr., BROMBERG, B.G., and PAYNE, L.E.: 'Application of the performance operator to aircraft automatic control', *J. Aero. Sci.*, 1948, **15**(9)

SHANNON, C.E.: 'The mathematical theory of communication', *BSTJ*, July & Oct. 1948, **27**

SOBCZYK, A.: 'Stabilization of carrier frequency servomechanisms', *JFI*, 1948, **246**, pp. 21–43, 95–121, 187–213

WARREN, H.E.: 'Precise turbine governor', *AIEE*, 1948, **67**, p. 571

WIENER, N.: *'Cybernetics: or control and communication in the animal and the machine'* (Wiley, New York, 1948)

1949

ANDRONOW, A.A., and CHAIKIN, C.E.: *'Theory of oscillations'* (Princeton University Press, 1949)

BENNER, A.H.: 'Phase-lead for A.C. servo mechanisms', *J. Applied Physics*, 1949, **20**, pp. 268–73

BROWN, G.S., and CAMPBELL, D.P.: 'Instrument engineering: its growth and promise in process-control problems', *Mechanical Engineering*, 1950, **72**, pp. 124–7, 136, discussion pp. 587–9

CHESTNUT, H., and MAYER, R.W.: 'Comparison of steady-state and transient performance of servomechanisms', *AIEE*, 1949, **68**, pp. 765–77

CURFMAN, H.J., and GARDINER, R.A.: 'Method for determining the frequency response characteristics of an element or system from the system transient output response to a known input function', *NACA/TN/1964* (1949)

EASTON, E.C., and THOMAS, C.H.: 'Graphical determination of transfer function loci for servomechanism components and systems', *AIEE*, 1949, **68**, pp. 307–18

HALL, A.C.: 'Damper stabilized instrument servomechanisms', *ibid.*, pp. 299–306

KAHN, D.A.: 'Analysis of relay servomechanisms', *ibid.*, pp. 1079–88

KLOEFFLER, R.G.: *'Industrial electronics and control'* (Wiley, New York, 1949)

KOOPMAN, R.J.W.: 'Operating characteristics of two-phase servo motors', *AIEE*, 1949, **68**, pp. 319–29

La VERNE, M.E., and BOKSENBOM, A.S.: 'Methods for determining frequency response of engines and control systems from transient data', *NACA/TN/1935*, 1949

LEBENBAUM, P.: 'Design of DC motors for use in automatic control systems', *AIEE*, 1949, **68**, pp. 1089–94

LYONS, D.J.: 'Present thoughts on the use of powered flying controls in aircraft', *J. Roy. Aero. Soc.*, 1949, **53**, pp. 253–92

MACK, C.: 'The calculation of the optimum parameters for a following system', *Phil. Mag.*, 1949, **40**, Series 7, pp. 922–8

McCANN, G.D., WELLS, C.H., and LOCANTHI, B.N.: 'Application of the California Inst. of Tech. electric analog computer to nonlinear mechanics and servomechanisms', *AIEE*, 1949, **68**, pp. 652–60

McDONALD, D.: 'Electromechanical lead networks for AC servomechanisms', *RSI*, 1949, **20**, pp. 775–9

MARCY, H.T., YACHTER, M., and ZANDERE, J.: 'Instrument inaccuracies in feedback control systems with particular reference to backlash', *AIEE*, 1949, **68**, pp. 778-88
MEREDITH, F.W.: 'The modern autopilot', *J. Roy. Aero. Soc.*, 1949, **50**, pp. 409-32
RAGAZZINI, J.R., and ZADEH, L.A.: 'Probability criterion for the design of servomechanisms', *J. App. Physics*, 1949, **20**, pp. 141-44
SHANNON, C.E., and WEAVER, W.: *The mathematical theory of communication* (University of Illinois Press, Urbana, 1949)
ZIEGLER, J.G., and NICHOLS, N.B.: 'Valve characteristics and process control', *Instruments*, 1949, **22**, p. 75

1950

BODE, H.W., and SHANNON, C.E.: 'A simplified derivation of linear least square smoothing and prediction theory', *IRE*, 1950, **33**, pp. 417-26
BROWN, G.S., and CAMPBELL, D.P.: 'Control systems', *Scientific American*, 1950, **187**, pp. 56-64
COSGRIFF, R.L.: 'Integral controller for use in carrier type servomechanisms', *AIEE*, 1950, **69**, pp. 379-83
CRUICKSHANK, A.J.O.: 'Servo control problems', *Trans. Inst. Engrs. & Shipblds in Scotland*, 1950, p. 266
DUERDOTH, W.T.: 'Some considerations in the design of negative-feedback amplifiers', *AIEE*, 1950, **97**(II), pp. 138-58
DUTILH, J.: 'Theorie des servomechanisms a relais', *Onde Elec.*, 1950, pp. 438-445
EDWARDS, C.M., and JOHNSON, E.C.: 'Electronic simulator for non-linear servomechanisms', *AIEE*, 1950, **69**, pp. 300-07
EVANS, W.R.: 'Control system synthesis by root locus method', *ibid.*, pp. 66-69
GOLDBERG, E.A.: 'Stabilization of D-C amplifiers', *RCA Review*, 1950, **11**, pp. 296-300
HALL, A.C.: 'Generalized analogue computer for flight simulation', *AIEE*, 1950, **69**, pp. 308-20
HARTREE, D.R.: *Calculating instruments and machines* (Cambridge University Press, 1950)
HAYES, K.A.: 'Servo-mechanisms: recent history and basic theory', *SIT*, 1950, **2**, pp. 2-13
HAZEBROCK, P., and VAN DER WAERDEN, B.L.: 'Theoretical considerations on the optimum adjustment of regulators', *ASME*, 1950, **72**, pp. 309-15
HAZEBROCK, P., and VAN DER WAERDEN, B.L.: 'The optimum adjustment of regulators', *ibid.*, pp. 317-22
HERST, R.: 'Mechanical and electrical factors in servomechanism design', *Elec. Mfg.*, 1950, **46**, pp. 76-79, 184, 186
HOPKIN, A.M.: 'A phase-plane approach to the design of saturating servomechanisms', *AIEE*, 1950, **70**, pp. 631-39
KIRBY, M.J.: 'Stability of servomechanisms with linearly varying elements', *AIEE*, 1950, **69**, pp. 662-68
KOCHENBURGER, R.J.: 'A frequency response method for analysing and synthesising contactor servomechanisms', *AIEE (T)*, 1950, **69**, pp. 270-83
KORN, G.A., and KORN, T.M.: 'Modern servomechanisms testers', *Elec. Eng.*, 1950, **69**, pp. 814-16
LEE, Y.W.: 'Application of statistical methods to communication problems', MIT Res. Lab. Electronics Tech. Rept. No. 181, 1950
LOZIER, L.C.: 'Carrier controlled relay servos', *Elec. Eng.*, 1950, **69**, pp. 1052-56
McDONALD, D.C.: 'Nonlinear techniques for improving servo performance', *Proc. Natl. Electronics Conf.*, 1950, **6**, pp. 400-21
McDONALD, D.C.: 'Improvements in characteristics of AC lead networks', *AIEE*, 1950, **69**, pp. 293-300
MILLER, K.S., and SCHWARZ, R.J.: 'Analysis of a sampling servo-mechanism', *J. Applied Physics*, 1950, **21**, p. 290
MORRIS, D.: 'A phase calibrated variable-frequency supply for the testing of servomechanisms', *IEE*, 1950, **97**(II), pp. 37-45
NEWTON, G.C.: 'Comparison of hydraulic and electric servomotors', *Proc. Natl. Conf. Industrial Hydraulics*, 1950, **3**, pp. 64-86

NOTTHOFF, A.P.: 'Phase lead for AC servo-systems with compensation for carrier frequency changes', *AIEE*, 1950, **69**, pp. 285-95
PORTER, A.: *'Introduction to servomechanisms'* (London, 1950)
PORTER, A., and STONEMAN, F.: 'A new approach to the design of pulse-monitored servo systems', *IEE*, 1950, **97**(II), pp. 597-611
REYNER, J.H.: *'Magnetic amplifiers'* (Stuart & Reynolds, London, 1950)
ROGERS, T.A., and HURTY, W.C.: 'Relay servomechanisms, the shunt motor servo with inertia load', *ASME*, 1950, **72**, pp. 1163-72
RUDENBERG, R.: *'Transient performance of electric power systems'* (McGraw-Hill, New York, 1950)
RUTHERFORD, C.I.: 'The practical application of frequency response analysis to automatic process control', *IMechE*, 1950, **162**, p. 334
SEACORD, C.L. Jr.: 'Application of frequency-response analysis to aircraft-autopilot stability', *J. Aero. Sciences*, 1950, **17**(8)
STOUT, T.M.: 'A note on control area', *J. Applied Physics*, 1950, **21**, pp. 1129-31
TIZARD, R.H., and HARRINGTON, B.G.V.: 'An electronic speed control for the towing carriage of a ship model testing tank', *IEE*, 1950, **97**(II), p. 651
TUSTIN, A.: 'Problems to be solved in the development of control systems', *SIT*, 1950, **2**(2), pp. 19-33
TUSTIN, A.: 'Progress in automatic control systems engineering', *Engineering*, 1950, **170**, pp. 360-63
UTLEY, A.M.: 'Stabilisation of closed loop control systems', *SIT*, 1950, **2**, pp. 14-18
WALTER, E.R., and REA, J.B.: 'Determination of frequency response from response to arbitrary input', *J. Aero. Sciences*, 1950, **17**, pp. 446-52
WEST, J.C.: 'D.C. generator amplifier', *Elec. Rev.*, 1950, **147**, p. 819
ZADEH, L.A.: 'Frequency analysis of variable networks', *IRE*, 1950, **38**, pp. 291-99
ZADEH, L.A., and RAGAZZINI, J.R.: 'An extension of Wiener's theory of prediction', *J. Applied Physics*, 1950, **21**, pp. 645-55

1951

'History of pre-Act response', *Taylor Technology*, 1951, **4**(1), pp. 16-20
'Report of feedback systems conference, December 6-7, 1951', *Electrical Engineering*, 1951, **70**, p. 736
'Self-control for machines: discussion at Cranfield', *Economist*, Aug. 1951, **161**, pp. 357-59
'Servomechanisms: Selected government Research Reports' (HMSO, London, 1951)
AARON, M.R.: 'Synthesis of feedback control systems by means of pole and zero location of the closed loop function', *AIEE Tech. Paper*, 1951, pp. 251-267
AHRENDT, W.R., and TAPLIN, J.F.: *'Automatic feedback control'* (McGraw-Hill, New York, 1951)
AIKMAN, A.R.: 'The frequency response approach to automatic control problems', *SIT*, 1951, **3**, pp. 2-16
ATTURA, G.M.: 'Effects of carrier shifts on derivative networks for AC servomechanisms', *AIEE*, 1951, **70**, pp. 612-18
BARTLETT, F.: 'Human control systems', *SIT*, 1951, **3**, pp. 134-42
BEHAR, M.F.: *'Handbook of measurement and control'* (Instrument Publishing, Pittsburgh, 1951)
BOLLAY, W.: 'Aerodynamic stability and automatic control', *J. Aero. Sciences*, 1951, **18**, pp. 569-624
BOWN, L.O.: 'Transfer function for a two-phase induction servo motor', *AIEE*, 1951, **70**, pp. 1890-93
CHESTNUT, H., and MAYER, R.W.: *'Servomechanisms and regulating system design'* Vol. I and II (Wiley, New York, 1951)
EVANS, W.R.: 'Application of root locus method to multi-coupled systems', Thesis, University of California, 1951
FARRINGTON, G.H.: *'Fundamentals of automatic control'* (Wiley, New York, 1951)
FETT, G.: *'Feedback control systems'* (Prentice Hall, New York, 1951)
FLECK, J.T., and ORDUNG, P.F.: 'Realization of a transfer ratio by means of a R-C ladder network', *IRE*, 1951, **39**, pp. 1969-74

GRAYBEAL, T.D.: 'Block diagram network transformation', *Electrical Engineering*, 1951, **70**, pp. 985-90
HARRIS, H., KIRBY, M.J., and VON ARY, E.F.: 'Servomechanism transient performance from decibel-log frequency plots', *AIEE Tech. Paper*, 1951, **51**, p. 269
HONNELL, P.M.: 'The generalized transmission matrix stability criterion', *AIEE*, 1951, **70**, pp. 292-98
HOPKIN, A.M.: 'Transient response of small two-phase servo-motors', *ibid.*, pp. 881-86
HOPKIN, A.M.: 'Compensation of saturating servos', *ibid.*, pp. 631-39
JACKSON, K.R.: 'Investigation of the effect of error plus error-rate equalization of shunt motor relay servomechanisms', Thesis, University of California, 1951
JOHNSON, R.L., and REA, J.B.: 'Importance of extending Nyquist servo analysis to include transient response', *J. Aero. Sciences*, 1951, **18**, pp. 43-49
KIRBY, M.J., and GUILIANELLI, R.M.: 'Stability of varying element servomechanisms with polynomial coefficients', *AIEE*, 1951, **70**, pp. 1447-51
LAWDEN, D.F.: 'A general theory of sampling servo systems', *IEE*, 1951, **98**(IV), pp. 31-36
LINVILLE, J.K.: 'Sampled data control systems studied through comparisons of sampling with amplitude modulation', *AIEE*, 1951, **70**, pp. 1779-88
MACMILLAN, R.H.: *An introduction to the theory of control in mechanical engineering* (Cambridge University Press, 1951)
MADDOCK, A.J.: 'Magnetic amplifiers', *SIT*, 1951, **3**, pp. 68-79
MOORE, J.R.: 'Combination open-cycle closed-cycle systems', *IRE*, 1951, **39**, pp. 1421-32
NIMS, P.T.: 'Some design criteria for automatic controls', *AIEE*, 1951, **70**, pp. 606-11
ORDUNG, P.F., *et al.*: 'Synthesis of paralleled three terminal R-C networks to provide complex zeros in the transfer function', *ibid.*, pp. 1861-67
PORTER, A.: 'The impact of servomechanism techniques on instrumentation', *ASME, Preprint* 51-SA-22 (1951)
SEBRING, R.B.: 'Proportionalizing dither increases sensitivity of hydraulic servo', *Machine Design*, 1951, **23**, pp. 125-28
SUTTON, G.G.: 'The technique of servo control applied to the design of an automatic spectrophotometer', *SIT*, 1951, **3**, pp. 157-69
TANNER, J.A.: 'A logarithmic plotting technique for the design of closed loop systems', *SIT*, 1951, **3**, pp. 17-82
WARNKE, G.F., and DISNEY, V.H.: 'A new parallel-type compensation network', *AIEE Misc. Paper*, 1951, pp. 52-95
WASS, C.A.A., and HAYMAN, E.G.: 'An approximate method of deriving the transient response of a linear system from the frequency response', RAE Tech. Note GW148, 1951
WHITELEY, A.L.: 'Servomechanisms: a review of progress', *IEE*, 1951, **98**(1), pp. 289-97; also in *Engineering*, 1951, **172**, pp. 825-56

1952

BARKER, R.H.: 'The pulse transfer-function and its application to sampling servo-systems', *IEE*, 1952, **99**, pp. 302-17
BRUNS, R.A.: 'Analogue computers for feedback control systems', *AIEE*, 1952, **71**, pp. 250-54
BURNS, D.O., and COOPER, C.W.: 'A harmonic-response-testing equipment for linear systems', IEE paper 1386, 1952
CHU, Y.: 'Feedback control systems with dead-time or distributed lag by root locus method', *AIEE*, 1952, **71**, pp. 291-96
CHU, Y.: 'Synthesis of feedback control systems by phase angle loci', *ibid.*, pp. 330-39
CUNNINGHAM, W.J.: 'Equivalences for the analysis of circuits with small non-linearities', *J. Applied Physics*, 1952, **23**, pp. 653-57
DRAPER, C.S., McKAY, W., and LEES, S.: *Instrument engineering* (McGraw-Hill, New York, 1952)
DUNCAN, W.J.: *The principles of the control and stability of aircraft* (Cambridge University Press, 1952)
DUSHKES, S.Z., and CAHN, S.L.: 'Analysis of some hydraulic components used in regulators and servomechanisms', *ASME*, 1952, **74**, pp. 595-601

GIMPEL, D.J., and CALVERT, J.F.: 'Signal component control', *AIEE*, 1952, **71**, pp. 339-43
GOLDFARB, L.C.: 'On some non-linear phenomena in regulating systems', *Automatiki i Telemek*, 1952, **8**, pp. 349-383; translation in National Bureau of Standards Report 1691, 1952
HAMER, H.: 'Linear circuit theory and servomechanism design', *Electrical Engineering*, 1952, **71**, pp. 614-15
HAUSER, A.A., Jr.: 'A generalised method for analyzing servomechanisms', *IRE*, 1952, **40**, pp. 197-202
HICK, W.E.: 'Why the human operator', *SIT*, 1952, **4**, pp. 67-77
JOHNSON, E.C.: 'Sinusoidal analysis of feedback control system containing non-linear elements', *AIEE*, 1952, **71**, pp. 169-81
JONES, A.L., and WHITE, J.S.: 'Analogue-computer simulation of an autopilot servo system having nonlinear response characteristics', NACA TN 2707, 1952
LANGHAM, E.M.: 'Electronics control applied to a modern textile beaming machine', *SIT*, 1952, **4**, pp. 104-12
LEWIS, J.B.: 'The use of nonlinear feedback to improve the transient response of a servomechanism', *AIEE*, 1952, **71**, pp. 449-53
McDONALD, D.: 'Multiple mode operation of servomechanisms', *RSI*, 1952, **23**, pp. 22-30
MILLIKEN, W.F., JR.: 'Dynamic stability and control research', 3rd Anglo-American Aeronautical Conference 1951 (Royal Aeronautical Society, London, 1952)
NEWTON, G.C., Jr.: Compensation of feedback control systems subject to saturation', *JFI*, 1952, **254**, pp. 281-86, 391-413
PESTARINI, J.: *'Metadyne statics'*, (Wiley, New York, 1952)
STOUT, T.M.: 'A block diagram approach to network analysis', *AIEE*, 1952, **71**, pp. 255-60
TUSTIN, A. (Ed.): *'Automatic and manual control: proceedings of the 1951 Cranfield Conference'* (Butterworths, London, 1952)
TUSTIN, A.: *'Direct current machines for control systems'* (E & FN Spon, London, 1952)
TUSTIN, A.: 'Specification and measurement of performance in servo systems', *IEE*, 1952, **99**(2), pp. 494-96
WEST, J.C.: 'A system utilizing coarse and fine position measuring elements simultaneously in a remote-position control servomechanism', *IEE*, 1952, **99**(2), pp. 135-43
WESTCOTT, J.H.: 'The frequency response method: its relationship to transient behaviour in control system design', *SIT*, 1952, **4**, pp. 113-24

1953

BANE, W.T.: 'Design charts for an on/off control system', *SIT*, 1953, **5**, pp. 52-71
BRIGGS, P.G., and DREWE, J.: 'Automatic control of alignment in multicolour printing', *SIT*, 1953, **5**, pp. 160-69
BUSHAW, D.W.: 'Differential equations with a discontinuous forcing term', Stevens Ints. Technol. Exp. Towing Tank Report 469, 1953
COHEN, G.H., and COON, G.A.: 'Theoretical consideration of retarded control', *ASME*, 1953, **75**, p. 827
CHU, Y.: 'Correlation between frequency and transient response of feedback control systems', *AIEE*, 1953, **72**(II), pp. 81-92
CROW, L.R.: *'Synchros, self-synchronous devices and electrical servomechanisms'* (Scientific Book Pub., Vincennes, Indiana, 1953)
D'OMBRAIN, G.L.: 'The training of the process control engineer and the presentation of control theory', *SIT*, 1953, **5**, pp. 90-104
FLUGGE-LOTZ, I.: *'Discontinuous automatic control'* (Princeton University Press, 1953)
GRAHAM, D., and LATHROP, R.C.: 'The synthesis of "optimum" transient response criteria and standard forms', *AIEE*, 1953, **72**(II), pp. 273-88
HAINES, J.E.: *'Automatic control of heating and air conditioning'* (McGraw-Hill, New York, 1953)
HOPKIN, H.R.: 'Automatic control of aircraft', *SIT*, 1953, **5**, pp. 72-89
JONES, R.W.: *'Electric control systems'* (Wiley, New York, 1953)
KAZDA, L.F.: 'Errors in relay servo systems', *AIEE*, 1953, **72**(II), pp. 323-28

KING, L.H.: 'Reduction of forced error in closed-loop systems', *IRE*, 1953, **41**, pp. 1037–42
MASON, S.J.: 'Feedback theory — some properties of signal flow graphs', *IRE*, 1953, **41**, pp. 1144–56
NEISWANDER, R.S., and MACNEAL, R.H.: 'Optimization of nonlinear control systems by means of nonlinear feedbacks', *AIEE*, 1953, **72**(II), pp. 262–72
NIXON, F.E.: *'Principles of automatic control'* (Prentice-Hall, Englewood Cliffs, NJ, 1953)
PERCIVAL, W.S.: 'The solution of passive electrical networks by means of mathematical trees', *IEE*, 1953, **100**, pp. 143–50
SIMPSON, R.F.C.: 'Automatic control problems in refinery fuel systems', *SIT*, 1953, **5**, pp. 31–37
STOUT, T.M.: 'Effects of friction in an optimum relay servomechanism', *AIEE*, 1953, **72**(II), pp. 329–36
STOUT, T.M.: 'Block diagram solutions for vacuum tube circuits', *ibid.*, pp. 561–67
THALER, R.J., and BROWN, R.G.: *'Servomechanism analysis'* (New York, 1953)
TUSTIN, A.: 'The nature of the load as a factor in the design of electrical servos', *SIT*, 1953, **5**, pp. 13–28
TUSTIN, A.: *'Mechanism of economic systems'* (Heinemann, London, 1953)
WALKER, J.R.D.: 'Frequency response methods applied to large servomechanisms', *SIT*, 1953, **5**, pp. 170–181
WALKER, R.C.: *'Relays for electronic and industrial control'* (London, 1953)
WEST, J.C.: *'Textbook of servomechanisms'* (EUP, London, 1953)

1954

ADKINS, B., et al.: *'Rotating amplifiers: the amplidyne, metadyne, magnicon, and the magnavolt and their uses in control systems'* (George Newnes, London, 1954)
AHRENDT, W.R.: *'Servomechanism practice'* (New York, 1954)
BELL, D.A.: 'General properties of electromagnetic amplifiers', *Wireless Engineer*, 1954, **31**, p. 310
BELLMAN, R.E., and DANSKIN, J.M.: 'Survey of the mathematical theory of time-lag, retarded control and hereditary processes', Rand Report R-256, 1954
BOGNER, I., and KAZDA, L.F.: 'An investigation of the switching criteria for higher order servomechanisms', *AIEE*, 1954, **73**, pp. 118–27
DIEBOLD, J.: *'Control system design notes'* (1954)
EVANS, W.R.: *'Control system dynamics'* (McGraw-Hill, New York, 1954)
FETT, G.H.: *'Feedback control systems'* (New York, 1954)
GEYGER, W.A.: *'Magnetic amplifier circuits'* (McGraw-Hill, New York, 1954)
IZAWA, K.: *'Introduction to automatic control'* (Ohm, 1954)
JONES, P.: 'Stability of feed-back systems using the dual Nyquist diagram', *IEEE*, 1954, CT-1, pp. 35–44
LA JOY, M.H.: *'Industrial automatic controls'* (McGraw-Hill, New York, 1954)
OPPELT, W.: *'Kleines Handbuch technisches Regelvorgänge'* (Chemie, 1954)
PETERS, J.: *'Einschwingvorgange, Gegenkopplung, Stabilitat'* (Springer, Berlin, 1954)
PRESTON, J.L.: 'Non-linear control of saturating third order servomechanisms', MIT Servo Lab. Tech. Memo 6897-TM-14, 1954
PROFOS, P.: *'Vektorielle Regeltheorie'* (Zurich, 1954)
SAY, M.G. (Ed.): *'Rotating amplifiers'* (Newnes, London, 1954)
SAY, M.G. (Ed.): *'Magnetic amplifiers and saturable reactors'* (Newnes, London, 1954)
SOROKA, W.W.: *'Analog methods in computation and simulation'* (McGraw-Hill, New York, 1954)
TAKAHASHI, Y.: *'The theory of automatic control'* (Iwanami, 1954) in Japanese
TRUXAL, J.G.: *'Feedback theory and control system synthesis'* (McGraw-Hill, New York, 1954)
TSIEN, H.S.: *'Engineering cybernetics'* (McGraw-Hill, New York, 1954)
WAX, N.: *'Selected papers on noise and stochastic processes'* (Dover, New York, 1954)
WESTCOTT, J.H.: 'The introduction of constraints into feedback systems designs', *IRE*, 1954, CT-1, pp. 39–49
YOUNG, A.J.: *'Process Control'* (Instruments Publishing, Pittsburgh, PA, 1954)
ZIEBOLZ, H., and PAYNTER, H.M.: 'Possibility of two-time scale computing system for control and simulation of dynamic systems', Nat. Electronics Conf. 9, 1954

1955

BRUNS, R.A., and SAUNDERS, R.M.: 'Analysis of feedback control systems, servomechanisms and automatic regulators' (McGraw-Hill, New York, 1955)

CHADWICK, J.H.: 'On the stabilization of roll', *Trans. Soc. Naval Arch. & Marine Eng.*, 1955, **63**, pp. 234–80

CHESTNUT, H., and MAYER, R.W.: 'Servomechanisms and regulating systems design' (McGraw-Hill, New York, 1955)

DRAPER, C.S.: 'Flight control', *J. Roy. Aero. Soc.*, 1955, pp. 451–77

KALMAN, R.E.: 'Analysis and design principles of second and higher order saturating servomechanisms', *AIEE*, 1955, **74**(II), pp. 294–310

NEWTON, G.C. Jr.: 'Design of control systems for minimum bandwidth', *AIEE*, 1955, **74**(II), pp. 161–68

PERCIVAL, W.S.: 'The graphs of active networks', *Proc. IEE*, 1955, **102**, pp. 270–78

PONTRYAGIN, L.S.: 'On the zeros of some elementary transcendental functions', *Amer. Math. Soc. Trans.*, 1955, **2**(1), pp. 95–110

STOLLARD, D.V.: 'Series method of calculating control system transient response from frequency response', *AIEE*, 1955, **74**(II), pp. 61–64

THALER, G.J.: 'Elements of servomechanisms' (McGraw-Hill, New York, 1955)

TRUXAL, J.G.: 'Automatic control system synthesis' (McGraw-Hill, New York, 1955)

TSYPKIN, Y.Z.: 'Theory of relay control systems' (State Press for Technical and Theoretical Literature, Moscow, 1955) in Russian; (Oldenbourg, Munich, 1958) in German

VAN VALKENBURG, M.E.: 'Network analysis' (Prentice-Hall, Englewood Cliffs, NJ, 1955)

YOUNG, A.J.: 'An introduction to process control systems design' (Longmans, London, 1955)

Index

ASME 53–4, 189
AT&T 61, 70–1, 76, 86
Admiralty Experiment Works 18
Admiralty Research Laboratory 11–2, 14, 18, 103, 120, 131–3
airborne interception 146–7
Alexanderson, E.F.W. 10
American Institute of Electrical Engineers 70, 189
amplidyne 8, 10–2, 21, 131, 140, 144–6, 149–50
amplifiers
  distortion in 73
  electronic 8, 21, 23
  frequency characteristics 90–1
  negative feedback 22, 43, 61, 70–81, 97, 167
  pneumatic 31–47
  positive feedback 74, 77, 88–9
  relay 31
  singing of 74
analogue simulation 198–9
anti-aircraft gun control 14, 115–6, 130–3
anti-hunt circuits 144–6
Askania Company 19, 62
Atanasoff, John V. 180
auto-correlation 177–8
auto-pilot 19
automatic curve followers 111
automatic follow-up *see* servomechanisms
automatic steering 18–20
automation 204–5

backlash 14, 102, 153–5, 199
  in gun drives 170, 196
Bailey Meter Company 3, 30, 39, 53
bandwidth 75, 77

Barber-Coleman Company 168
Barkhausen, H. 84
Bawdsey Research Station 117–8
Beck, Charles 8
Behar, M.E. 21, 38, 49, 54, 61, 62
Bell Telephone Laboratories 61-2, 73, 81, 84, 90, 149, 170, 174, 180–1
Bell System 71
bellows, in instruments
  helical spring 29, 31, 40, 42
  Bourdon tube 29, 34
Belsey, F.H. 155, 196
Bendix Corporation 18–9
Bergman, P.G. 180
Bigelow, John 173–8, 180
Black, Harold Stephen 43, 70–81, 87, 90–1, 97, 108, 131, 139, 164
Blackett, P.M.S. 111
block diagrams 190
Blumlein, A.D. 147
Bode diagram 86, 138–9, 190–1
Bode, Hendrik 76, 84–6, 91, 164, 171, 194
boiler control, automatic 3, 22
Bourdon tube *see* bellows
Bristol Company 1, 29, 30, 37, 39, 44
Bristol, Edgar H. 31
Bristol, William H. 34
British Thomson-Houston Company 5, 9–10, 20–1, 119, 131, 149, 150
Brown Brothers 18
Brown, Gordon S. 108, 111–2, 138–40, 146, 150, 172, 192, 199, 204
Brown Instrument Company 39, 53, 189, 199
Brownian motion 175
Bush, Vannevar 97–104, 108, 175–76, 186
Bushaw, D.W. 200–1

Butterworth, S. 153
by-pass valves 36, 37

Caldwell, Samuel H. 121, 146, 171–3
Callendar recorder 30
Cambridge Scientific Instrument
    Company 30
Campbell, Duncan P. 140
Carrick Engineering Company 36
Carson, John 75–6, 99
Characteristic equations
    roots of 56, 134
Chestnut, Harold 194, 196
Churchill, Winston 116, 119, 124–5
Clarridge, Ralph E. 47–8, 50
Clausen, H. 12
Coales, John F. 119, 154
Cockcroft, J.D. 149
communications 2, 62
    carrier systems 72–3, 76–81
    cross-talk 77
    telephone 70–3
    wireless 86–9
computing mechanisms
    analogue 98–106
    Cinema Integraph 104–6, 109
    digital 203
    Differential Analyzer 55–60, 103–4, 173
    network analyser 97–9
    POLYTHEMUS 62
    Product Integraph 98–103
Cooke, P.A. 19
Concordia, C. 198–9
constant temperature ovens 88
control system design *see* frequency
    response methods, *and* transient
    response methods
controllability 61
controller classification, Mitereff's 51
Cossor, (A.C.) Company 118, 149, 151
Craik, K.T.W. 167
Cranfield Conference (1951) 200
cybernetics 181, 204–5

Daniell, P.J. 136, 154, 170, 175, 180, 190
data transmission systems
    Autosyn 12
    Magslip 12, 116–8
    selsyn 7, 10, 20, 196
    step-by-step 11
    synchros 77
definite correction servomechanisms
    30–31, 38, 109, 156, 196
Denny (William) & Brothers 18

digital computers *see* computing
    mechanisms
Differential Analyzer *see* computing
    mechanisms
directors *see* predictors
dither 14, 34
divided reset 155, 196
Dixon, A.M. 40
Dow Chemical Company 38, 49, 53
Draper, C.S. 138, 153
Drysdale, C.V. 12

Eckman, Donald P. 189, 199
electro-mechanical systems 21
    oscillators 89
    telephone repeaters 71–2
electronic valves
    triode 71
    vacuum tube repeaters 71–2, 73–4
Elliott Brothers 3
engineering education 91–2, 186–8
    Northampton Polytechnic, 187–8
    servomechanism courses at MIT 111–2
    University of Pittsburgh 92
errors in servomechanisms
    steady state 139, 190
    velocity error 12–4
Evans, Walter R. 60, 196–8

Fairchild, C.O. 189
feed-forward control 14
Ferranti, de V.Z. 187
Ferrell, E.B. 190
filters 72, 76
fire control 10–4
    anti-aircraft 115–20
    predictors 11, 14, 116–7
    section of NDRC 121
floating control 3, 38
Flugge-Lotz, I. 200
Ford, J.M. 12
Forrester, Jay W. 140
Fourier transform 82
Foxboro Company 31–4, 37, 39–46, 49
    Potentiometric Stabilog 44
    Stabilog controller 39–46, 49
Frahm, H. 15
frequency response methods 6, 50, 53, 61,
    73–4, 138, 143, 190–1, 194, 201
    Bode diagram 86, 138–9, 190–1
    gain margin 51, 191
    inverse Nyquist 150–1, 194–5
    M-circles 192
    N-circles 192

Nichols diagram 183
Nyquist method 84, 86, 90, 142–4, 150, 154
Frequency Response Symposium (1953) 200
friction 14, 34, 43, 152
Fry, Thornton, C. 146, 179
Frymoyer, W.W. 40

gain margin 51, 191
galvanometer 29–30, 101
Garner, H.M. 19
Gates, S.B. 19
General Electric Company 3–4, 7–9, 10, 12, 20–21, 23, 90, 92, 98, 125, 131, 144, 194
Getting, Ivan A. 121, 125, 142
Gibson Island Conferences 62
Godet, Sidney 144–5
Graham, Robert E. 190
Gray, Truman, S. 105, 140
Grebe, J.J. 38, 48–9, 53
Gulliksen, F.H. 5–6
gun control 10–4, 115–6, 130–3
  aided laying of 165–7
  anti-aircraft 115–6, 130–3
gyroscope 11–2, 15, 18–9

Hall, A.C. 111, 130, 139–43, 146, 150, 172–73, 178, 192, 199
harmonic response 134–6, 166, 176
Harris, Herbert 90, 140–1, 150–1, 194
Hartree Douglas 55, 60, 111
Hawkins, J.T. 34–5
Haus, F. 19
Hayes, K.A. 187
Hazen, Harold Locke 23, 91, 97–112, 121, 131, 146, 155–6, 164, 167–8
heat treatment furnaces 31
Herwald, S.W. 189, 199
Honeywell Company 19, 50
Hull, A.W. 4
human operator 1, 31, 50, 166–7
hunting 4, 7, 22, 34–6, 51, 98, 133
Hurewicz, Wittold 157, 190
hydraulic control 12–4, 19, 36
ICI 55, 60
IEE 187–9
IFAC 203
IFF (Identification Friend and Foe) 147
Inglis, C.C. 190
Instrument Society of America 62
intruments, industrial
  manufacturers 28–9, 1, 61

recording *see* recorders
sales 28–9
temperature measuring 28–30
Inverse Nyquist diagram 150–1, 194–5
IRE 189
Ivanoff, A. 49, 51, 59, 189

Jackson, Dugald C. 97
Jackson and Moreland, Consulting Engineers 97–8
James, Hubert M. 143, 153
Jofeh, L. 151, 154
Johnson, Fritzhof 23, 92
Johnson, W.S. 31

Kalman, R.E. 203
Kent, (George) Company 3, 39
Kolmogorov, A.N. 181
Küpfmüller, K. 84

Laplace Transform
  use in design 141, 192, 201
Lawry, Clint 143–5
lead mechanisms 172
Leeds, Morris E. 30
Leeds & Northrup Company 30–1, 39, 50, 91
  Micromax recorder 30
  Speedamax recorder 30
Lee-Wiener network 173
Levinson, Norman 174–5, 178
Lindemann, F.A. (Lord Cherwell) 116, 119, 149
Liston, J. 8
Ludbrook, L.C. 6, 149

M9 Director 149, 170–1
M-circles 192
MacColl, L.A. 84, 155, 191–2
McCann, G.D. 199–200
Magslip 12, 116–8
Manchster University 55, 60
manual tracking 165–8
Marcy, H.T. 150–1, 194
Mason, Clesson E. 40–3, 49, 50, 54–5, 91, 189
Mason-Neilan Company 46
mechanical integrators 99
Melvill Jones, B. 19
Meredith, F.W. 19
Metadyne 10, 21, 131, 148–9
Metropolitan Vickers Company 10, 118, 131–3, 154, 170
Mikhailov, A.V. 84, 196

minimum phase 76, 85, 172, 191
Minneapolis-Honeywell Regulator
   Company 50
Minorsky, Nicholas 15, 18, 48, 91, 107–8, 154
MIT 60, 90, 92, 97–8
   Division of Industrial Cooperation 140
   Electrical Engineering Department 172
   Radiation Laboratory 124, 126, 143, 149, 178, 190
   Servomechanisms Laboratory 111, 140
Mitereff, S.D. 48–50, 51
Modelling, mathematical 23, 51–2, 53, 55
Morristown trials 70, 76
multivariable systems 202–3

N-circles 192
National Science Foundation 186
Naval Research Laboratory 116, 180–1
NDRC 121–5, 139, 156, 168–71
negative feedback
   reduces distortion 74–6, 79–80
   reduces noise disturbance 79–80
   amplifier 70–81
Neilan Company 37
Newton, George C. 111, 140
Nichols chart 193
Nichols, Nathaniel B. 60–1, 143, 153, 189
Nieman, C.W. 103, 108
non-linear systems 52, 72, 82, 89, 153–5, 200–1
Northampton Polytechnic 187–8
Nyquist criterion
   extension of 191–2
Nyquist diagram 84, 86, 90, 142–4, 150, 154
Nyquist, Harry 5, 75–6, 81–4, 86, 91, 108, 136, 139, 164, 194

operational amplifiers 180
operational calculus 91, 99, 175–6
Oppelt, W. 20
Osbon, W.O. 199
oscillators frequency regulation of 88–90
OSRD 121–5, 190

pasteurisation 36
performance criteria
   M-circles 142
   overshoot 142
   rms error 142–3
   square of error 176
   Whiteley's Standard Forms 150–2
performance evaluation 54, 107

Pestarini, J. 10, 131
phase lead 6, 172–3
phase margin 51, 191
phase-plane 155–6
phase shift networks 50–1
Philbrick, G.A. 54–5, 62, 180, 199
Phillips, R.S. 142, 153, 180
photoelectric cell 21, 30, 105–6, 108, 110–11
Pittsburgh, University of 92
pneumatic control system 19
   flapper-nozzle amplifiers 31–4, 37, 40–3, 46–7, 91
   relay 31
Poincaré, H. 155
Poitras, Edward J. 169–70, 179
POLYTHEMUS 199
Porter, Arthur 55, 60, 103, 111, 148–9, 165, 168–9, 186–7
Post, Wiley 18–9
pre-act 47–9, 55
predictors (directors) 11, 14, 116–7, 156–7
   BTL10 170–1
   dynamic tester for 168–70
   errors in 168–70
   Kerrison 171
   M7 168
   M9 149, 170–1
   performance 125, 179
   tracking errors 178–9
   Wiener 176–80
printing press control 9
process control 2
   theory of 49–55
   PD control 6, 46–9
   PID control 48–9, 56–7
   PI control 165, 199
process controllers
   Ampliset Free-Vane 44
   Damplifier 44
   Dubl-Response 43
   Fulscope 44, 49
   narrow band 31, 37, 39
   on-off 31–4
   Potentiometric Stabilog 44
   Stabilog 39–46, 49
   tuning of 54, 58–9, 60–1, 189
   wide band 39–46
Profos, P. 197
proportional band 37
pulse data system *see* sampled data systems
pyrometers 29–30

radar 116, 119

airborne 117
automatic following 126, 148–9, 154
centermetric 146–7
GLI 118–9
GLII 118–9
GLIII 118–9, 154
GCI (Ground Controlled Interception) 117
Home Chain 117
RDF (Radio Direction Finding) 116, 119
  see also radar
recorders 1, 28–30
  Callendar 30
  Celectray 30
  Gawatcon 30
  Micromax 30
  potentiometric 30–1, 38
  Speedomax 30
  thread recorder 30
regeneration theory 81–4
regulators 20–1
  carbon pile 3, 9
  frequency 3, 98
  speed 8
  Tirrill 2–4
  voltage 2–6
relays 6–7, 200–1
  control by 102, 109, 155–6
remote control, power operated 12, 14, 180–3
Republic Flow Meters Company 53
reset action 5, 38, 43, 46
Rhodes, Thomas J. 189
Ridenour, Louis N. 190
Rogers, H.W. 8–9
Root Locus 60, 196–8
Rosenblueth, Arturo 181
Routh Criteria 23, 53, 77, 83, 145, 199
Royal Aircraft Establishment 19, 167

sampled data systems 126, 196
Scientific Advisory Committees 124–5
SCR-584 143–6, 149, 181, 193
searchlight control 12–4, 126, 130–1, 149
secrecy in wartime 124–5
sectional drives 7–10, 77
Selsyn 7, 10, 20, 196
servomechanisms 2, 10, 12, 18–20, 30–31, 46, 62, 101–3
  classification of 192, 195
  definite correction 109, 30–1, 38, 156, 196
  figure-of-merit 107
  Hazen's definition of 108

high-speed 105
lag in 102–3
mechanical 99
relay 102, 109
velocity lag 130–1, 133
Servo-Panel 120–3, 133–4, 146, 188–90
Shannon, Claude, E. 171, 180
Sheffield University 136, 170
Siemens Company 3, 19
simulators
  Differential Analyser 55–60, 134
  electronic 55
  Network Analyser 97–9
  POLYTHEMUS 62
  at TRE 148
Smith, Ed. E. 49, 61–2, 189
Smoot Engineering Company 3, 37
Society of Instrument Technologists 62, 187
Spencer, Hugh H. 98
Sperry, Elmer 2, 11, 15
Sperry Gyroscope Company 18, 130–1, 140–1, 168–70, 173
Spirules 198–9
stability 5–6, 10, 19–20, 56, 190
  aircraft 19
  characteristic equation, roots of 56, 134
  closed loop 50–2
  Nyquist's criterion 83–4
  Routh-Hurwitz criteria 23, 53, 77, 83, 145, 199
stabilization, ship 12, 15–8
Standard Forms (Whiteley) 195
state-space approach 203
statistical analysis of flight path 174–7
Stein, I. Melville 50
Stewart, Duncan J. 168
Stibitz, George 156–7, 168–9, 176–80
stochastic systems 178–9, 202
systems engineering 125–6, 164

Tagliabue C J Company 30, 44
  Damplifier 44
Taplin, John 90, 139
Taylor, G.I. 175
Taylor Instrument Companies 30, 43–4, 47, 49, 50
  Dubl-Response 43
  Fulscope 44, 49
Telecommunications Research Establishment (TRE) 146–8
terminology (nomenclature) 53–4
thermocouples 29–30
thyratron 4–5, 8, 21, 157

250 *Index*

time constants 36–7
time delays 36–7, 50–1, 57
time lags 54, 133
time series 178–80
Tizard Committee 116, 124–5
Tizard, Henry 116–7
tachogenerator 8
torque amplifier 103–4, 107
transfer function 141–3, 192
transfer locus 142
transient response methods 134, 138, 190, 194
Tustin, Arnold 80, 92, 126, 131, 133–6, 153–5, 165–6, 170, 187, 200, 204

United Shoe Machinery Company 169
Uttley, J.A. 167, 187

valves, process control
  Evenaction 43
  friction in 43
  motor driven (electric) 37
  pneumatic (diaphragm) 40
  solenoid 37
Van der Pol, B. 136
velocity feedback 46, 145

Velodyne 147–8
Vickers Company 11, 119, 171

Walker, A.M. 190
Wannamaker, W.H. 199
Ward-Leonard systems 131, 149
Waterbury Tool Company 140
Watson-Watt, Robert 116
Weaver, Warren 121, 167–73, 179–80
Weiss, H.K. 138, 155, 168–71
Western Electric Company 72
Westinghouse Company 3, 7–9, 21, 92, 126, 189, 199
White, E.C.L. 147
Whiteley, A.L. 5–6, 9–10, 20, 131, 148–54, 187, 194–95
Wiener-Hopf equation 174–5
Wiener-Kintchine equation 176
Wiener, Norbert 85, 97, 104–6, 171–81, 204
Williams, Frederick C. 111, 145–8
Wilson-Maeulen Company 44
Wimperis, H.E. 116–7, 119

Z-transform 157
Ziegler, J.G. 44, 41–7, 49, 51, 60–1, 189